BIOZONE

BIOLOGY

FOR NGSS

MW00700682

THIS BOOK IS THE PROPERTY OF

Name _____

Class _____

School _____

BIOLOGY
FOR NGSS

About the Authors

Jillian Mellanby *Editor*

Jill began her science career with a degree in biochemistry and, after a short spell in research labs, became a science teacher both in the UK and then New Zealand. She spent many years managing the Royal Society of New Zealand's academic publishing programme of eight science journals which allowed her to hone her project management and editorial skills. She was also a part of the Expert Advice writing team at the Royal Society of New Zealand, producing science pieces for a public audience. She joined the BIOZONE team in late 2021, as editor.

Kent Pryor *Author*

Kent has a BSc from Massey University, majoring in zoology and ecology and taught secondary school biology and chemistry for 9 years before joining BIOZONE as an author in 2009.

Sarah Gaze *Author*

Sarah has 16 years experience as a Science and Chemistry teacher, recently completing MEd. (1st class hons) with a focus on curriculum, science, and climate change education. She has a background in educational resource development, academic writing, and art. Sarah joined the BIOZONE team, at the start of 2022.

Lissa Bainbridge-Smith *Author*

Lissa graduated with a Masters in Science (hons) from the University of Waikato. After graduation she worked in industry in a research and development capacity for eight years. Lissa joined BIOZONE in 2006 and is hands-on developing new curricula. Lissa has also taught science theory and practical skills to international and ESL students.

Cover photograph

Pit viper snake *Trimeresurus Insularis*
The beautifully colored, but venomous, pit viper snake is found across the lesser Sunda Islands of Indonesia. Most individuals of this species are green in color, the striking blue coloration is relatively uncommon. The species is highly aggressive, hunting lizards, frogs, small birds, and small mammals at dusk or in the evening.

PHOTO: https://stock.adobe.com
Photo ID: 295055844

Acknowledgements:
BIOZONE wishes to thank and acknowledge the team for their efforts and contributions to the production of this title.

ISBN 978-1-98-856692-4

Third Edition 2022

Copyright © 2022 Richard Allan
Published by BIOZONE International Ltd

Second printing
Printed by Replika Press, (INDIA)
Using FSC paper.

Next Generation Science Standards (NGSS) is a registered trademark of Achieve. Neither Achieve nor the lead states and partners that developed the Next Generation Science Standards were involved in the production of this product and do not endorse it.

Purchases of this book may be made direct from the publisher:

BIOZONE Corporation
USA and Canada
FREE phone: 1-855-246-4555
FREE fax: 1-855-935-3555
Email: sales@biozone.com
Web: www.biozone.com

Contents

CODING: Activity is marked: ☐ to be done ☑ when completed ● Practical investigation

Contents

CODING: Activity is marked: 🔲 to be done   ☑ when completed   ● Practical investigation

Contents

CODING: Activity is marked: ▣ to be done   ☑ when completed   ● Practical investigation

Contents

CODING: **Activity** is marked: ▣ to be done ☑ when completed ● Practical investigation

Using This Book

Each chapter begins with a broad **anchoring phenomenon**. This is something you may have seen, heard about, or experienced but may not necessarily be able to explain. Activities that present everyday and investigative phenomena make up most of this book. As you work through the activities in each chapter, you will build a deeper understanding of scientific concepts. You can then test your understanding in the review activity at the end of the chapter.

Structure of a chapter

Chapter introduction
Identifies the activities relating to the guiding questions.

Summing Up
Find out what you know about the ideas, connections, and skills you have explored in the chapter.

Introductory activity
The first activity acts as an **anchoring phenomenon**. It introduces a phenomenon that can be explained by the rest of the activities in the chapter.

Introductory activity revisited
Once you have completed the activities in the chapter, you should be able to explain the anchoring phenomenon more fully.

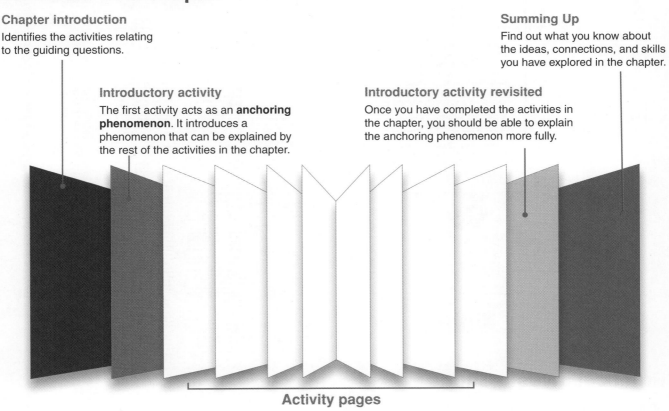

Activity pages

Chapter Introductions

The chapter number is identified for easy navigation.

Mark the check boxes to indicate the outcomes you should complete. Check them off when you have finished.

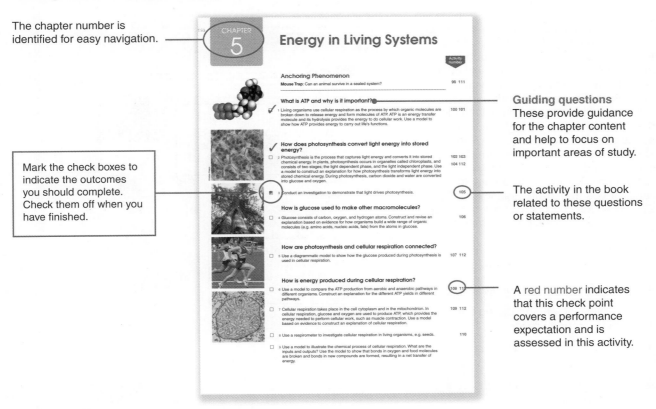

Guiding questions
These provide guidance for the chapter content and help to focus on important areas of study.

The activity in the book related to these questions or statements.

A red number indicates that this check point covers a performance expectation and is assessed in this activity.

Glossary Terms

Building communication skills and scientific literacy is an important feature of any science course. By speaking with, listening to your peers and teachers, and writing answers, you naturally practice and develop communication skills. To help develop **scientific literacy** we have included a **glossary** at the back of this worktext (pages 393-400). The glossary provides a definition in English and also in Spanish. Refer to the glossary to help you understand the meaning of a key term. It is easy to see which key terms are in the glossary, the terms have been **bolded** within an activity (see below). Note: Key terms are only bolded the first time they appear within an activity.

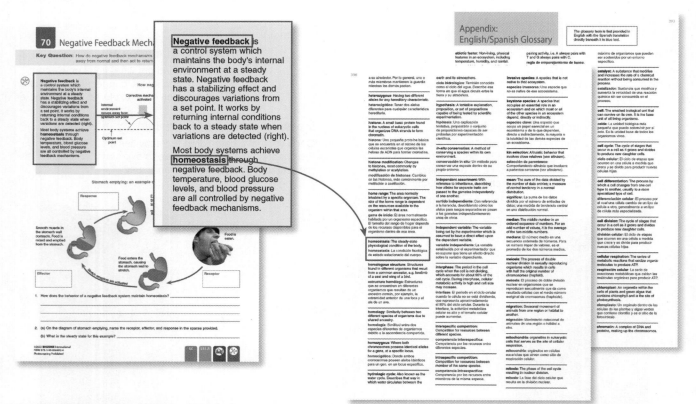

Practical Investigations

An important part of physical science involves carrying out investigations and carefully observing and recording what occurs during them. Throughout the book, you will notice green investigation panels (like the one shown right). Each investigation has been designed using simple equipment found around the home or in most high school laboratories. The investigations provide opportunities for you to investigate phenomena for yourself. The investigations have different purposes depending on where they occur within the chapter. Some provide stimulus material or ask questions to encourage you to think about a particular phenomenon before you study it in detail. Others build on work you have already carried out and provide a more complex scenario for you to explain. Equipment lists are provided as an appendix at the back of the book. The investigations will help you develop:

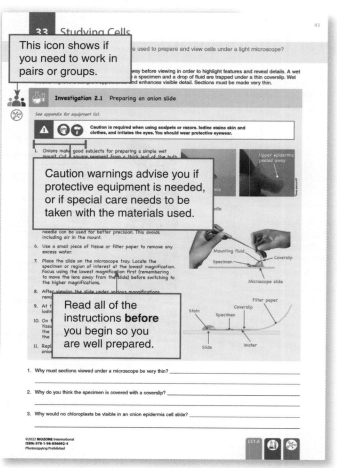

▸ Skills in observation

▸ Skills in critical analysis and problem solving

▸ Skills in mathematics and numeracy

▸ Skills in collecting and analyzing data and maintaining accurate records

▸ Skills in working independently and collaboratively as part of a group

▸ Skills in communicating and contributing to group discussions

Using the Tab System

The tab system is a useful way to quickly identify the Disciplinary Core ideas, Crosscutting concepts, and Science and Engineering Practices embedded within each activity. The tabs also indicate whether or not the activity is supported online on BIOZONE's **Research Hub**.

The **orange** Disciplinary Core Idea (DCI) tabs indicate the core ideas that are covered in the activity. These are covered in the introduction to each chapter, under the guiding questions. The code itself is just a guide for your teacher.

The gray hub tab indicates that the activity is supported online at the **BIOZONE RESOURCE HUB**. Online support may include videos, animations, games, simulations, articles, 3D models, and computer models.

The **green** Crosscutting Concepts tabs indicate activities that share the same crosscutting concepts. You will become familiar with the concepts that connect all areas of science.

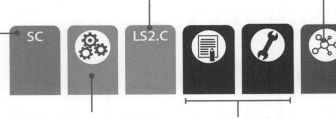

The ETS icon indicates an engineering design DCI is included in the activity.

The **blue** science and engineering practices tabs use picture codes to identify the Science and Engineering Practices (SEPs) relevant to the activity. You will use science and engineering practices in the course of completing the activities.

Science and Engineering Practices

Asking questions (for science) and defining problems (for engineering)
Asking scientific questions about observations or content in texts helps to define problems and draw valid conclusions.

Developing and using models
Models can be used to represent a system or a part of a system. Using models can help to visualize a structure, process, or design and understand how it works. Models can also be used to improve a design.

Planning and carrying out investigations
Planning and carrying out investigations is an important part of independent research. Investigations allow ideas and models to be tested and refined.

Analyzing and interpreting data
Once data is collected, it must be analyzed to reveal any patterns or relationships. Tables and graphs are just two of the many ways to display and analyze data for trends.

Using mathematics and computational thinking
Mathematics is a tool for understanding scientific data. Converting or transforming data helps to see relationships more easily while statistical analysis can help determine the significance of the results.

Constructing explanations (for science) and designing solutions (for engineering)
Constructing explanations for observations and phenomena is a dynamic process and may involve drawing on existing knowledge as well as generating new ideas.

Engaging in argument from evidence
Scientific argument based on evidence is how new ideas gain acceptance in science. Logical reasoning based on evidence is required when considering the merit of new claims or explanations of phenomena.

Obtaining, evaluating, and communicating information
Evaluating information for scientific accuracy or bias is important in determining its validity and reliability. Communicating information includes reports, graphics, oral presentation, and models.

Crosscutting Concepts

P   **Patterns**
We see patterns everywhere in science. These guide how we organize and classify events and organisms and prompt us to ask questions about the factors that create and influence them.

CE   **Cause and effect**
A major part of science is investigating and explaining causal relationships. The mechanisms by which they occur can be tested in one context and used to explain and predict events in new contexts.

SPQ   **Scale, proportion, and quantity**
Different things are relevant at different scales. Changes in scale, proportion, or quantity affect the structure or performance of a system.

SSM   **Systems and system models**
Making a model of a system (e.g. physical, mathematical) provides a way to understand and test ideas.

EM   **Energy and matter**
Energy flows and matter cycles. Tracking these fluxes helps us understand how systems function.

SF   **Structure and function**
The structure of a substance or object determines many of its properties and functions.

SC   **Stability and change**
Science often deals with constructing explanations of how things change or how they remain stable.

X

Using BIOZONE's Resource Hub

▶ BIOZONE's Resource Hub provides links to online content that supports the activities in the book. From this page, you can also check for any corrections or clarifications to the book since printing.

▶ The Resource Hub provides a range of different resources to help explain or support the activity in the work text. They provide great support to help your understanding of a topic.

www.BIOZONEhub.com

Then enter the code in the text field   **NBI3-6924**

Or scan this QR code

Search for an activity here.

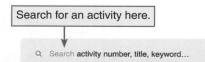

Q Search activity number, title, keyword…

BIOLOGY for NGSS

BIOZONE's Resource Hub provides links to online content that supports the activities in the book. From this page, you can also check for any errata or clarifications to the book or model answers since printing.

The external websites are, for the most part, narrowly focused animations and video clips directly relevant to some aspect of the activity on which they are cited. They provide great support to help your understanding.

Chapter title

→ **Chapter 1 - Science Practices**

Click on an activity title to go directly to the resources available for that activity.

- 1. How Do We Do Science
- 2. Systems and Systems Models
- 3. Observations, Hypotheses, and Assumptions
- 4. Accuracy and Precision
- 5. Working with Numbers
- 6. Tallies, Percentages, and Rates
- 7. Fractions and Ratios
- 8. Dealing with Large Numbers
- 9. Apparatus and Measurement
- 10. Types of Data
- 11. Variables and Controls
- 12. A Case Study: Catalase Activity
- 13. Recording Results

- 14. Practicing Data Manipulations
- 15. Constructing Tables
- 16. Which Graph to Use?
- 17. Drawing Line Graphs
- 18. Interpreting Line Graphs
- 19. Drawing Scatter Graphs
- 20. Correlation or Causation?
- 21. Drawing Bar Graphs
- 22. Drawing Histograms
- 23. Mean, Median, and Mode

View resources →

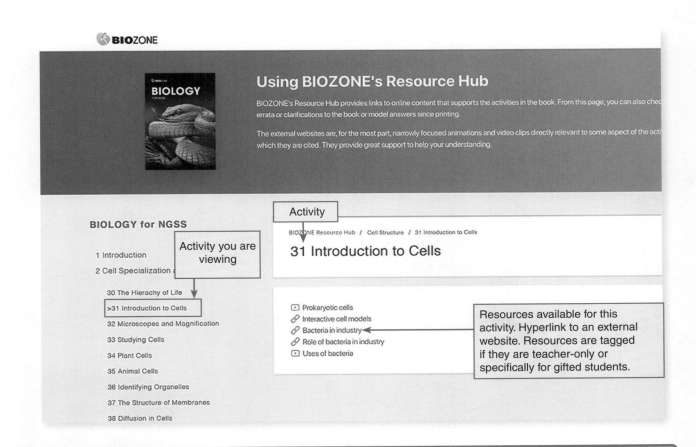

The Resource Hub icons

 Games
 Simulations
 Weblinks
 Slideshow
 3D Models
 PDF
 Spreadsheet
 Video
 Reference

Explore videos

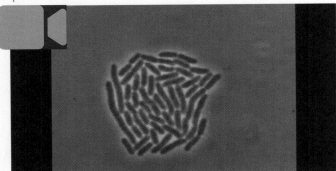

Explore spreadsheet modeling

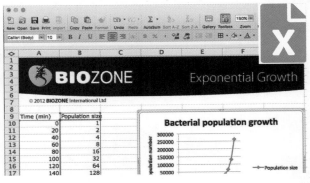

Explore web based resources

Explore 3D models

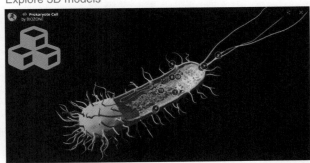

Life Sciences: A Flow of Ideas

This concept map shows the broad areas of content covered within each performance expectation of **Biology for NGSS.** The dark blue boxes indicate the book sections, each of which has its own concept map. The blue ovals are the chapters in each section. We have placed some major connections between topics, but you can make more of your own.

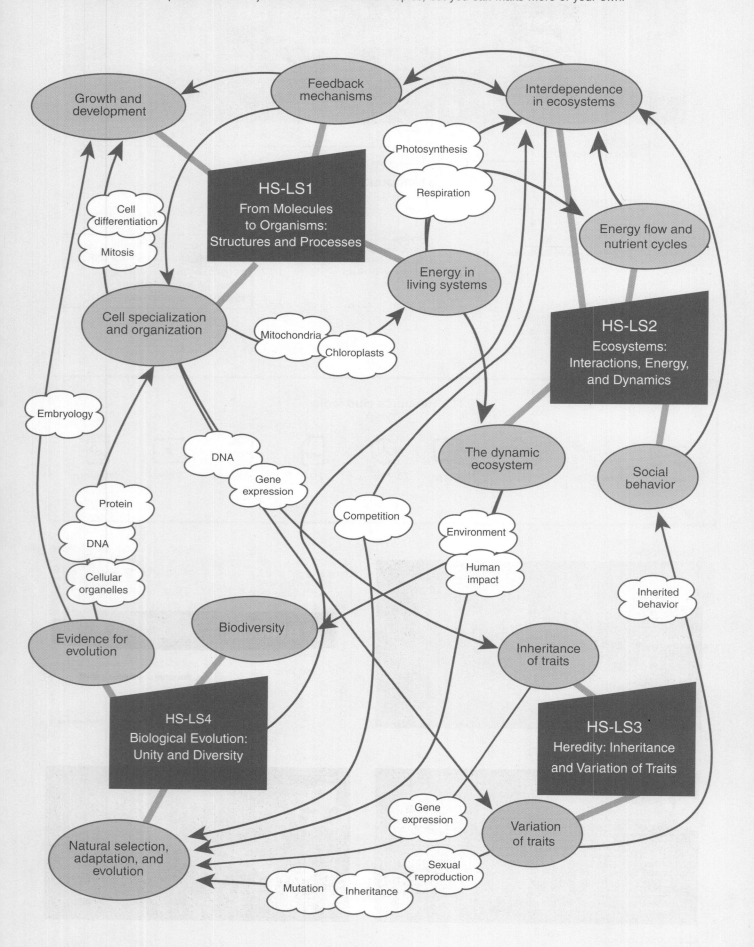

CHAPTER 1

Science Practices

Activity number

Science and engineering practices
Background in activities noted. Covered in following chapters in context.

Asking questions and defining problems

☐ 1 Demonstrate an understanding of science as inquiry. Appreciate that unexpected results may lead to new hypotheses and to new discoveries. **1**

☐ 2 Ask and evaluate questions that can be investigated with the resources available. Ask questions that arise from observation or examining models or theories, or to find out more information, determine relationships, or refine a model. **3 5**

Developing and using models

☐ 3 Develop and use models to describe systems or their components and how they work. **2**

☐ 4 Make accurate biological drawings to record important features of specimens. **26 27**

Planning and carrying out investigations

☐ 5 Plan and conduct investigations to test a hypothesis based on observations. Identify any assumptions in the design of your investigation. **12 28**

☐ 6 Consider and evaluate the accuracy and precision of the data that you collect. **4 9**

☐ 7 Use appropriate tools to collect and record data. Understand what is meant by quantitative, qualitative, and ranked data. **10 11 13**

☐ 8 Understand that variables are factors that can change or be changed in an experiment. Make and test hypotheses about the effect on a dependent variable when an independent variable is manipulated. Understand and use controls appropriately. **11 28**

Analyzing and interpreting data

☐ 9 Use graphs appropriate to the data to visualize data and identify trends. **16 17 19 21 22**

☐ 10 Summarize data and describe its features using descriptive statistics. **23 24 25**

☐ 11 Apply concepts of statistics and probability to answer questions and solve problems. **79 217**

☐ 12 Demonstrate an ability to use mathematics and computational tools to analyze, represent, and model data. Recognize and use appropriate units in calculations. **5 18 23**

☐ 13 Demonstrate an ability to apply ratios, rates, percentages, and unit conversions. **5 6 7 8**

Construct explanations and design solutions

☐ 14 Explain results based on evidence and applying scientific ideas and principles. **20 28**

Engage in argument from evidence

☐ 15 Use evidence to defend and evaluate claims and explanations about science. **28 59**

Obtain, evaluate, and communicate information

☐ 16 Evaluate the validity and reliability of designs, methods, claims, and evidence. **1 28**

1 How Do We Do Science?

Key Question: How does science, the rigorous, dynamic process of observation, investigation, and analysis, help us build an understanding of the world we live in?

▶ Science is a way of understanding the world we live in: how it formed, the rules it obeys and how it changes over time. Science distinguishes itself from other ways of understanding by using empirical standards, logical arguments, and skeptical review. Science allows our understanding to change over time as our knowledge increases.

▶ It is important to realize that science is a human endeavor and requires creativity and imagination. New research and ways of thinking can be based on the well-argued idea of a single person.

▶ Science influences, and is influenced by, society and technology. As society's beliefs and desires change, what is or can be researched is also affected. As technology advances, what is or can be researched changes. Scientific discoveries advance technology and can change society's beliefs.

▶ Science can never answer questions with absolute certainty. It can be confident of certain outcomes, but only within the limits of the data. Science might help us predict with 99.9% certainty that a system will behave in a certain way, but that still means there's one chance in a thousand it won't.

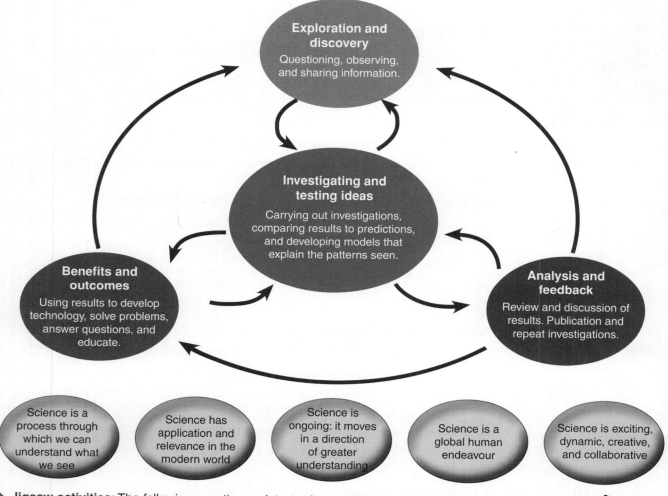

Exploration and discovery
Questioning, observing, and sharing information.

Investigating and testing ideas
Carrying out investigations, comparing results to predictions, and developing models that explain the patterns seen.

Benefits and outcomes
Using results to develop technology, solve problems, answer questions, and educate.

Analysis and feedback
Review and discussion of results. Publication and repeat investigations.

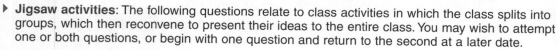

Science is a process through which we can understand what we see

Science has application and relevance in the modern world

Science is ongoing: it moves in a direction of greater understanding

Science is a global human endeavour

Science is exciting, dynamic, creative, and collaborative

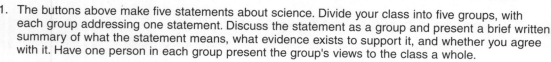

▶ **Jigsaw activities**: The following questions relate to class activities in which the class splits into groups, which then reconvene to present their ideas to the entire class. You may wish to attempt one or both questions, or begin with one question and return to the second at a later date.

1. The buttons above make five statements about science. Divide your class into five groups, with each group addressing one statement. Discuss the statement as a group and present a brief written summary of what the statement means, what evidence exists to support it, and whether you agree with it. Have one person in each group present the group's views to the class a whole.

2. The work of scientists in many disciplines has contributed to a deeper understanding of the processes involved in evolution. Have each person in the class choose a different example from the timeline opposite (in larger classes you may want to work in pairs). Find out about the work of the scientist or scientists involved, its significance to our understanding of evolution, and who their influences were. Summarize your findings in a report and present it to the class. Collate all the information to add more information to the timeline. How many people appear on the timeline now?

©2022 BIOZONE International
ISBN: 978-1-98-856692-4
Photocopying Prohibited

▶ Although Charles Darwin is largely credited with the development of the theory of evolution by natural selection, his ideas did not develop in isolation, but within the context of the work of others before him. The modern synthesis (below) has a long history, with contributors from all fields of science. Evolution by natural selection is one of the best substantiated theories in the history of science, supported by evidence from many disciplines, including paleontology, geology, genetics, and developmental biology.

▶ The diagram below summarizes just some of the important players in the story of evolutionary biology. This is not to say they were collaborators or always agreed. Some of the work, such as Haeckel's work on embryology, was flawed, and even untruthful.

▶ However, the work of many has contributed to a deeper understanding of evolutionary processes. This understanding continues to increase in the light of increasingly sophisticated molecular techniques and the collaboration of scientists internationally.

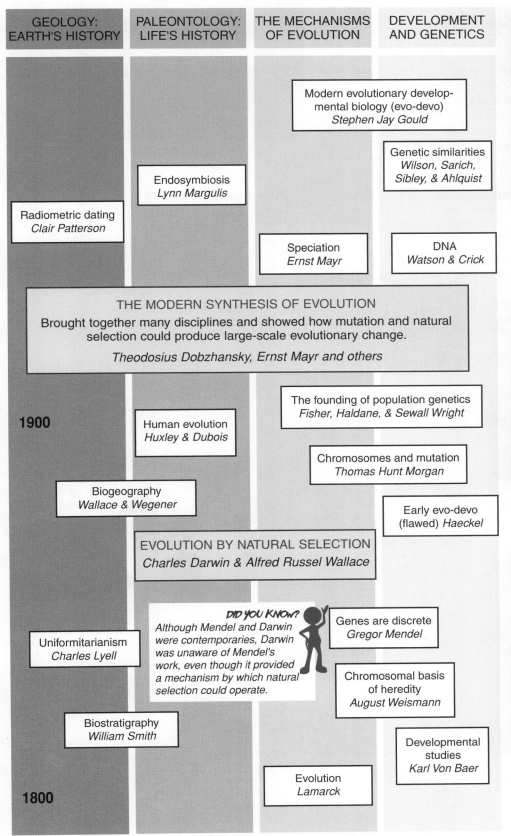

GEOLOGY: EARTH'S HISTORY | PALEONTOLOGY: LIFE'S HISTORY | THE MECHANISMS OF EVOLUTION | DEVELOPMENT AND GENETICS

Modern evolutionary developmental biology (evo-devo)
Stephen Jay Gould

Genetic similarities
Wilson, Sarich, Sibley, & Ahlquist

Endosymbiosis
Lynn Margulis

Radiometric dating
Clair Patterson

Speciation
Ernst Mayr

DNA
Watson & Crick

THE MODERN SYNTHESIS OF EVOLUTION
Brought together many disciplines and showed how mutation and natural selection could produce large-scale evolutionary change.

Theodosius Dobzhansky, Ernst Mayr and others

1900

Human evolution
Huxley & Dubois

The founding of population genetics
Fisher, Haldane, & Sewall Wright

Chromosomes and mutation
Thomas Hunt Morgan

Biogeography
Wallace & Wegener

Early evo-devo
(flawed) *Haeckel*

EVOLUTION BY NATURAL SELECTION
Charles Darwin & Alfred Russel Wallace

DID YOU KNOW?
Although Mendel and Darwin were contemporaries, Darwin was unaware of Mendel's work, even though it provided a mechanism by which natural selection could operate.

Genes are discrete
Gregor Mendel

Uniformitarianism
Charles Lyell

Chromosomal basis of heredity
August Weismann

Biostratigraphy
William Smith

Developmental studies
Karl Von Baer

Evolution
Lamarck

1800

Stephen Jay Gould (1941-2002)

Gould's work on the genetic triggers for development reinstated the credibility of embryological work. Today, "evo-devo" is providing some of the strongest evidence for how novel forms can rapidly arise.

Mayr (1904-2005)

Mayr worked on how species arise. He collaborated with Dobzhansky to formulate the modern evolutionary synthesis.

Darwin (1809-1882)

Darwin and Wallace independently proposed the theory of evolution by natural selection. Both amassed large amounts of evidence to support their theory.

©2022 **BIOZONE** International
ISBN: 978-1-98-856692-4
Photocopying Prohibited

2 Systems and System Models

Key Question: What are models and why do we use them in science?

▸ A system is a set of interrelated components that work together. Energy flow in ecosystems such as the one on the image on the right, gene regulation, interactions between organ systems, and feedback mechanisms are all examples of systems studied in biology.

▸ Scientists often used **models** to learn about biological systems. A model is a representation of a system and is useful for breaking a complex system down into smaller parts that can be studied more easily. Often, only part of a system is modelled. As scientists gather more information about a system, more **data** can be put into the model so that eventually it represents the real system more closely.

Modeling data

There are many different ways to model data. Often, seeing data presented in different ways can help to understand it better. Some common examples of models are shown here.

Visual models

Visual models can include drawings, such as these plant cells on the right.

Three dimensional models can be made out of materials such as modeling clay and sticks, like this model of a water molecule (below).

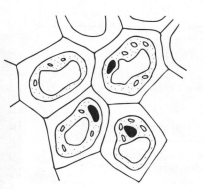

Mathematical models

Displaying data in a **graph** or as a mathematical equation, as shown below for logistic growth, often helps us to see relationships between different parts of a system.

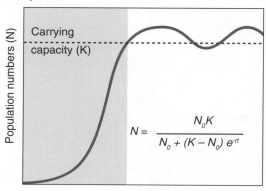

$$N = \frac{N_0 K}{N_0 + (K - N_0)\, e^{-rt}}$$

(graph axes: Population numbers (N) vs Time; Carrying capacity (K))

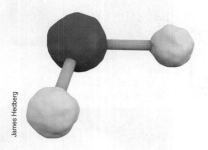

James Hedberg

Analogy

An analogy is a comparison between two things. Sometimes, comparing a biological system to an everyday object can help us to understand it better. For example, the heart pumps blood in blood vessels in much the same way a fire truck pumps water from a fire hydrant through a hose. Similarly, ATP is like a fully charged battery in a phone.

… a charged phone battery

ATP is like…

1. What is a system? _____

_____

2. (a) What is a model? _____

_____

(b) Why do scientists often study one part of a system rather than the whole system? _____

_____

_____

_____

SSM

©2022 BIOZONE International
ISBN: 978-1-98-856692-4
Photocopying Prohibited

3 Observations, Hypotheses, and Assumptions

Key Question: What is the importance of making observations, producing hypotheses, and recognizing assumptions?

Observations and hypotheses

▸ An **observation** is watching or recording what is happening. Observation is the basis for forming hypotheses and making **predictions**. An observation may generate a number of hypotheses (tentative explanations for what we see). Each **hypothesis** will lead to one or more predictions, which can be tested by investigation.

▸ A hypothesis is often written as a statement to include the prediction: "If X is true, then if I do Y (the experiment), I expect Z (the prediction)". Hypotheses are accepted, changed, or rejected on the basis of investigations. A hypothesis should have a sound theoretical basis and should be testable.

Observation 1:

• Some caterpillar species are brightly colored and appear to be highly visible to predators, such as insectivorous birds. Predators appear to avoid these caterpillars.

• These caterpillars are often found in groups, rather than as solitary animals.

Observation 2:

• Some caterpillar species have excellent camouflage. When alerted to danger they are difficult to see because they blend into the background.

• These caterpillars are usually found alone.

Assumptions

Any investigation requires you to make **assumptions** about the system you are working with. Assumptions are features of the system you are studying that you assume to be true, but that you do not (or cannot) test. Some assumptions about the examples above include:

• Insect eating birds have color vision.
• Caterpillars that look bright to us, also appear bright to insectivorous birds.
• Birds can learn about the taste of prey by eating them.

Read the two observations about the caterpillars above and then answer the following questions:

1. Generate a hypothesis to explain the observation that some caterpillars are brightly colored and highly visible, while others are camouflaged and blend into their surroundings:

   _____

   _____

2. Describe one of the assumptions being made in your hypothesis: _____

   _____

   _____

3. Generate a prediction about the behavior of insect eating birds towards caterpillars: _____

   _____

   _____

   _____

4 Accuracy and Precision

Key Question: What do accuracy and precision mean, how are they different, and why are they important when taking measurements?

The terms **accuracy** and **precision** are often used when talking about measurements.

▶ Accuracy refers to how close a measured value is to its true value, i.e. the correctness of the measurement.

▶ Precision refers to the closeness of repeated measurements to each other, i.e. the ability to be exact. For example, a digital device such as a pH meter (right) will give very precise measurements, but its accuracy depends on correct calibration.

Using the analogy of a target, repeated measurements are compared to arrows shot at a target. This analogy is useful when distinguishing between accuracy and precision.

Accurate but imprecise	Inaccurate and imprecise	Precise but inaccurate	Accurate and precise

The measurements are all close to the true value but quite spread apart.	The measurements are all far apart and not close to the true value.	The measurements are all clustered close together but not close to the true value.	The measurements are all close to the true value and also clustered close together.
Analogy: The arrows are all close to the bullseye.	**Analogy**: The arrows are spread around the target.	**Analogy**: The arrows are all clustered close together but not near the bullseye.	**Analogy**: The arrows are clustered close together near the bullseye.

Significant figures

Significant figures (sf) are the digits of a number that carry meaning contributing to its precision. They communicate how well you could actually measure the data.

For example, you might measure the height of 100 people to the nearest cm. When you calculate their mean height, the answer is 175.0215 cm. If you reported this number, it implies that your measurement technique was accurate to 4 decimal places. You would have to round the result to the number of significant figures you had accurately measured. In this instance the answer is 175 cm.

Non-zero numbers (1-9) are always **significant**.

All zeros between non-zero numbers are always **significant**.

$$0.005704510$$

Zeros to the left of the first non-zero digit after a decimal point are not significant.

Zeros at the end of number where there is a decimal place are **significant** (e.g. 4600.0 has five sf).
BUT
Zeros at the end of a number where there is no decimal point are not significant (e.g. 4600 has two sf).

1. Why are precise but inaccurate measurements not helpful in a biological investigation? _____

_____

_____

_____

2. State the number of significant figures in the following examples:

(a) 3.15985 _____   (d) 1000.0 _____

(b) 0.0012 _____   (e) 42.3006 _____

(c) 1000 _____   (f) 120 _____

©2022 BIOZONE International
ISBN: 978-1-98-856692-4
Photocopying Prohibited

5 Working With Numbers

Key Question: How is mathematical notation used, and how does converting and manipulating numbers make them easier to understand?

Commonly used mathematical symbols

In mathematics, universal symbols are used to represent mathematical concepts. They save time and space when writing. Some commonly used symbols are shown below.

=   Equal to

<   The value on the left is less than the value on the right

<<   The value on the left is much less than the value on the right

>   The value on the left is greater than the value on the right

>>   The value on the left is much greater than the value on the right

∝   Proportional to. A ∝ B means that A = a constant x B

~   Approximately equal to

Decimal and standard form

▶ Decimal form (also called ordinary form) is the longhand way of writing a number (e.g. 15,000,000). Very large or very small numbers can take up too much space if written in decimal form and are often expressed in a condensed standard form. For example, 15,000,000 is written as 1.5×10^7 in standard form.

▶ In standard form a number is always written as $A \times 10^n$, where A is a number between 1 and 10, and n (the exponent) indicates how many places to move the decimal point. n can be positive or negative.

▶ For the example above, A = 1.5 and n = 7 because the decimal point moved seven places (see below).

$$1{\overset{\frown}{5}}\,0{\overset{\frown}{0}}0{\overset{\frown}{0}}\,0{\overset{\frown}{0}}0{\overset{\frown}{0}} = 1.5 \times 10^7$$

▶ Small numbers can also be written in standard form. The exponent (n) will be negative. For example, 0.00101 is written as 1.01×10^{-3}.

$$0.\,0{\overset{\frown}{0}}1{\overset{\frown}{0}}1 = 1.01 \times 10^{-3}$$

▶ Converting can make calculations easier. Work through the following example to solve $4.5 \times 10^4 + 6.45 \times 10^5$.

1. Convert $4.5 \times 10^4 + 6.45 \times 10^5$ to decimal form:

   _____

2. Add the two numbers together: _____

3. Convert to standard form: _____

Estimates

▶ When carrying out calculations, typing the wrong number into your calculator can put your answer out by several orders of magnitude. An estimate is a way of roughly calculating what answer you should get, and helps you decide if your final calculation is correct.

▶ Numbers are often rounded to help make estimation easier. The rounding rule is, if the next digit is 5 or more, round up. If the next digit is 4 or less, it stays as it is.

▶ For example, to estimate 6.8 x 704 you would round the numbers to 7 x 700 = 4900. The actual answer is 4787, so the estimate tells us the answer (4787) is probably right.

Use the following examples to practise estimating:

4. 43.2 x 1044: _____

5. 3.4 x 72 ÷ 15: _____

6. 658 ÷ 22: _____

Conversion factors and expressing units

▶ Measurements can be converted from one set of units to another by using a conversion factor. This is a numerical factor that multiplies or divides one unit to convert it into another.

▶ Conversion factors are commonly used to convert non-SI units to SI units (e.g. converting pounds to kilograms). Note that mL and cm^3 are equivalent, as are L and dm^3.

In the space below, convert 5.6 cm^3 to mm^3 (1 cm^3 = 1000 mm^3):

7. _____

   _____

▶ The value of a variable must be written with its units where possible. SI units or their derivations should be used in recording measurements: volume in cm^3 (mL) or dm^3 (L), mass in kilograms (kg) or grams (g), length in meters (m), time in seconds (s). To denote 'per', you can use a solidus (/) or a negative exponent, e.g. per second is written as /s or s^{-1} and per meter squared is written as /m^2 or m^{-2}.

▶ For example the rate of oxygen consumption should be expressed as:

Oxygen consumption (mL/g/s) *or*
Oxygen consumption mL g^{-1} s^{-1})

6 Tallies, Percentages, and Rates

Key Question: How is unprocessed (raw) data manipulated or transformed to make it easier to understand and to identify important features?

▶ The **data** collected by measuring or counting in the field or laboratory is called **raw data**. Raw data often needs to be processed into a form that makes it easier to identify its important features, e.g. trends, and make meaningful comparisons between samples or treatments. Basic calculations, such as totals (the sum of all data values for a variable), are commonly used to compare treatments. Some common methods of processing data include creating tally charts, and calculating percentages and rates. These are explained below.

Tally Chart
Records the number of times a value occurs in a data set

HEIGHT (cm)	TALLY	TOTAL
0-0.99	III	3
1-1.99	++++ I	6
2-2.99	++++ ++++	10
3-3.99	++++ ++++ II	12
4-4.99	III	3
5-5.99	II	2

- A useful first step in analysis; a neatly constructed tally chart doubles as a simple histogram.
- Cross out each value on the list as you tally it, to prevent double entries.

Percentages
Expressed as a fraction of 100

Men	Body mass (kg)	Lean body mass (kg)	% lean body mass
Athlete	70	60	85.7
Lean	68	56	82.3
Normal weight	83	65	78.3
Overweight	96	62	64.6
Obese	125	65	52.0

- Percentages express what proportion of data fall into any one category, e.g. for pie charts.
- Allows meaningful comparison between different samples.
- Useful to monitor change, e.g. % increase from one year to the next.

Rates
Expressed as a measure per unit time

Time (minutes)	Cumulative sweat loss (mL)	Rate of sweat loss (mL/min)
0	0	0
10	50	5
20	130	8
30	220	9
60	560	11.3

- Rates show how a variable changes over a standard time period, e.g. one second, one minute, or one hour.
- Rates allow meaningful comparison of data that may have been recorded over different time periods.

Example: Height of 6 day old seedlings.

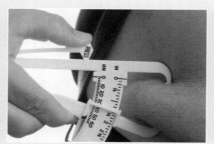

Example: Percentage of lean body mass in men.

Example: Rate of sweat loss during exercise in cyclists.

1. What is raw data? _____

_____

2. Why is it useful to process raw data and express it differently, e.g. as a rate or a percentage? _____

_____

_____

3. Identify the best data transformation in each of the following examples:

(a) Comparing harvest (in kg) of different grain crops from a farm: _____

(b) Comparing amount of water loss from different plant species: _____

7 Fractions and Ratios

Key Question: How are fractions and ratios used to provide a meaningful comparison of sample data where the sample sizes are different?

Fractions

- Fractions express how many parts of a whole are present.

- Fractions are expressed as two numbers separated by a solidus (/). For example, 1/2.

- The top number is the numerator. The bottom number is the denominator. The denominator can not be zero.

Simplifying fractions

- Fractions are often written in their simplest form (the top and bottom numbers cannot be any smaller, while still being whole numbers). Simplifying makes working with fractions easier.

- To simplify a fraction, the numerator and denominator are divided by the highest common number that divides into both numbers equally.

- For example, in a class of 10 students, two had blonde hair. This fraction is 2/10. To simplify this fraction 2 and 10 are divided by the highest common factor (2).

$$2 \div 2 = 1 \text{ and } 10 \div 2 = 5$$

- The simplified fraction is 1/5.

Adding fractions

- To add fractions, the denominators must be the same. If the denominators are the same the numerators are simply added, e.g. 5/12 + 3/12 = 8/12

- When the denominators are different one (or both) fractions must be multiplied to give a common denominator, e.g. 4/10 + 1/2. By multiplying 1/2 by 5 the fraction becomes 5/10. The fractions can now be added together (4/10 + 5/10 = 9/10).

Ratios

- Ratios give the relative amount of two or more quantities, i.e. it shows how much of one thing there is relative to another.

- Ratios provide an easy way to identify patterns.

- Ratios do not require units.

- Ratios are usually expressed as $a : b$.

- In the example below, there are 3 blue squares and 1 gray square. The ratio would be written as 3:1.

Calculating ratios

- Ratios are calculated by dividing all the values by the smallest number.

- Ratios are often used in Mendelian genetics to calculate phenotype (appearance) ratios. Some examples for pea plants are given below.

882 inflated pod   *299 constricted pod*

To obtain the ratio, divide both numbers by 299.
299 ÷ 299 = 1
882 ÷ 299 = 2.95
The ratio = 2.95 : 1

| 495 | 152 | 158 | 55 |
| round yellow | wrinkled yellow | round green | wrinkled green |

For the example above of pea seed shape and color, all of the values were divided by 55. The ratio obtained was:
9 : 2.8 : 2.9 : 1

1. (a) A student prepared a slide of the cells of an onion root tip and counted the cells at various stages in the cell cycle. The results are presented in the table (right). Calculate the ratio of cells in each stage (show your working):

_____

_____

Cell cycle stage	No. of cells counted	No. of cells calculated
Interphase	140	
Prophase	70	
Telophase	15	
Metaphase	10	
Anaphase	5	
Total	**240**	**4800**

   (b) Assuming the same ratio applies in all the slides examined in the class, calculate the number of cells in each phase for a cell total count of 4800.

2. Simplify the following fractions:

   (a) 3/9 : _____   (b) 84/90: _____   (c) 11/121: _____

3. In a class, 5/20 students had blue eyes. In another class, 5/12 students had blue eyes. What fraction of students had blue eyes in both classes combined?

_____

8 | Dealing with Large Numbers

Key Question: How does using logarithms or log-linear (semi-log) graphs make large scale changes in numerical data more manageable?

▶ In biology, numerical data indicating scale can often decrease or increase exponentially. Examples include the exponential growth of populations, exponential decay of radioisotopes, and the pH scale.

▶ Exponential changes in numbers are defined by a function. A function is simply a rule that allows us to calculate an output for any given input. Exponential functions are common in biology and may involve very large numbers.

▶ Log transformations of exponential numbers can make them easier to handle.

Exponential function

▶ Exponential growth occurs at an increasingly rapid rate in proportion to the growing total number or size.

▶ In an exponential function, the base number is fixed (constant) and the exponent is variable.

▶ The equation for an exponential function is $y = c^x$.

▶ Exponential growth and decay (reduction) are possible.

▶ Exponential changes in numbers are easy to identify because the curve has a J-shape appearance due to its increasing steepness over time.

▶ An example of exponential growth is the growth of a microbial population in an unlimiting, optimal growth environment.

Log transformations

▶ A log transformation makes very large numbers easier to work with. The log of a number is the exponent to which a fixed value (the base) is raised to get that number. So $\log_{10}(1000) = 3$ because $10^3 = 1000$.

▶ Both $\log_{10}$ (common logs) and $\log_e$ (natural logs or *ln*) are commonly used.

▶ Log transformations are useful for **data** where there is an exponential increase or decrease in numbers. In this case, the transformation will produce a straight line plot.

▶ To find the $\log_{10}$ of 32, using a calculator, key in log 32 = . The answer should be 1.51.

▶ Alternatively, the untransformed data can be plotted directly on a log-linear scale (as below). This is not difficult. You just need to remember that the log axis runs in exponential cycles. The paper makes the log for you.

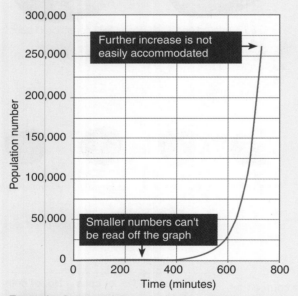

Example: Cell growth in a yeast culture where growth is not limited by lack of nutrients or build up of toxins.

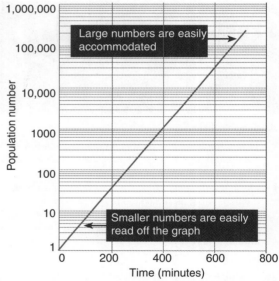

Example: The same yeast cell growth plotted on a log-linear scale. The y axis present 6 exponential cycles.

1. Why is it useful to plot exponential growth using semi-log paper? _____

_____

_____

2. What would you do to show yeast exponential growth as a straight line plot on normal graph paper?

_____

3. Log transformations are often used when a value of interest ranges over several orders of magnitude. Can you think of another example of data from the natural world where the data collected might show this behavior?

_____

©2022 BIOZONE International
ISBN: 978-1-98-856692-4
Photocopying Prohibited

9 Apparatus and Measurement

Key Question: Why must the apparatus used in experimental work be appropriate for the experiment or analysis and be used correctly?

Selecting the correct equipment

It is important that you choose equipment that is appropriate for the type of measurement you want to take. For example, if you wanted to accurately weigh out 5.65 g of sucrose, you need a balance that accurately weighs to two decimal places. A balance that weighs to only one decimal place would not allow you to make an accurate enough measurement.

Study the glassware (right). Which would you use if you wanted to measure 225 mL? The graduated cylinder has graduations every 10 mL, whereas the beaker has graduations every 50 mL. It would be more accurate to measure 225 mL in a graduated cylinder.

Percentage errors

Percentage error is a way of mathematically expressing how far out your result is from the ideal result. The equation for measuring percentage error is:

$$\frac{\text{experimental value - ideal value}}{\text{ideal value}} \times 100$$

For example, to determine the **accuracy** of a 5 mL pipette, dispense 5 mL of water from the pipette and weigh the dispensed volume on a balance. The mass (g) = volume (mL). The volume is 4.98 mL.

$$\frac{\text{experimental value (\textbf{4.98}) - ideal value (\textbf{5.0})}}{\text{ideal value (\textbf{5.0})}} \times 100$$

The percentage error = –0.4% (the negative sign tells you the pipette is dispensing **less** than it should).

Recognizing potential sources of error

It is important to know how to use equipment correctly to reduce errors. A spectrophotometer measures the amount of light absorbed by a solution at a certain wavelength. This information can be used to determine the concentration of the absorbing molecule (e.g. density of bacteria in a culture). The more concentrated the solution, the more light is absorbed. Incorrect use of the spectrophotometer can alter the results. Common mistakes include incorrect calibration, errors in sample preparation, and errors in sample measurement.

A cuvette (left) is a small clear tube designed to hold spectrophotometer samples. Inaccurate readings occur when:

- The cuvette is dirty or scratched (light is absorbed giving a falsely high reading).

- Some cuvettes have a frosted side to aid alignment. If the cuvette is aligned incorrectly, the frosted side absorbs light, giving a false reading.

- Not enough sample is in the cuvette and the beam passes over, rather than through the sample, giving a lower absorbance reading.

1. Assume that you have the following measuring devices available: 50 mL beaker, 50 mL graduated cylinder, 25 mL graduated cylinder, 10 mL pipette, 10 mL beaker. What would you use to accurately measure:

   (a) 21 mL: _____ (b) 48 mL: _____ (c) 9 mL: _____

2. Calculate the percentage error for the following situations (show your working):

   (a) A 1 mL pipette delivers a measured volume of 0.98 mL: _____

   _____

   (b) A 10 mL pipette delivers a measured volume of 9.98 mL: _____

   _____

10 Types of Data

Key Question: What types of data may be collected during an investigation?

Data is information collected during an investigation. Data may be quantitative, qualitative, or ranked. When planning a biological investigation, it is important to consider the type of data that will be collected. It is best to collect quantitative or numerical data, because it is easier to analyze it objectively (without bias).

Types of Data

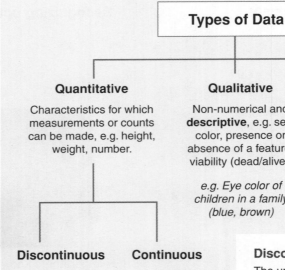

Quantitative

Characteristics for which measurements or counts can be made, e.g. height, weight, number.

Qualitative

Non-numerical and **descriptive**, e.g. sex, color, presence or absence of a feature, viability (dead/alive).

e.g. Eye color of children in a family (blue, brown)

Ranked

Data which can be ranked on a scale that represents an order, e.g. abundance (abundant, common, rare); color (dark, medium, pale).

e.g. Birth order in a family (1, 2, 3)

Discontinuous

e.g. Number of children in a family (3, 0, 4)

Continuous

e.g. Height of children in a family (1.5 m, 0.8 m)

Discontinuous or discrete data:

The unit of measurement cannot be split up (e.g. can't have half a child).

Continuous data:

The unit of measurement can be a part number (e.g. 5.25 kg).

A: Skin color

B: Eggs per nest

C: Bacterial colony diameter

1. For each of the photographic examples A-C above, classify the data as quantitative, ranked, or qualitative:

   (a) Skin color: _____

   (b) Number of eggs per nest: _____

   (c) Bacterial colony diameter: _____

2. Why is it best to collect quantitative data where possible in biological studies? _____

   _____

   _____

   _____

   _____

3. Give an example of data that could not be collected quantitatively, and explain your answer: _____

   _____

   _____

   _____

   _____

©2022 BIOZONE International
ISBN: 978-1-98-856692-4
Photocopying Prohibited

11 Variables and Controls

Key Question: What are dependent, independent, or controlled variables, and how are they used in an experiment?

Types of variables

A **variable** is a factor that can be changed during an experiment, e.g. temperature. Investigations often look at how changing one variable affects another.

There are several types of variables:

- Independent
- Dependent
- Controlled

Only one variable should be changed at a time. Any changes seen are a result of the changed variable.

Remember! The **dependent variable** is "dependent" on the **independent variable**.

Example: *When heating water, the temperature of the water depends on the time it is heated for. Temperature (dependent variable) depends on time (independent variable).*

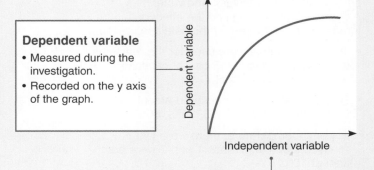

Dependent variable
- Measured during the investigation.
- Recorded on the y axis of the graph.

Controlled variable
- Factors that are kept the same.

Independent variable
- Set by the experimenter, it is the variable that is changed.
- Recorded on the graph's x axis.

Experimental controls

▶ A **control** is the standard or reference treatment in an experiment. Controls make sure that the results of an experiment are due to the variable being tested, e.g. nutrient level, and not due to another factor such as the equipment not working correctly.

▶ A control is identical to the original experiment except that it lacks the altered variable. The control undergoes the same preparation, experimental conditions, **observations**, measurements, and analysis as the test group.

▶ If the control works as expected, it means the experiment has run correctly, and the results are due to the effect of the variable being tested.

Test plant
(nutrient added)

Control plant
(no nutrient added)

An experiment was designed to test the effect of a nutrient on plant growth. The control plant had no nutrient added to it. Its growth sets the baseline for the experiment. Any growth in the test plant greater than that seen in the control plant is due to the added nutrient.

1. What is the difference between a dependent variable and an independent variable? _____

_____

_____

_____

2. Why do we control the variables we are not investigating? _____

_____

_____

3. What is the purpose of the experimental control? _____

_____

12 A Case Study: Catalase Activity

Key Question: How does manipulating one variable (the independent variable) affect the response of the dependent variable?

Investigation: catalase activity

Catalase is an enzyme that converts hydrogen peroxide (H_2O_2) to oxygen and water. An experiment investigated the effect of temperature on the rate of the catalase reaction.

- 10 cm³ test tubes were used for the reactions, each tube contained 0.5 cm³ of catalase enzyme and 4 cm³ of H_2O_2.

- Reaction rates were measured at four temperatures (10°C, 20°C, 30°C, 60°C).

- For each temperature, there were two reaction tubes, e.g. tubes 1 and 2 were both kept at 10°C.

- The height of oxygen bubbles present after one minute of reaction was used as a measure of the reaction rate. A faster reaction rate produced more bubbles than a slower reaction rate.

- The entire experiment was repeated on two separate days.

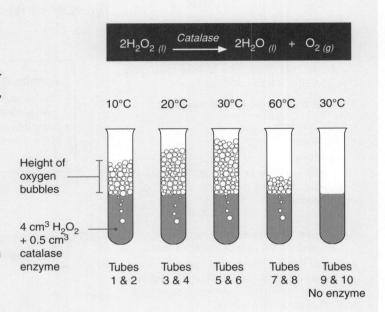

$$2H_2O_{2\ (l)} \xrightarrow{\text{Catalase}} 2H_2O_{\ (l)} + O_{2\ (g)}$$

10°C   20°C   30°C   60°C   30°C

Height of oxygen bubbles

4 cm³ H_2O_2 + 0.5 cm³ catalase enzyme

Tubes 1 & 2   Tubes 3 & 4   Tubes 5 & 6   Tubes 7 & 8   Tubes 9 & 10 No enzyme

1. Write a suitable aim for this experiment: _____

_____

_____

2. Write a hypothesis for this experiment: _____

_____

_____

_____

3. (a) What is the independent variable in this experiment? _____

   (b) What is the range of values for the independent variable? _____

   (c) Name the unit for the independent variable: _____

   (d) List the equipment needed to set the independent variable, and describe how it was used: _____

_____

_____

4. (a) What is the dependent variable in this experiment? _____

   (b) Name the unit for the dependent variable: _____

   (c) List the equipment needed to measure the dependent variable, and describe how it was used: _____

_____

_____

_____

5. Which tubes are the control for this experiment? _____

©2022 BIOZONE International
ISBN: 978-1-98-856692-4
Photocopying Prohibited

13 Recording Results

Key Question: How does accurately recording results (using tables or data loggers) make it easier to understand and analyze your data later?

Ways to record data

▶ Recording your results accurately is very important in any type of scientific investigation. If you have recorded your results accurately and in an organized way, it makes analyzing and understanding your **data** easier. Log books and dataloggers are two methods by which data can be recorded.

Log books

A log book records your ideas and results throughout your scientific investigation. It also provides proof that you have carried out the work.

- An A4 lined exercise book is a good choice for a log book. It gives enough space to write ideas and record results and provides space to paste in photos or extra material such as printouts.
- Each entry must have the date recorded.
- Make sure that you can read what you write at a later date. A log book entry is meaningless if it is incomplete or cannot be read.

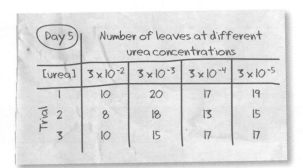

Day 5	Number of leaves at different urea concentrations			
[urea]	3×10^{-2}	3×10^{-3}	3×10^{-4}	3×10^{-5}
Trial 1	10	20	17	19
Trial 2	8	18	13	15
Trial 3	10	15	17	17

If you are not using a datalogger, a **table** (above) is often a good way to record and present your results as you collect them. Tables can also be useful for showing calculated values (such as rates and means). Recording data in a table as your experiment proceeds, lets you identify any trends early on and change experimental conditions if necessary.

Dataloggers

A datalogger (also called a data recorder) is an electronic device that automatically records data over time.

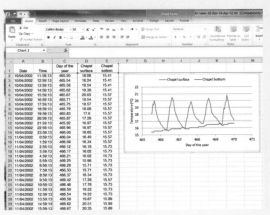

- Dataloggers have a variety of sensors to measure different physical properties. Common sensors include light, temperature, pH, conductivity, and humidity.
- Dataloggers can be used in both field or laboratory experiments, and can be left to collect data without the experimenter being present.
- Information collected by the datalogger can be downloaded to a computer (below) so that the data can be accessed and analyzed.

1. Why is it important to accurately record your results? _____

_____

_____

2. Why must log book entries be well organized? _____

_____

_____

3. (a) What is a datalogger? _____

_____

    (b) What are the advantages of using a datalogger over manually recording results? _____

_____

_____

14 Practicing Data Manipulations

Key Question: How are percentages, rates, and frequencies used to manipulate raw data?

1. Complete the transformations for each of the tables on the right. The first value, and the working, is provided for each example.

(a) TABLE: Incidence of red clover in different areas:

Working: 124 ÷ 159 = 0.78 = 78%

> This is the number of red clover out of the total.

Incidence of red and white clover in different areas

Clover plant type	Frost free area		Frost prone area		Totals
	Number	%	Number	%	
Red	124	78	26		
White	35		115		
Total	159				

(b) TABLE: Plant water loss using a bubble potometer:

Working: (9.0 − 8.0) ÷ 5 min = 0.2

> This is the distance the bubble moved over the first 5 minutes. Note that there is no **data** entry possible for the first reading (0 min) because no difference can be calculated.

Plant water loss using a bubble potometer

Time (min)	Pipette arm reading (cm³)	Plant water loss (cm³/min)
0	9.0	–
5	8.0	0.20
10	7.2	
15	6.2	
20	4.9	

(c) TABLE: Frequency of size classes in a sample of eels:

Working: (7 ÷ 270) x 100 = 2.6 %

> This is the number of individuals out of the total that appear in the size class 0-50 mm. The relative frequency is rounded to one decimal place.

Frequency of size classes in a sample of eels

Size class (mm)	Frequency	Relative frequency (%)
0-50	7	2.6
50-99	23	
100-149	59	
150-199	98	
200-249	50	
250-299	30	
300-349	3	
Total	270	

(d) TABLE: Body composition in women:

Working: (38 ÷ 50) x 100 = 76 %

> This is lean body mass. The percentage lean body mass is calculated by dividing lean body mass by total body mass. It is multiplied by 100 to convert it into a percentage.

Body mass composition in women

Women	Body mass (kg)	Lean body mass (kg)	% lean body mass
Athlete	50	38	76
Lean	56	41	
Normal weight	65	46	
Overweight	80	48	
Obese	95	52	

©2022 BIOZONE International
ISBN: 978-1-98-856692-4
Photocopying Prohibited

15 Constructing Tables

Key Question: What is the purpose of recording data in an organized table during an experiment?

▸ **Tables** are used to record **data** during an investigation. Your log book should present neatly tabulated data (right).

▸ Tables allow a large amount of information to be condensed, and can provide a summary of the results.

▸ Presenting data in tables allows you to organize your data in a way that allows you to more easily see the relationships and trends.

▸ Columns can be provided to display the results of any data transformations such as rates. Basic **descriptive statistics** (such as **mean** or standard deviation) may also be included.

▸ Complex data sets tend to be graphed rather than tabulated.

Features of tables

Tables should have an accurate, descriptive title. Number tables consecutively through a report.

Heading and subheadings identify each set of data and show units of measurement.

Table 1: Length and growth of the third internode of bean plants receiving three different hormone treatments.

Independent variable in the left column.

Treatment	Sample size	Mean rate of internode growth (mm/day)	Mean internode length (mm)	Mean mass of tissue added (g/day)
Control	50	0.60	32.3	0.36
Hormone 1	46	1.52	41.6	0.51
Hormone 2	98	0.82	38.4	0.56
Hormone 3	85	2.06	50.2	0.68

Control values should be placed at the beginning of the table.

Each row should show a different experimental treatment, organism, sampling site etc.

Columns for comparison should be placed alongside each other. Show values only to the level of significance allowable by your measuring technique.

Organize the columns so that each category of like numbers or attributes is listed vertically.

1. What are two advantages of using a table format for data presentation?

(a) _____

_____

(b) _____

_____

2. Why might you tabulate data before you presented it in a graph? _____

_____

_____

_____

16 Which Graph to Use?

Key Question: How does the type of data you collected affect the type of graph you should choose to display your data?

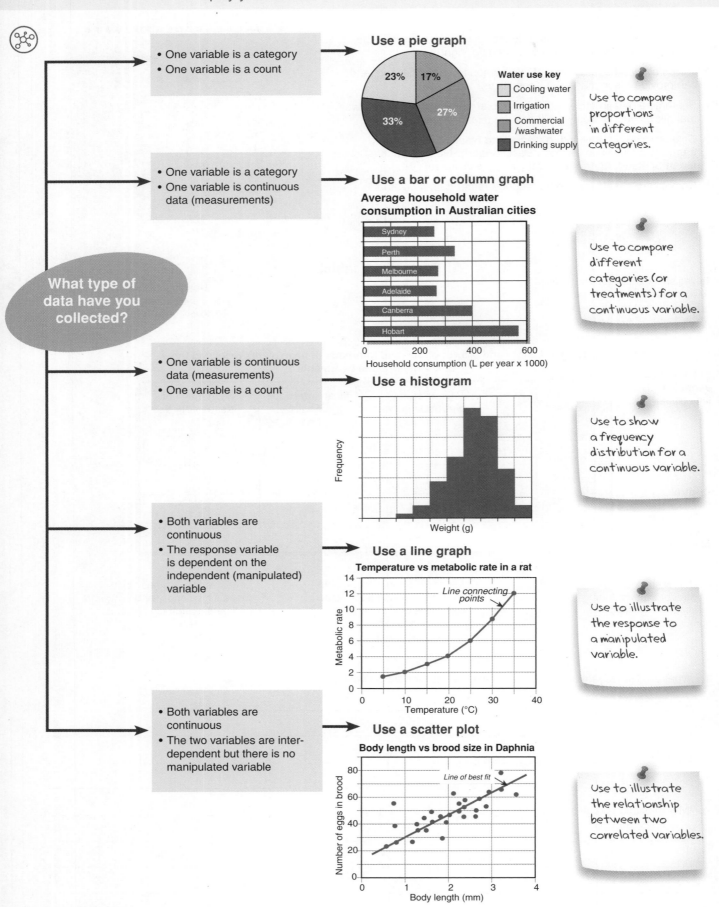

What type of data have you collected?

- One variable is a category
- One variable is a count

Use a pie graph

Water use key
- Cooling water
- Irrigation
- Commercial /washwater
- Drinking supply

17% | 27% | 33% | 23%

Use to compare proportions in different categories.

- One variable is a category
- One variable is continuous data (measurements)

Use a bar or column graph

Average household water consumption in Australian cities

Sydney, Perth, Melbourne, Adelaide, Canberra, Hobart

Household consumption (L per year x 1000)

Use to compare different categories (or treatments) for a continuous variable.

- One variable is continuous data (measurements)
- One variable is a count

Use a histogram

Frequency vs Weight (g)

Use to show a frequency distribution for a continuous variable.

- Both variables are continuous
- The response variable is dependent on the independent (manipulated) variable

Use a line graph

Temperature vs metabolic rate in a rat

Line connecting points

Metabolic rate vs Temperature (°C)

Use to illustrate the response to a manipulated variable.

- Both variables are continuous
- The two variables are inter-dependent but there is no manipulated variable

Use a scatter plot

Body length vs brood size in Daphnia

Line of best fit

Number of eggs in brood vs Body length (mm)

Use to illustrate the relationship between two correlated variables.

©2022 BIOZONE International
ISBN: 978-1-98-856692-4
Photocopying Prohibited

17 Drawing Line Graphs

Key Question: What kind of data is plotted on line graphs, and how do they show the relationship between the independent variable and the dependent variable?

Graphs provide a way to visually see **data** trends. Line graphs are used when one variable (the **independent variable**) affects another, the **dependent variable**. Important features of line graphs are:

▶ The data must be continuous for both variables.

▶ The dependent variable is usually a biological response.

▶ The independent variable is often time or the experimental treatment.

▶ The relationship between two variables can be represented as a continuum and the data points are plotted accurately and connected directly (point to point).

▶ Line graphs may be drawn with a measure of error. The data are presented as points (the calculated **means**), with bars above and below, indicating a measure of variability or spread in the data, e.g. standard deviation.

▶ More than one curve can be plotted per set of axes. If the two data sets use the same measurement units and a similar range of values for the dependent variable, one scale on the y axis is used. If the two data sets use different units and/or have a very different range of values for the dependent variable, two scales for the y axis are used (see right). Distinguish between the two curves with a key.

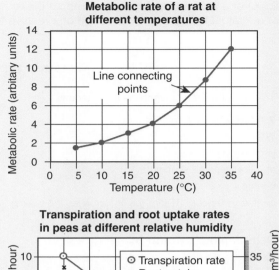

Metabolic rate of a rat at different temperatures

Line connecting points

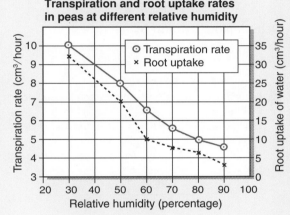

Transpiration and root uptake rates in peas at different relative humidity

⊙ Transpiration rate
× Root uptake

1. The results (shown right) were collected in a study investigating the effect of temperature on the activity of an enzyme.

   (a) Using the results provided, plot a line graph on the grid below:

   (b) Estimate the rate of reaction at 15°C: _____

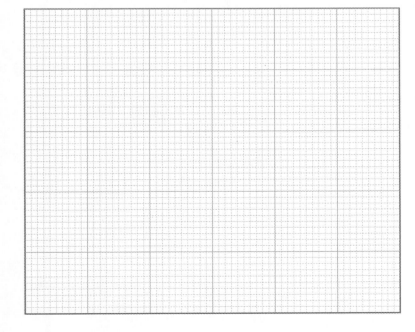

Lab Notebook

An enzyme's activity at different temperatures

Temperature (°C)	Rate of reaction (mg of product formed per minute)
10	1.0
20	2.1
30	3.2
35	3.7
40	4.1
45	3.7
50	2.7
60	0

©2022 BIOZONE International
ISBN: 978-1-98-856692-4
Photocopying Prohibited

18 Interpreting Line Graphs

Key Question: What is the equation for a straight line? What is the significance of a line with a positive, negative, or zero slope?

▸ The equation for a linear (straight) line on a graph is y = mx + c. The equation can be used to calculate the gradient (slope) of a straight line and tells us about the relationship between x and y (how fast y is changing relative to x). For a straight line, the rate of change of y relative to x is always constant.

Measuring gradients and intercepts

The equation for a straight line is written as:

y = mx + c

Where :

y = the y-axis value

m = the slope (or gradient)

x = the x-axis value

c = the y intercept (where the line crosses the y-axis).

Determining "m" and "c"

To find "c" just find where the line crosses the y-axis.

To find "m":

1. Choose any two points on the line.

2. Draw a right-angled triangle between the two points on the line.

3. Use the scale on each axis to find the triangle's vertical length and horizontal length.

4. Calculate the gradient of the line using the following equation:

$$\frac{\text{change in y}}{\text{change in x}}$$

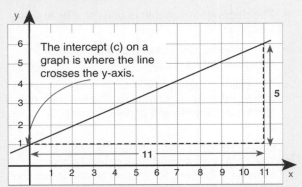

For the example above:

c = 1

m = 0.45 (5 ÷11)

Once c and m have been determined you can choose any value for x and find the corresponding value for y.

For example, when x = 9, the equation would be:

y = 9 x 0.45 + 1

y = 5.05

A line may have a positive, negative, or zero slope

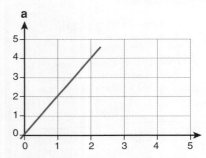

a

Positive gradients: the line slopes upward to the right (y is increasing as x increases).

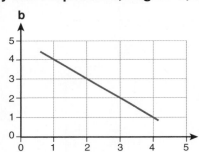

b

Negative gradients: the line slopes downward to the right (y is decreasing as x increases).

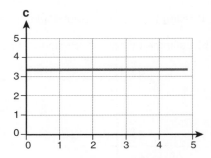

c

Zero gradients: the line is horizontal (y does not change as x increases).

1. For the graph (right):

(a) Identify the value of c: _____

(b) Calculate the value of m: _____

(c) Determine y if x = 2: _____

_____

(d) Describe the slope of the line: _____

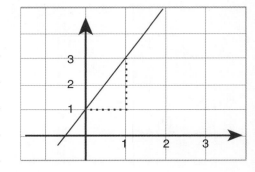

©2022 BIOZONE International
ISBN: 978-1-98-856692-4

19 Drawing Scatter Graphs

Key Question: How does a scatter graph show continuous data where there is a relationship between two interdependent variables?

Scatter graphs are used to display continuous **data** where there is a relationship between two interdependent **variables**.

▸ The data must be continuous for both variables.

▸ There is no independent (manipulated) variable, but the variables are often correlated, i.e. they vary together in some predictable way.

▸ Scatter graphs are useful for determining the relationship between two variables.

▸ The points on the graph should not be connected, but a line of best fit is often drawn through the points to show the relationship between the variables.

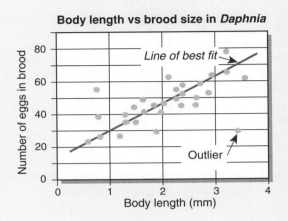

Body length vs brood size in *Daphnia*

1. In the example below, metabolic measurements were taken from seven Antarctic fish *Pagothenia borchgrevinski*. The fish are affected by a gill disease, which increases the thickness of the gas exchange surfaces and affects oxygen uptake. The results of oxygen consumption of fish with varying amounts of affected gill (at rest and swimming) are tabulated below.

(a) Plot the data on the grid (bottom right) to show the relationship between oxygen consumption and the amount of gill affected by disease. Use different symbols or colors for each set of data (at rest and swimming), and use only one scale for oxygen consumption.

(b) Draw a line of best fit through each set of points.

2. Describe the relationship between the amount of gill affected and oxygen consumption in the fish:

(a) For the "at rest" data set:

_____

_____

_____

_____

_____

(b) For the swimming data set:

_____

_____

_____

_____

_____

_____

_____

Oxygen consumption of fish with affected gills

Fish number	Percentage of gill affected	Oxygen consumption (cm³/g/hour)	
		At rest	Swimming
1	0	0.05	0.29
2	95	0.04	0.11
3	60	0.04	0.14
4	30	0.05	0.22
5	90	0.05	0.08
6	65	0.04	0.18
7	45	0.04	0.20

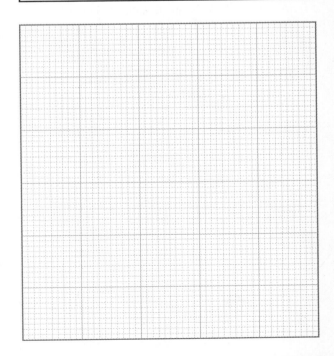

©2022 BIOZONE International
ISBN: 978-1-98-856692-4

20 Correlation or Causation?

Key Question: What does correlation mean, and why can you not assume a correlation is the result of causation?

Correlation does not imply causation

▶ You may come across the phrase "correlation does not necessarily imply causation". This means that even when there is a strong correlation between variables (they vary together in a predictable way), you cannot assume that change in one variable caused change in the other.

▶ **Example**: When data from the organic food association and the office of special education programmes is plotted (below), there is a strong correlation between the increase in organic food and rates of diagnosed autism. However, it is unlikely that eating organic food causes autism, so we can not assume a causative effect here.

Drawing the line of best fit

Some simple guidelines need to be followed when drawing a line of best fit on your scatter plot.

▶ Your line should follow the trend of the data points.

▶ Roughly half of your data points should be above the line of best fit, and half below.

▶ The line of best fit does not necessarily pass through any particular point.

▶ A line of best fit should pivot around the point representing the mean of the x and y variables.

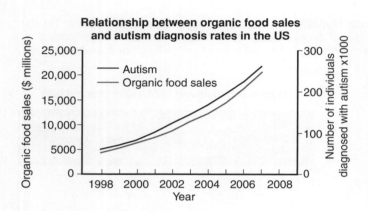

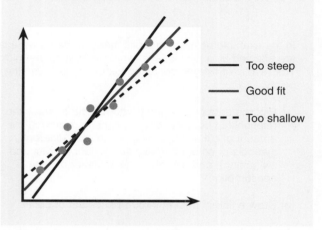

1. What does the phrase "correlation does not imply causation" mean? _____

_____

_____

2. A student measured the hand span and foot length measurements of 21 adults and plotted the data as a scatter graph (right).

   (a) Draw a line of best fit through the data:

   (b) Describe the results: _____

   _____

   _____

   _____

   _____

   (c) Using your line of best fit as a guide, comment on the correlation between hand span and foot length:

   _____

   _____

   _____

   _____

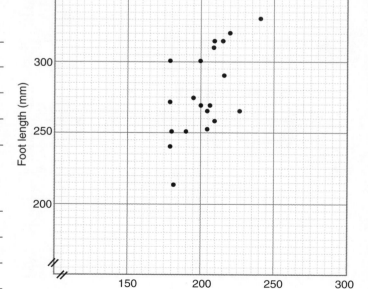

©2022 BIOZONE International
ISBN: 978-1-98-856692-4
Photocopying Prohibited

21 Drawing Bar Graphs

Key Question: What kind of data is shown on bar graphs?

Bar graphs are appropriate for **data** that is non-numerical and discrete for at least one **variable**.

▸ There are no dependent or **independent variables**.

▸ Data is collected for discontinuous, non-numerical categories (e.g. place, color, and species), so the bars do not touch.

▸ Multiple sets of data can be displayed side by side for direct comparison.

▸ Axes may be reversed, i.e. the bars can be vertical or horizontal. When they are vertical, these **graphs** are called column graphs.

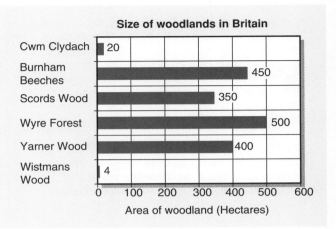

1. Counts of eight mollusk species were made from a series of quadrat samples at two sites on a rocky shore. The summary data are presented on the right.

   (a) Tabulate the **mean** (average) numbers per square meter at each site in the table (below).

   (b) Plot a bar graph of the tabulated data on the grid below. For each species, plot the data from both sites side by side using different colors to distinguish the two sites.

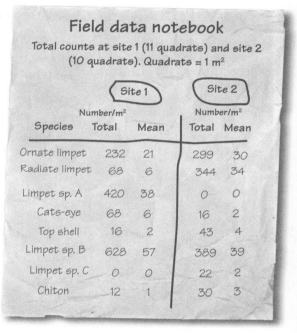

Average abundance of 8 mollusk species from two sites along a rocky shore

Species	Mean (no./m²)	
	Site 1	Site 2

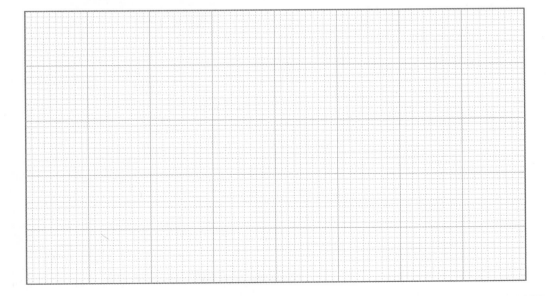

22 Drawing Histograms

Key Question: What kind of data is shown on a histogram?

Histograms are plots of continuous **data** and are often used to represent frequency distributions, where the y-axis shows the number of times a particular measurement or value was obtained. For this reason, they are often called frequency histograms. Important features of histograms include:

▸ The data are numerical and continuous, e.g. height or weight, so the bars touch.

▸ The x-axis usually records the class interval. The y-axis usually records the number of individuals in each class interval (frequency).

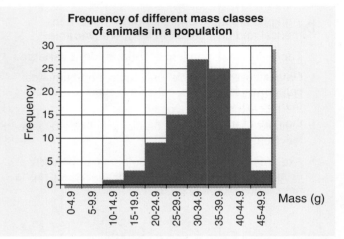

1. The weight data provided below were recorded from 95 individuals (male and female), older than 17 years.

(a) Create a tally chart (frequency table) in the table provided (right). An example of the tally for the weight grouping 55-59.9 kg has been completed for you. Note that the **raw data** values, once they are recorded as counts on the tally chart, are crossed off the data set in the notebook. It is important to do this in order to prevent data entry errors.

(b) Plot a frequency histogram of the tallied data on the grid below.

Weight (kg)	Tally	Total
45-49.9		
50-54.9		
55-59.9	THL II	7
60-64.9		
65-69.9		
70-74.9		
75-79.9		
80-84.9		
85-89.9		
90-94.9		
95-99.9		
100-104.9		
105-109.9		

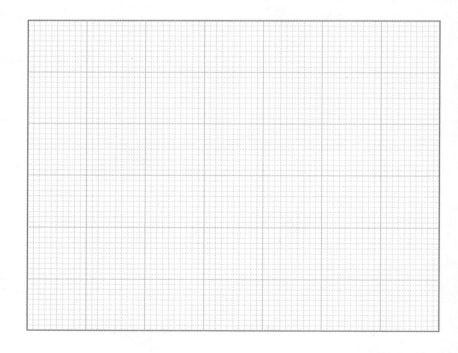

Lab notebook

Weight (in kg) of 95 individuals

63.4	81.2	65
56.5	83.3	75.6
84	95	76.8
81.5	105.5	67.8
73.4	82	68.3
56	73.5	63.5
60.4	75.2	58
83.5	63	58.5
82	70.4	50
61	82.2	92
55.2	87.8	91.5
48	86.5	88.3
53.5	85.5	81
63.8	87	72
69	98	66.5
82.8	71	61.5
68.5	76	66
67.2	72.5	65.5
82.5	61	67.4
83	60.5	73
78.4	67	67
76.5	86	71
83.4	85	70.5
77.5	93.5	65.5
77	62	68
87	62.5	90
89	63	83.5
93.4	60.	73
83	71.5	66
80	73.8	57.5
76	77.5	76
56	74	

©2022 BIOZONE International
ISBN: 978-1-98-856692-4
Photocopying Prohibited

23 Mean, Median, and Mode

Key Question: What are descriptive statistics and how are they used to summarize a data set and describe its basic features?

Descriptive statistics

▶ When we describe a set of **data**, it is usual to give a measure of central tendency. This is a single value identifying the central position within that set of data.

▶ **Descriptive statistics**, such as **mean**, **median**, and **mode**, are all valid measures of central tendency depending on the type of data and its distribution. They help to summarize features of the data, so are often called summary statistics.

▶ The appropriate statistic for different types of data variables and their distributions is described below.

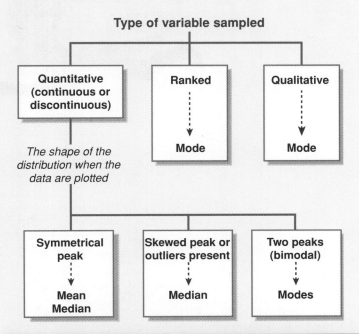

Distribution of data

Variability in continuous data is often displayed as a frequency distribution. There are several types of distribution.

- Normal distribution (A): Data has a symmetrical spread about the mean. It has a classical bell shape when plotted.
- Skewed data (B): Data is not centered around the middle but has a "tail" to the left or right.
- Bimodal data (C): Data which has two peaks.

The shape of the distribution will determine which statistic (mean, median, or mode) should be used to describe the central tendency of the sample data.

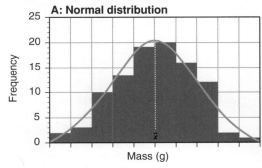

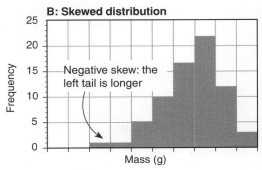

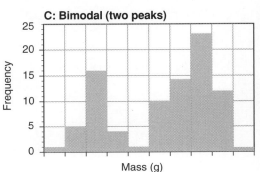

Statistic	Definition and when to use it	How to calculate it
Mean	• The average of all data entries. • Measure of central tendency for normally distributed data.	• Add up all the data entries. • Divide by the total number of data entries.
Median	• The middle value when data entries are placed in rank order. • A good measure of central tendency for skewed distributions.	• Arrange the data in increasing rank order. • Identify the middle value. • For an even number of entries, find the mid point of the two middle values.
Mode	• The most common data value. • Suitable for bimodal distributions and qualitative (categorical) data.	• Identify the category with the highest number of data entries using a tally chart or a bar graph.

1. The birth weights of 60 newborn babies are provided (right). Create a tally chart (frequency table) of the weights in the table provided below. Choose an appropriate grouping of weights.

Weight (kg)	Tally	Total

NEED HELP? See Activity 22

Birth weights (Kg)

3.740	2.660	4.170	3.970	3.570
3.830	3.375	4.400	3.840	3.620
3.530	3.840	3.770	4.710	3.260
3.095	3.630	3.400	4.050	3.315
3.630	3.810	3.825	4.560	3.230
1.560	2.640	3.130	3.350	3.790
3.910	3.955	3.400	3.380	2.620
4.180	2.980	3.260	3.690	3.030
3.570	3.350	4.100	1.495	
2.660	3.780	3.220	3.260	
3.150	3.260	3.135	3.430	
3.400	4.510	3.090	3.510	
3.380	3.800	3.830	3.230	

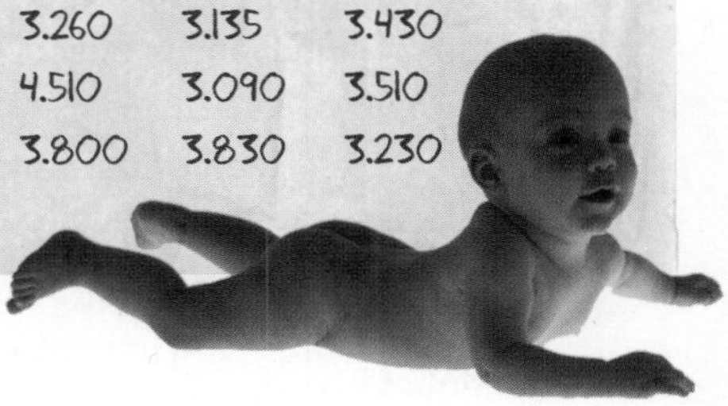

2. (a) On the graph paper (right) draw a frequency histogram for the birth weight data.

   (b) What type of distribution does the data have?

   _____

   _____

   _____

   (c) Predict whether mean, median, or mode would be the best measure of central tendency for the data:

   _____

   (d) Explain your reason for your answer in (c):

   _____

   _____

   _____

   (e) Calculate the mean, median, and mode for the birth weight data:

   Mean: _____

   Median: _____

   Mode: _____

   (f) What do you notice about the results in (e)? _____

   _____

   _____

   (g) Explain the reason for this: _____

   _____

   _____

24 What is Standard Deviation?

Key Question: What does standard deviation measure, what is its purpose, and how is it calculated?

▸ While it is important to know the mean of a data set, it is also important to know how well the mean represents the data set as a whole. This is evaluated using a simple measure of the spread in the data called standard deviation.

▸ In general, if the standard deviation is small, the mean will more accurately represent the data than if it is large.

Standard deviation

- Standard deviation is usually presented as $\bar{x} \pm s$. In normally distributed data, 68% of all data values will lie within one standard deviation (s) of the mean ($\bar{x}$) and 95% of all data values will lie within two standard deviations of the mean (right).
- Different sets of data can have the same mean and range, yet a different data distribution. In both the data sets below, 68% of the values lie within the range $\bar{x} \pm 1s$ and 95% of the values lie within $\bar{x} \pm 2s$. However, in B, the data values are more tightly clustered around the mean.
- Standard deviation is easily calculated using a spreadsheet. Data should be entered as columns. In a free cell, type the formula for standard deviation (this varies depending on the program) and select the cells containing the data values, enclosing them in parentheses.

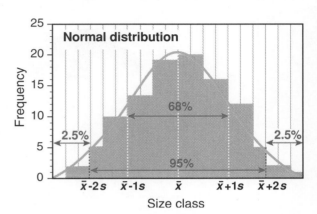

Histogram A has a larger standard deviation; the values are spread widely around the mean.

Both plots show a normal distribution with a symmetrical spread of values about the mean.

Histogram B has a smaller standard deviation; the values are clustered more tightly around the mean.

Calculating s

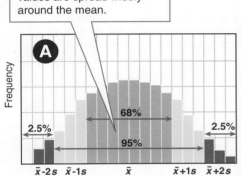

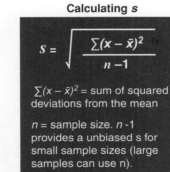

$$S = \sqrt{\frac{\sum(x - \bar{x})^2}{n-1}}$$

$\sum(x - \bar{x})^2$ = sum of squared deviations from the mean

n = sample size. $n-1$ provides a unbiased s for small sample sizes (large samples can use n).

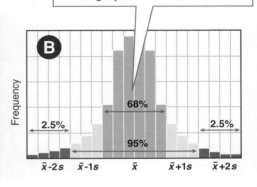

1. Two data sets have the same mean. The first data set has a much larger standard deviation than the second data set. What does this tell you about the spread of data around the mean in each case? Which data set is most reliable?

   _____

   _____

   _____

2. The data on the right shows the heights for 29 male swimmers.

   (a) Calculate the mean for the data: _____

   (b) Use manual calculation, a calculator, or a spreadsheet to calculate the standard deviation (*s*) for the data:

   _____

   (c) State the mean ± 1s: _____

   (d) What percentage of values are within 1s of the mean? _____

   (e) What does this tell you about the spread of the data? _____

   _____

Raw data: Height (cm)					
178	177	188	176	186	175
180	181	178	178	176	175
180	185	185	175	189	174
178	186	176	185	177	176
176	188	180	186	177	

25 Detecting Bias in Samples

Key Question: What is sampling bias, and how can it be detected and eliminated?

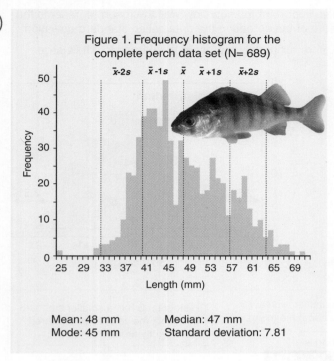

Figure 1. Frequency histogram for the complete perch data set (N= 689)

Mean: 48 mm   Median: 47 mm
Mode: 45 mm   Standard deviation: 7.81

Bias is the selection for or against one particular group and can influence the findings of an investigation. Bias can occur when sampling is not random and certain members of a population are under- or over-represented. Small sample sizes can also bias results. Bias can be reduced by random sampling (sampling in which all members of the population have an equal chance of being selected). Using appropriate collection methods will also reduce bias.

- This exercise illustrates how random sampling, large sample size, and sampling bias affect our statistical assessment of variation in a population. In this exercise, perch were collected and their body lengths (mm) were measured. Data are presented as a frequency histogram and with **descriptive statistics** (**mean**, **median**, **mode** and standard deviation).
- Figure 1 shows the results for the complete data set. The sample set was large (N= 689) and the perch were randomly sampled. The data are close to having a normal distribution.
- Figures 2 and 3 show results for two smaller sample sets drawn from the same population. The data collected in Figure 2 were obtained by random sampling but the sample was relatively small (N = 30). The person gathering the data displayed in Figure 3 used a net with a large mesh size to collect the perch.

1. (a) Compare the results for the two small data sets (Figures 2 and 3). How close are the mean and median to each other in each sample set?

    _____

    _____

    _____

    _____

    (b) Compare the standard deviation for each sample set:

    _____

    _____

    _____

    (c) Describe how each of the smaller sample sets compares to the large sample set (Figure 1):

    _____

    _____

    _____

    _____

    (d) Why do you think the two smaller sample sets look so different from each other?

    _____

    _____

    _____

    _____

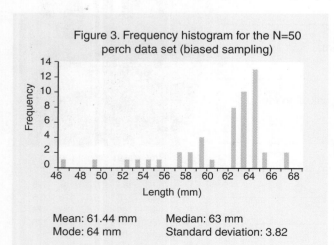

Figure 2. Frequency histogram for the N=30 perch data set (random sampling)

Mean: 49.23 mm   Median: 49.5 mm
Mode: 38 mm   Standard deviation: 11.37

Figure 3. Frequency histogram for the N=50 perch data set (biased sampling)

Mean: 61.44 mm   Median: 63 mm
Mode: 64 mm   Standard deviation: 3.82

©2022 BIOZONE International
ISBN: 978-1-98-856692-4
Photocopying Prohibited

26 Biological Drawings

Key Question: What is the purpose of a good biological drawing when studying a specimen?

▸ Drawing is a very important skill to have in biology. Drawings record what a specimen looks like and give you an opportunity to record its important features. Often, drawing something will help you remember its features at a later date, e.g. in a test.

▸ **Biological drawings** require you to pay attention to detail. It is very important that you draw what you actually see, and not what you think you should see.

▸ Biological drawings should include as much detail as you need to distinguish different structures and types of tissue, but avoid unnecessary detail which can make your drawing confusing.

▸ Attention should be given to the symmetry and proportions of your specimen. Accurate labeling, a statement of magnification or scale, the view (section type), and type of stain used (if applicable) should all be noted on your drawing.

▸ Some key points for making good biological drawing are described on the example below. The drawing of *Drosophila* (right) is well executed but lacks the information required to make it a good biological drawing.

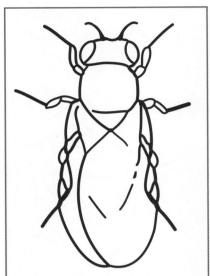

This drawing of *Drosophila* is a fair representation of the animal, but has no labels, title, or scale.

All drawings must include a title. Underline the title if it is a scientific name. ⟶

Place your drawing on the left of the page. This will leave room to place all the labels to the right of the drawing.

If you need to represent depth, use stippling (dotting). Do not use shading as this can smudge and obscure detail.

Use simple, narrow lines to make your drawings.

Use a sharp pencil for drawing. Make your drawing on plain white paper.

Your drawing must include a scale or magnification to indicate the size of your subject.

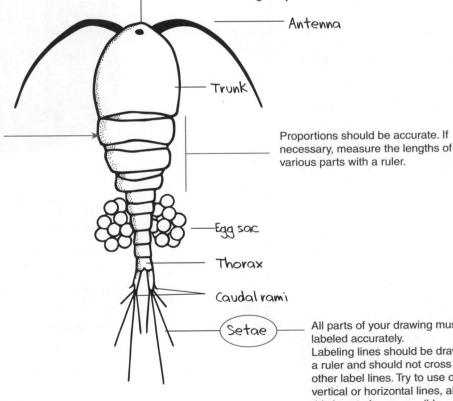

Copepod

Single eye

Antenna

Trunk

Proportions should be accurate. If necessary, measure the lengths of various parts with a ruler.

Egg sac

Thorax

Caudal rami

Setae

All parts of your drawing must be labeled accurately.
Labeling lines should be drawn with a ruler and should not cross over other label lines. Try to use only vertical or horizontal lines, although this is not always possible.

Scale
0.2 mm

©2022 BIOZONE International
ISBN: 978-1-98-856692-4
Photocopying Prohibited

Annotated diagrams

An annotated diagram is a diagram that includes a series of explanatory notes. These provide important or useful information about your subject.

Transverse section through collenchyma of <u>Helianthus</u> stem. Magnification (x 450)

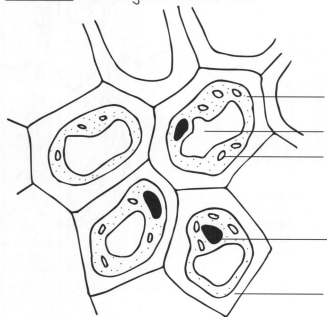

Cytoplasm - Solution of dissolved substances, enzymes, and organelles.

Vacuole containing cell sap.

Chloroplast - Organelles containing chlorophyll where photosynthesis occurs.

Nucleus - A large organelle containing most of the cell's DNA.

Primary wall with secondary thickening.

Plan diagrams

Plan diagrams are drawings made of samples viewed under a microscope at low or medium power. They are used to show the distribution of the different tissue types in a sample without any cellular detail. The tissues are identified, but no detail about the cells within them is included.

The example here shows a plan diagram produced after viewing a light micrograph of a transverse section through a dicot stem.

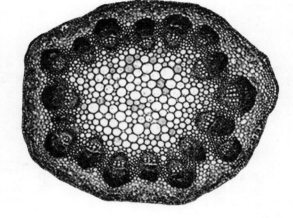

WBS

Light micrograph of a transverse section through a dicot stem.

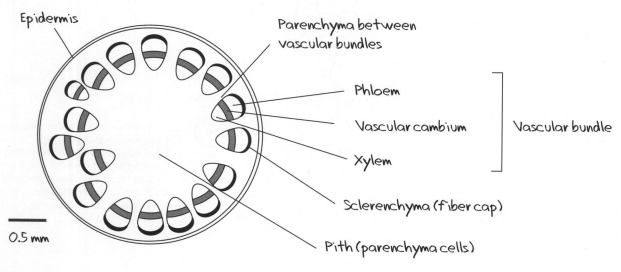

Epidermis

Parenchyma between vascular bundles

Phloem

Vascular cambium

Xylem

Vascular bundle

Sclerenchyma (fiber cap)

Pith (parenchyma cells)

0.5 mm

27 Practicing Biological Drawings

Key Question: What kind of detail is needed when making accurate and useful biological drawings?

Above: Use relaxed viewing when drawing at the microscope. Use one eye (the left for right handers) to view and the right eye to look at your drawing.

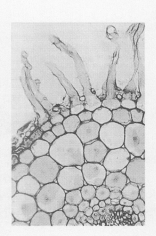

Above: Light micrograph Transverse section (TS) through a *Ranunculus* root. Right: A biological drawing of the same section.

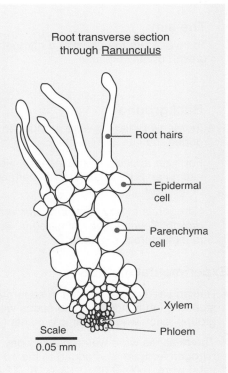

Root transverse section through <u>Ranunculus</u>

Root hairs

Epidermal cell

Parenchyma cell

Xylem

Phloem

Scale
0.05 mm

1. The image below is a labeled photomicrograph (x50) showing a partial section through a dicot root. Use this image to construct a plan diagram:

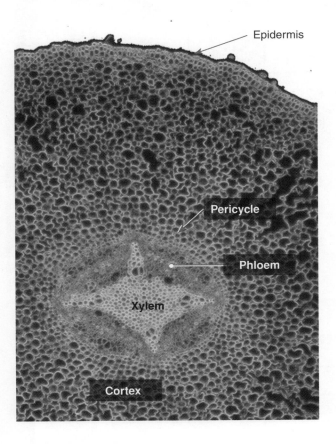

Epidermis

Pericycle

Phloem

Xylem

Cortex

28 Analyzing Experimental Data

Key Question: What is the effect of fertilizer on the growth of radishes?

The aim

To investigate the effect of a nitrogen fertilizer on the growth of radish plants.

Background

Inorganic fertilizers were introduced to crop farming during the late 19th century.

Fertilizer addition increased crop yields. An estimated 50% of crop yield is attributable to the use of fertilizer.

Nitrogen is a very important element for plant growth. Several types of nitrogen fertilizers are manufactured, e.g. urea.

Radishes

Hypothesis

If plants need nitrogen to grow, radish growth will increase with increasing nitrogen concentration.

Experimental method

▶ Radish seeds were planted in separate identical pots (5 cm x 5 cm wide x 10 cm deep) and kept together in standard lab conditions. The seeds were planted into a commercial soil mixture and divided randomly into six groups, each with five sample plants (a total of 30 plants in six treatments).

▶ The radishes were watered every day at 10 am and 3 pm with 500 mL per treatment per watering. Water soluble nitrogen fertilizer was added to the 10 am watering on the 1st, 11th, and 21st days. The fertilizer concentrations used were: 0.00, 0.06, 0.12, 0.18, 0.24, and 0.30 g, and each treatment received a different concentration.

▶ The plants were grown for 30 days before being removed from the pots, washed, and the radish root weighed. The results are presented below.

Fertilizer concentration (g/L)	Sample number				
	1	2	3	4	5
0	80.1	83.2	82.0	79.1	84.1
0.06	109.2	110.3	108.2	107.9	110.7
0.12	117.9	118.9	118.3	119.1	117.2
0.18	128.3	127.3	127.7	126.8	DNG*
0.24	23.6	140.3	139.6	137.9	141.1
0.30	122.3	121.1	122.6	121.3	123.1

*DNG = did not germinate

† Based on data from M S Jilani, et al Journal Agricultural Research

1. Identify the independent variable for the experiment and its range: _____

_____

2. Identify the dependent variable for the experiment: _____

3. What is the sample size for each concentration of fertilizer? _____

4. (a) One of the radishes recorded in the table on the previous page did not grow as expected and produced an extreme value. Record the outlying value here:

   _____

   (b) Why should this value not be included in future calculations? _____

   _____

   _____

5. Use Table 1 below to record the raw data from the experiment. You will need to include column and row headings and a title, and complete some simple calculations. Some headings have been entered for you.

Table 1: _____

	Mass of radish root (g)					Total mass	Mean mass

6. The students decided to collect more data by counting the number of leaves on each radish plant at day 30. The data are presented in Table 2.
   Use the space below to calculate the mean, median and mode for the leaf data. Add these data to table 2.

Table 2: Number of leaves on radish plants under six different fertilizer concentrations.

| Fertilizer concentration (g/L) | Number of leaves at day 30 | | | | | Mean | Median | Mode |
| | Sample (n) | | | | | | | |
	1	2	3	4	5			
0	9	9	10	8	7			
0.06	15	16	15	16	16			
0.12	16	17	17	17	16			
0.18	18	18	19	18	DNG*			
0.24	6	19	19	18	18			
0.30	18	17	18	19	19			

* DNG: Did not germinate

©2022 **BIOZONE** International
ISBN: 978-1-98-856692-4
Photocopying Prohibited

7. Use the grid below to draw a line graph of the experimental results. Plot your calculated mean mass data from Table 1, and remember to include a title and correctly labelled axes.

8. Which fertilizer concentration appeared to produce the best growth of radish mass? _____

9. Which fertilizer concentration appeared to produce the best growth of leaves? _____

10. Write a short conclusion for the entire experiment: _____

11. (a) What assumptions were made in the design of this experiment? _____

(b) Can you suggest ways in which the design could be improved? How could you evaluate whether the mass differences between the fertilizer concentrations were significant?

From Molecules to Organisms: Structures and Processes

Concepts and connections
Use arrows to make connections between related concepts in this section of the book

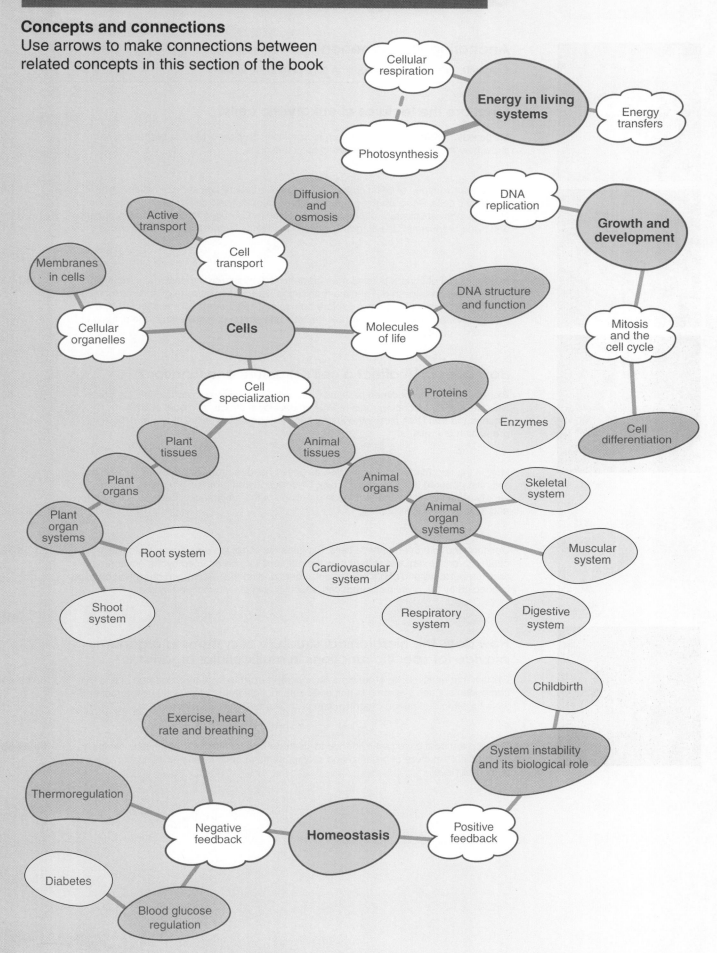

CHAPTER 2

Cell Specialization and Organization

Activity number

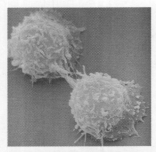

Anchoring Phenomenon

Frogsicle: How does the wood frog survive freezing in winter?

17 32

What are the features of eukaryotic cells?

☐ 1 Multicellular organisms are built from hierarchical structures. Models of these systems show how they interact to carry out specific functions.

30

☐ 2 There are two types of cells: prokaryotic and eukaryotic. Plant and animal cells are eukaryotic cells. Describe the general features of eukaryotic cells. Use a microscope to view the internal structure of plant and animal cells. Calculate the size of different cells and living organisms. Use diagrams of cells to identify features on real SEM images of cells.

31-36

☐ 3 Build a model of a cell's plasma membrane to show its three dimensional structure. Explore how proteins in the plasma membrane control the passage of specific molecules across the plasma membrane. Explore how role of specialized cells is affected by their size and shape, and investigate how the size and shape of a cell affects the diffusion of molecules into its interior.

37-46

How does DNA affect a cell's structure and function?

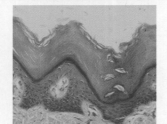

☐ 4 Extract DNA and associated proteins from cells to observe its macroscopic structure. Use the A-T, C-G bonding rules of DNA to develop a structural model of DNA. Understand that DNA stores the instructions for controlling cells actions and for the production of proteins.

47-50 66

☐ 5 How is the information stored in DNA as genes used to produce proteins? Explain how scientists worked out the genetic code. Understand that proteins are made up of a variety of amino acids and that the interactions of the amino acids determine the shape and function of the protein.

51-54

☐ 6 Describe the role of proteins in cells. Explain how some proteins act as biological catalysts (enzymes). Investigate how their actions can be affected by the environment (e.g. temperature and pH). Develop an understanding that a cell's structure and function is affected by the proteins it produces, which is ultimately controlled by the cell's DNA.

55-59

How does the hierarchical structure of systems in organisms provide for specific functions in multicellular organisms?

☐ 7 Explain that multicellular organisms have complex organ systems, made up of many components. Each system has a specific role, but different organ systems interact and work together so that the organism can carry out essential life functions.

30 60-64

☐ 8 Develop a model based on evidence to illustrate how multicellular organisms exhibit a hierarchical structure of organization. Components at each level of organization are part of the next level.

60-64 66

Photos: EII

EII

29 Frogsiscle

Key Question: How does the wood frog survive freezing in winter?

Freezing cells and tissues

▶ Try this: take a vegetable like a lettuce or carrot and put it in the freezer overnight. Thaw it out the next day. What happens? If you have a fridge with the cooling panel exposed at the back, or an open ice box for making ice cubes, you may have come across this problem. Sometimes, fruit or vegetables get pushed to the back of the fridge and part of them freezes. When they thaw out, they become limp, and sometimes a bit slimy.

▶ This is because the cells in them have frozen, usually rapidly. When this happens, the water in the vegetable forms ice crystals which damage the cells and tissues and cause them to lose their structure.

▶ If this happens in a living organism, the result is normally death.

▶ Wood frogs (*Rana sylvatica*) are able to survive freezing down to -6°C. How do they do this?

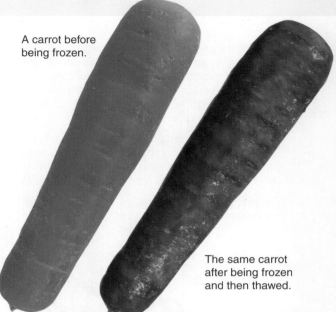

A carrot before being frozen.

The same carrot after being frozen and then thawed.

Mechanisms to survive freezing:

▶ When tissues freeze, initially ice forms outside the cells, and water is drawn out of the cell. Studies of the wood frog have found that it survives freezing by:

• Increasing the amount of glucose in the blood around the organs by up to 100 times the normal amount.

• Increasing urea in the extracellular spaces (spaces outside the cells).

• Having proteins that operate at lower temperatures than similar proteins in other frogs.

• Proteins in the cells' plasma membranes allow glucose and urea to enter the cell as freezing proceeds.

• Both breathing and the heartbeat stop.

• Antifreeze proteins bind to ice crystals.

1. What happens to cells when tissues freeze? _____

_____

_____

_____

2. For each of the mechanisms to survive freezing listed above, write down how you think they might help the frog survive.

(a) Increasing the amount of glucose in the blood: _____

_____

(b) Increasing urea in the extracellular spaces: _____

_____

(c) Having proteins that operate at lower temperatures than similar proteins: _____

_____

(d) Proteins in the cells' plasma membranes allow glucose and urea to enter the cell, as freezing proceeds:

_____

_____

©2022 **BIOZONE** International
ISBN: 978-1-98-856692-4
Photocopying Prohibited

30 The Hierarchy of Life

Key Question: How are the cells of organisms organized so that they work together in a coordinated way?

All multicellular organisms are organized in a hierarchy of structural levels, where each level builds on the one below it. It is traditional to start with the simplest components (parts) and build from there. Higher levels of organization are more complex than lower levels.

Hierarchical organization enables specialization so that individual components perform a specific function or set of related functions. Specialization enables organisms to function more efficiently.

The diagram below explains this hierarchical organization for a human.

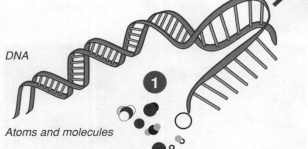

The cellular level

Cells are the basic structural and functional units of an organism. Cells are specialized to carry out specific functions, e.g. cardiac (heart) muscle cells (below).

DNA

1

Atoms and molecules

The chemical level

All the chemicals essential for maintaining life, e.g. water, ions, fats, carbohydrates, amino acids, **proteins**, and nucleic acids.

2

The organelle level

Molecules associate together to form the organelles and structural components of cells, e.g. the nucleus (above).

3

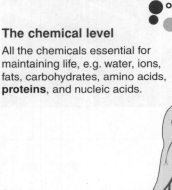

7 The organism

The cooperating organ systems make up the organism, e.g. a human.

The tissue level

4

Groups of cells with related functions form tissues, e.g. cardiac (heart) muscle (above). The cells of tissue often have a similar origin.

6 The system level

Groups of organs with a common function form an **organ system**, e.g. cardiovascular system (right).

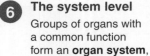

5

The organ level

An organ is made up of two or more types of tissues to carry out a particular function. Organs have a definite form and structure, e.g. heart (left).

31 Introduction to Cells

Key Question: What are the distinguishing features of living organisms, prokaryotic cells, and eukaryotic cells?

▸ The **cell** is the smallest unit of life. Cells are often called the building blocks of life.

▸ Cells are either **prokaryotic cells** or **eukaryotic cells**. Within each of these groups, cells may vary greatly in their size, shape, and functional role.

Prokaryotic cells

▸ Prokaryotic cells are bacterial cells.

▸ Prokaryotic cells lack a membrane-bound nucleus or any membrane-bound **organelles**.

▸ They are small (generally 0.5-10 µm) single cells (unicellular).

▸ They are relatively basic cells and have very little cellular organization (their **DNA**, ribosomes, and **enzymes** are free floating within the cell cytoplasm).

▸ Single, circular chromosome of naked DNA.

▸ Prokaryotes have a cell wall, but it is different from the cell walls that some eukaryotes have.

Eukaryotic cells

▸ Eukaryotic cells have a membrane-bound nucleus, and other membrane-bound organelles.

▸ Plant cells, animals cells, fungal cells, and protists are all eukaryotic cells.

▸ Eukaryotic cells are large (30-150 µm). They may exist as single cells or as part of a multicellular organism.

▸ Multiple linear chromosomes consisting of DNA and associated proteins.

▸ They are more complex than prokaryotic cells. They have more structure and internal organization.

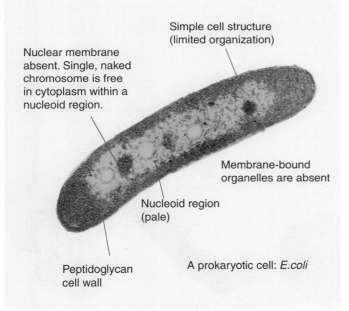

Nuclear membrane absent. Single, naked chromosome is free in cytoplasm within a nucleoid region.

Simple cell structure (limited organization)

Membrane-bound organelles are absent

Nucleoid region (pale)

Peptidoglycan cell wall

A prokaryotic cell: *E.coli*

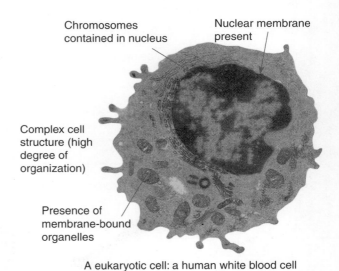

Chromosomes contained in nucleus

Nuclear membrane present

Complex cell structure (high degree of organization)

Presence of membrane-bound organelles

A eukaryotic cell: a human white blood cell

1. What are the main features of a prokaryotic cell? _____

_____

_____

2. (a) What are the main features of a eukaryotic cell? _____

_____

_____

(b) Name examples of eukaryotic cells: _____

_____

LS1.A

The cell theory

Cells are the fundamental unit of life. Study the images below:

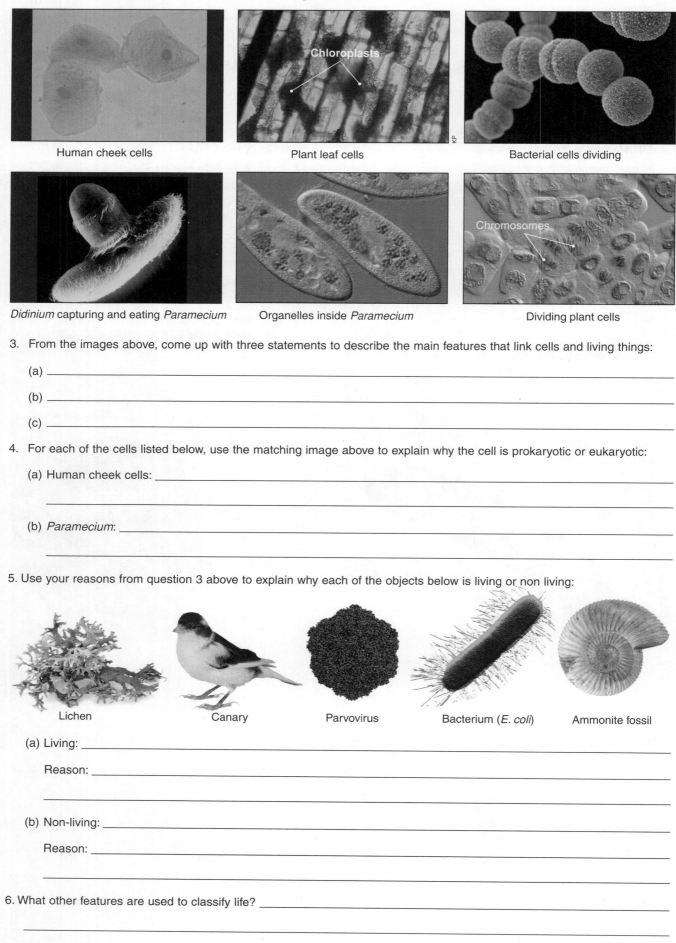

Human cheek cells

Plant leaf cells

Chloroplasts

Bacterial cells dividing

Didinium capturing and eating *Paramecium*

Organelles inside *Paramecium*

Chromosomes

Dividing plant cells

3. From the images above, come up with three statements to describe the main features that link cells and living things:

(a) _____

(b) _____

(c) _____

4. For each of the cells listed below, use the matching image above to explain why the cell is prokaryotic or eukaryotic:

(a) Human cheek cells: _____

_____

(b) *Paramecium*: _____

_____

5. Use your reasons from question 3 above to explain why each of the objects below is living or non living:

Lichen

Canary

Parvovirus

Bacterium (*E. coli*)

Ammonite fossil

(a) Living: _____

Reason: _____

_____

(b) Non-living: _____

Reason: _____

_____

6. What other features are used to classify life? _____

_____

©2022 **BIOZONE** International
ISBN: 978-1-98-856692-4
Photocopying Prohibited

32 Microscopes and Magnification

Key Question: What are the important features of a light microscope, and how do you calculate the magnification of the image they produce?

▸ The light (or optical) microscope (LM) is an important tool in biology. Light microscopy involves illuminating a sample and passing the light that is transmitted or reflected through lenses to give a magnified view of the sample.

▸ High power, compound light microscopes use visible light and a combination of lenses to magnify objects up to several hundred times. Bright field microscopy (below) is the simplest, and involves illuminating the specimen from below and viewing from above. Specimens must be thin and mostly transparent so that light can pass through. No detail will be seen in specimens that are thick or opaque.

▸ The wavelength of light limits the resolution of light microscopes to around 0.2 µm. Objects closer than this will not be distinguished as separate. Electron microscopes provide higher resolutions as they use a shorter wavelength electron beam, rather than light.

Structure of a typical compound light microscope

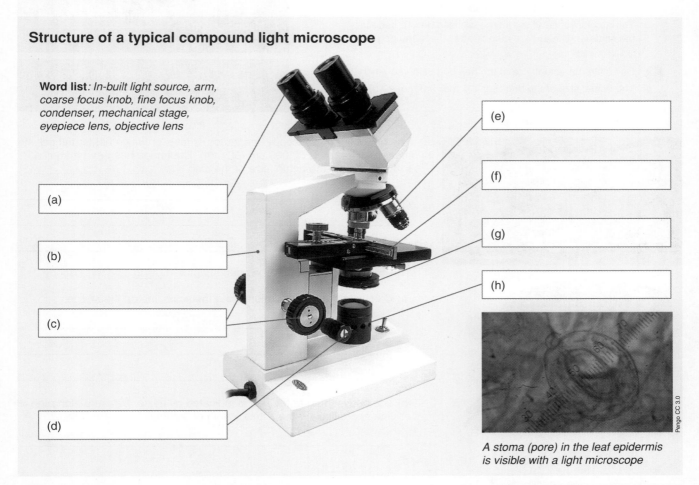

Word list: *In-built light source, arm, coarse focus knob, fine focus knob, condenser, mechanical stage, eyepiece lens, objective lens*

(a)

(b)

(c)

(d)

(e)

(f)

(g)

(h)

A stoma (pore) in the leaf epidermis is visible with a light microscope

Pengo CC 3.0

What is magnification?

Magnification refers to the number of times larger an object appears compared to its actual size.

Magnification is calculated as follows:

$$\text{Objective lens power} \times \text{Eyepiece lens power}$$

What is resolution?

Resolution is the ability to distinguish between close together but separate objects. Examples of high and low resolution for separating two objects viewed under the same magnification are shown on the right. Factors such as dirty lenses can reduce resolution.

High resolution

Low resolution

1. Label the image above of the compound light microscope (a) to (h). Use words from the list supplied.

2. Determine the magnification of a microscope using:

   (a) 15 X eyepiece and 40 X objective lens: _____

   (b) 10 X eyepiece and 60 X objective lens: _____

©2022 **BIOZONE** International
ISBN: 978-1-98-856692-4
Photocopying Prohibited

LS1.A

How do we calculate linear magnification?

▸ Magnification is how much larger an object appears compared to its actual size. It can be calculated from the ratio of image height to object height.

▸ If the ratio is greater than one, the image is enlarged. If it is less than one, it is reduced. In calculating magnification, all measurements should be in the same units.

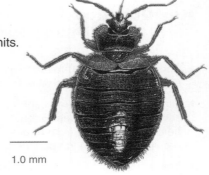

1.0 mm

Worked example

1 Measure the body length of the bed bug image (right). Your measurement should be 40 mm (not including the body hairs and antennae).

2 Measure the length of the scale line marked 1.0 mm. You will find it is 10 mm long. The magnification of the scale line can be calculated using equation 1.

The magnification of the scale line is **10** (10 mm ÷ 1 mm)

The magnification of the image will also be x10 because the scale and image are magnified to the same degree.

3 Calculate the actual size of the bed bug using equation 2. The actual size of the bed bug is **4 mm** (40 mm ÷ 10)

Microscopy equations

$$1.\ \text{Magnification} = \frac{\text{measured size of object}}{\text{actual size of object}}$$

$$2.\ \text{Actual object size} = \frac{\text{size of the image}}{\text{magnification}}$$

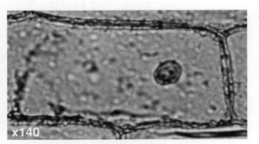

x140

3. The bright field microscopy image (left) shows an onion epidermal cell. Its measured length is 52,000 µm (52 mm). The image has been magnified x140. Calculate the actual size of the cell:

_____

_____

_____

4. The image of the flea (left) has been captured using light microscopy.

    (a) Calculate the magnification using the scale line on the image:

_____

_____

_____

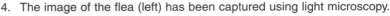

0.5 mm

    (b) The body length of the flea is indicated by a line. Measure along the line and calculate the actual length of the flea:

_____

_____

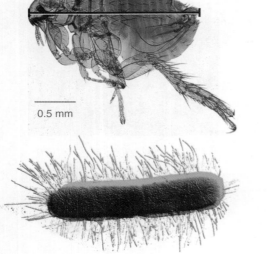

5. The image size of the *E.coli* cell (left) is 43 mm, and its actual size is 2 µm. Using this information, calculate the magnification of the image:

_____

_____

6. Explain why a higher magnification is not particularly useful if the resolution is poor: _____

_____

_____

7. When focusing a specimen it necessary to focus on the lowest magnification first, before switching to higher magnifications. Why do you think this is important?

_____

_____

©2022 **BIOZONE** International
ISBN: 978-1-98-856692-4
Photocopying Prohibited

33 Studying Cells

Key Question: What techniques are used to prepare and view cells under a light microscope?

 ▶ Specimens are usually prepared in some way before viewing in order to highlight features and reveal details. A wet mount is a temporary preparation in which a specimen and a drop of fluid are trapped under a thin coverslip. Wet mounts improve a sample's appearance and enhances visible detail. Sections must be made very thin.

  Investigation 2.1 Preparing an onion slide

See appendix for equipment list.

⚠ 👁 👕 **Caution is required when using scalpels or razors. Iodine stains skin and clothes, and irritates the eyes. You should wear protective eyewear.**

1. Onions make good subjects for preparing a simple wet mount. Cut a square segment from a thick leaf of the bulb using a razor or scalpel.

2. Bend the segment towards the upper epidermis (upper cell layer) until the lower epidermis and inner leaf tissue (the parenchyma) snaps, leaving the upper epidermis attached.

3. Carefully peel off the parenchyma from one side of the snapped leaf and then the other, leaving a peel of just the upper epidermis.

4. Place the peel in the centre of a clean glass microscope slide and cover it with a drop of water.

5. Carefully lower a coverslip over the peel. A mounted needle can be used for better precision. This avoids including air in the mount.

6. Use a small piece of tissue or filter paper to remove any excess water.

7. Place the slide on the microscope tray. Locate the specimen or region of interest at the lowest magnification. Focus using the lowest magnification first (remembering to move the lens away from the slide) before switching to the higher magnifications.

8. After viewing the slide under various magnifications, remove the slide and place it on the bench.

9. At the edge of the coverslip, place a small drop of iodine stain.

10. On the opposite side of the coverslip use a piece of tissue or filter paper to draw the water out from under the coverslip. The iodine will be drawn under the coverslip.

11. Replace the slide on the microscope and view the stained onion peel.

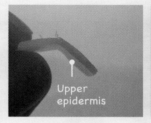
Upper epidermis

Upper epidermis peeled away

These photos KP

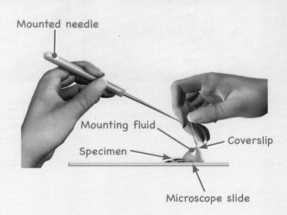
Mounted needle
Mounting fluid
Coverslip
Specimen
Microscope slide

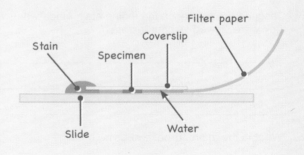
Filter paper
Coverslip
Stain
Specimen
Slide
Water

1. Why must sections viewed under a microscope be very thin? _____

_____

2. Why do you think the specimen is covered with a coverslip? _____

_____

3. Why would no chloroplasts be visible in an onion epidermis cell slide? _____

_____

LS1.A

Stains and their uses

▶ Staining material for viewing under a microscope can make it easier to distinguish particular cell structures.

▶ Stains and dyes can be used to highlight specific components or structures. Stains contain chemicals that interact with molecules in the cell. Some stains bind to a particular molecule, making it easier to see where those molecules are. Others cause a change in a target molecule, which changes the color, making them more visible.

▶ Most stains are non-viable, and are used on dead specimens, but harmless, viable stains can be applied to living material.

Some commonly used stains		
Stain	Final color	Used for
Iodine solution	Blue-black	Starch
Crystal violet	Purple	Gram staining
Aniline sulfate	Yellow	Lignin
Methylene blue	Blue	Nuclei
Hematoxylin and eosin (H&E)	H=dark blue/ violet E=red/pink	H=Nuclei E=Proteins

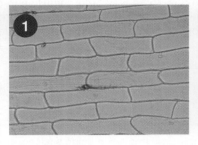 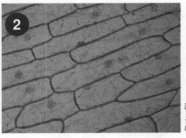

The light micrographs 1 and 2 (above) show how the use of a stain can enhance certain structures. The left image (1) is unstained and only the cell wall is easily visible. Adding iodine (2) makes the cell wall and nuclei stand out.

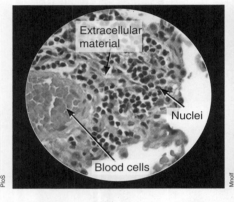

H&E stain is a common stain for animal tissues. Nuclei stain dark blue, whereas proteins, extracellular material, and red blood cells stain pink or red.

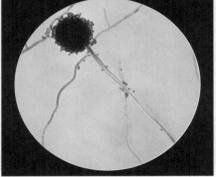

Viable stains do not immediately harm living cells. Trypan blue is a vital stain that stains dead cells blue but is excluded by live cells. It is also used to study fungi.

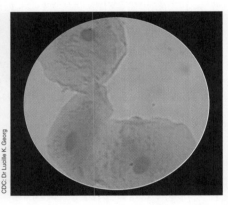

Methylene blue is a common temporary stain for animal cells, such as these cheek cells. It stains DNA, making the nuclei more visible.

4. Why is it necessary to focus on the lowest magnification first, before switching to higher magnifications?

_____

_____

5. Describe the difference the iodine stain made when viewing the onion cells under the microscope, compared to when they were viewed without the stain:

_____

_____

_____

6. What is the main purpose of using a stain? _____

_____

7. What is the difference between a viable and non-viable stain? _____

_____

_____

8. Identify a stain that would be appropriate for distinguishing each of the following:

(a) Live vs dead cells: _____ (c) Lignin in a plant root section: _____

(b) Red blood cells in a tissue preparation: _____ (d) Nuclei in cheek cells: _____

©2022 **BIOZONE** International
ISBN: 978-1-98-856692-4
Photocopying Prohibited

34 Plant Cells

Key Question: What are the general and specific features of a plant cell?

What is an organelle?

▶ The word **organelle** means "small organ". Therefore, organelles are the cell's "organs" and carry out the cell's work.

▶ Organelles represent one level of organization in a multicellular organism. One component (the cell) is made up of many smaller parts (organelles).

▶ Eukaryotic cells contain many different types of organelles. Each type of organelle has a specific role in the cell to help it function.

▶ Plant cells have several types of membrane-bound organelles called plastids. These make and store food and pigments. Some of the organelles found in a plant cell are shown below.

Features of a plant cell

▶ Plant cells are **eukaryotic cells**. Features that identify plant cells as eukaryotic cells include:

▶ A membrane-bound nucleus.

▶ Membrane-bound organelles, e.g. nucleus, mitochondria, endoplasmic reticulum.

▶ Features that can be used to identify a plant cell include the presence of:

- Cellulose cell wall.
- Chloroplasts and other plastids.
- Large vacuole (often centrally located).

A generalized plant cell

Chloroplast

A specialized plastid containing the green pigment, chlorophyll. Chloroplasts are the site for photosynthesis. Photosynthesis uses light energy to convert carbon dioxide to glucose.

Cellulose cell wall

A semi-rigid structure that lies outside the plasma membrane. It has several roles, including protecting the cell and providing shape. Many materials pass freely through the cell wall.

Plasma membrane

Located inside the cell wall in plants. It controls the movement of materials into and out of the cell.

Large central vacuole:

Plant vacuoles contain cell sap. Sap is a watery solution containing dissolved food material, ions, waste products, and pigments. Functions include storage, waste disposal, and growth.

Mitochondrion

Mitochondria are the cell's energy producers. They use the chemical energy in glucose to make ATP (the cell's usable energy).

Endoplasmic reticulum (ER)

A network of tubes and flattened sacs continuous with the nuclear membrane. There are two types of ER. Rough ER has ribosomes attached. Smooth ER has no ribosomes (so it appears smooth).

Nuclear pore

Nuclear membrane

Nucleus

Most of a plant cell's DNA is here.

Ribosomes

These small structures make proteins by joining amino acids.

Cytoplasm

A watery solution containing dissolved materials, enzymes, and the cell organelles.

Amyloplast

Specialized plastid that makes and stores starch (a glucose polymer).

Golgi apparatus

A structure made up of membranous sacs. It stores, modifies, and packages proteins.

SF LS1.A

1. Use the diagram of a plant cell on the previous page to become familiar with the features of a plant cell. Use your knowledge to label the ten structures in the transmission electron micrograph (TEM) of the cell below.

   Use the following list of terms to help you: *nuclear membrane, cytoplasm, endoplasmic reticulum, mitochondrion, starch granule, nucleus, vacuole, plasma membrane, cell wall, chloroplast.*

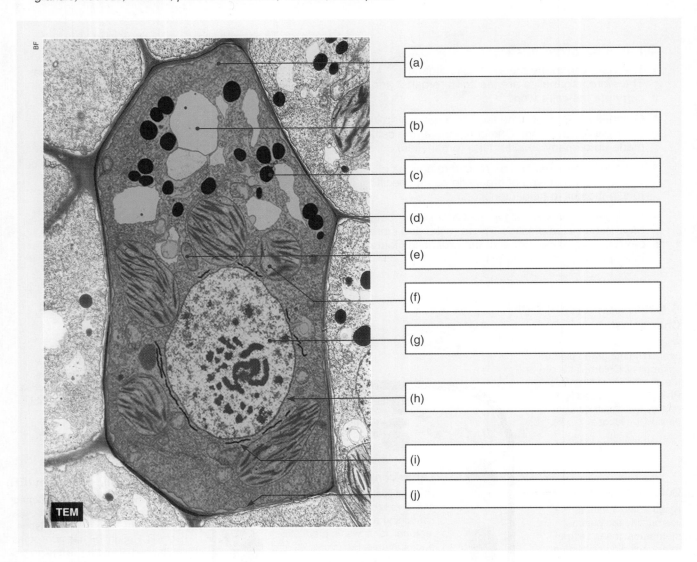

(a)

(b)

(c)

(d)

(e)

(f)

(g)

(h)

(i)

(j)

2. Which features identify this plant cell as eukaryotic? _____

_____

_____

3. (a) What is an organelle? _____

   _____

   _____

   (b) Why are there so many different types of organelles in eukaryotic cells, e.g. plant and animal cells? _____

   _____

   _____

   _____

4. (a) Name the organelle where photosynthesis occurs: _____

   (b) How many of these organelles are present in the labeled cell above? _____

   (c) What is the pigment in these organelles that captures light for photosynthesis? _____

©2022 **BIOZONE** International
ISBN: 978-1-98-856692-4
Photocopying Prohibited

35 Animal Cells

Key Question: What are the general and specific features of animal cells? How are they different from plant cells?

Animal cells are **eukaryotic cells**. Features that identify them as eukaryotic cells include:

▸ A membrane-bound nucleus.

▸ Membrane-bound **organelles**.

DID YOU KNOW?

Animal cells lack the rigid cell wall found in plant cells, so their shape is more irregular and they can sometimes move about or change shape.

Features of an animal cell

▸ Animal cells have many of the same structures and organelles that plant cells have, but several features help to identify them, including:

- No cell wall.
- Often have an irregular shape.
- No chloroplasts or other plastids.
- No large vacuoles (if any).
- They have centrioles (not found in the cells of most plants).

A generalized animal cell

Mitochondrion

Organelles involved in the production of ATP (usable energy).

Smooth endoplasmic reticulum (smooth ER)

Its main role is to make lipids and phospholipids.

Plasma membrane

The cell boundary. The membrane is a semi-fluid, phospholipid bilayer with embedded proteins. It separates the cell from its external environment and controls the movement of substances into and out of the cell.

Small vacuole

Not always present.

Ribosomes

These make proteins. They can be found floating free in the cytoplasm or attached to the surface of rough ER.

Golgi apparatus

The flattened, disc-shaped sacs of the Golgi are stacked one on top of each other, very near, and sometimes connected to, the ER. Vesicles bud off from the Golgi and transport protein products away.

Microvilli

Small finger-like extensions which increase the cell's surface area. Not all animal cells have these.

Lysosome

Sac-like organelles containing enzymes that break down foreign material, cell debris, and worn-out organelles.

Rough endoplasmic reticulum (rough ER)

These have ribosomes attached to the surface. Proteins are made here.

Nuclear pore

This is a hole in the nuclear membrane. It allows molecules to pass between the nucleus and the rest of the cell.

Nucleus

A large organelle containing most of the cell's DNA. Within the nucleus, is a denser structure called the nucleolus (*n*).

Cytoplasm

Centrioles

Paired cylindrical structures contained within the centrosome (an organelle that organizes the cell's microtubules). The centrioles form the spindle fibers involved in nuclear division. They are made of protein microtubules and are always at 90° to each other.

©2022 **BIOZONE** International
ISBN: 978-1-98-856692-4

SF LS1.A

1. Study the diagram of an animal cell on the previous page to become familiar with the features of an animal cell. Use your knowledge to identify and label the structures in the transmission electron micrograph (TEM) of the cell below.

    Use the following list of terms to help you: *cytoplasm, plasma membrane, rough endoplasmic reticulum, mitochondrion, nucleus, centriole, Golgi apparatus, lysosome.*

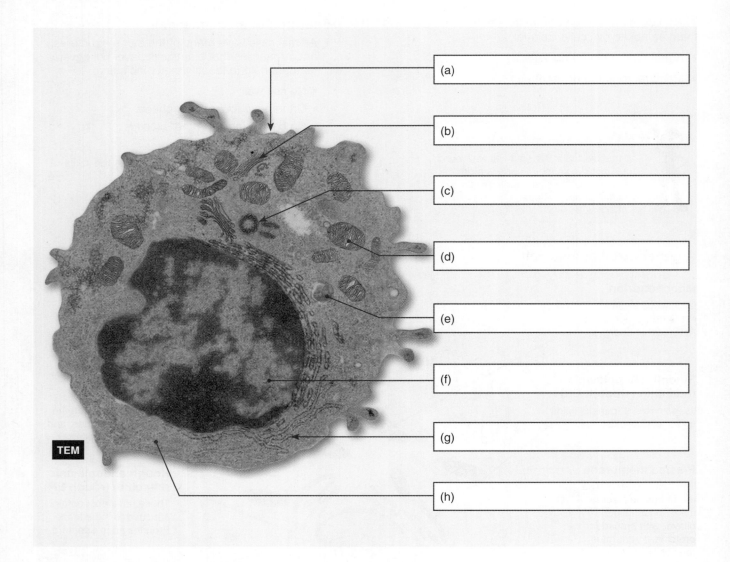

(a)

(b)

(c)

(d)

(e)

(f)

(g)

(h)

TEM

2. Name the features on the cell above that identify it as an animal cell: _____
    _____

3. (a) Where is the plasma membrane located in an animal cell? _____

    _____

    (b) All organelles have a specialized function. What is the function of the plasma membrane? _____

    _____

    _____

4. (a) Name the largest organelle visible in the animal cell above: _____

    (b) What important material does this organelle contain? _____

    _____

5. Which organelle is shown clearly in transverse (cross section) view and longitudinal view?

    _____

36 Identifying Organelles

Key Question: What features of cells can be identified using electron microscopes?

The photographs on the left were taken using a transmission electron microscope (TEM). They show the ultrastructure of some **organelles**. Use the information on the previous pages to identify the organelles and help answer the following questions.

1. (a) Identify this organelle (arrowed): _____

   (b) Describe the function of this organelle: _____

   _____

   _____

2. (a) Name the circled organelle: _____

   (b) Which kind of cell(s) would this organelle be found in? _____

   _____

   (c) Describe the function of this organelle: _____

   _____

   _____

   _____

   _____

3. (a) Name the large, circular organelle: _____

   (b) Which kind of cell(s) would this organelle be found in? _____

   _____

   (c) Describe the function of this organelle: _____

   _____

   _____

   _____

4. (a) Name the ribbon-like organelle in this photograph (arrowed): _____

   _____

   (b) Which kind of cell(s) would this organelle be found in? _____

   _____

   (c) What are the small dark granular structures attached to the organelle?

   _____

5. (a) Name this large circular structure (arrowed): _____

   (b) Which kind of cell(s) would this organelle be found in? _____

   _____

   (c) Describe the function of this organelle: _____

   _____

   _____

   _____

©2022 **BIOZONE** International
ISBN: 978-1-98-856692-4

SF   LS1.A

37 The Structure of Membranes

Key Question: What are the key components of plasma membranes?

The fluid mosaic model of membrane structure

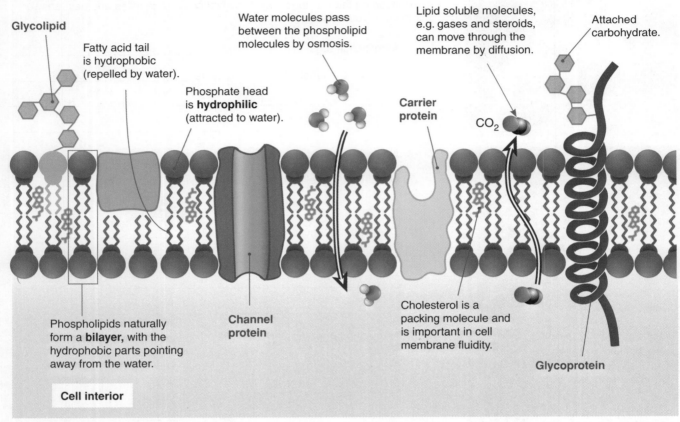

Glycolipid

Fatty acid tail is hydrophobic (repelled by water).

Water molecules pass between the phospholipid molecules by osmosis.

Phosphate head is **hydrophilic** (attracted to water).

Lipid soluble molecules, e.g. gases and steroids, can move through the membrane by diffusion.

Carrier protein

CO_2

Attached carbohydrate.

Phospholipids naturally form a **bilayer,** with the hydrophobic parts pointing away from the water.

Channel protein

Cholesterol is a packing molecule and is important in cell membrane fluidity.

Glycoprotein

Cell interior

The fluid-mosaic model of membrane structure (above) describes a phospholipid bilayer with **proteins** of different types moving freely within it. The double layer of lipids is quite fluid. It is a dynamic structure and is actively involved in cellular activities.

1. List the important components of the plasma membrane: _____

   _____

   _____

   _____

2. Identify which kind of molecule on the diagram:

   (a) Can move through the plasma membrane by diffusion: _____

   _____

   (b) Forms a channel through the membrane: _____

   _____

3. List the types of proteins pictured in the diagram: _____

   _____

   _____

4. (a) On the diagram (right) label the hydrophobic and hydrophilic ends of the phospholipid.

   (b) Which end is attracted to water? _____

©2022 **BIOZONE** International
ISBN: 978-1-98-856692-4
Photocopying Prohibited

Modeling the plasma membrane

▶ Plasma membranes are often shown as two dimensional structures (as shown on the previous page). Even when drawn to represent a three dimensional structure, the nature of the plasma membrane may not be obvious. In this part of the activity, you will build a simple, three dimensional plasma membrane.

1. Cut out the plasma membrane along the red lines. Cut out the solid black circles. Fold along the black lines. Use clear tape to stick the sides together to produce a 3D, slightly curved box.

2. Cut out the three proteins along the red lines. Fold along the black lines and use clear tape to produce three cylinders.

3. Cut out both carbohydrate chains. Fold over the black squares. Stick one to the black square on the end of the glycoprotein. Stick the other to the black square on the plasma membrane surface to produce a glycolipid.

4. Slide the two transmembrane proteins into the channels created by cutting out the circles from the plasma membrane.

5. Slide the peripheral protein about halfway into the final hole. This completes your plasma membrane model.

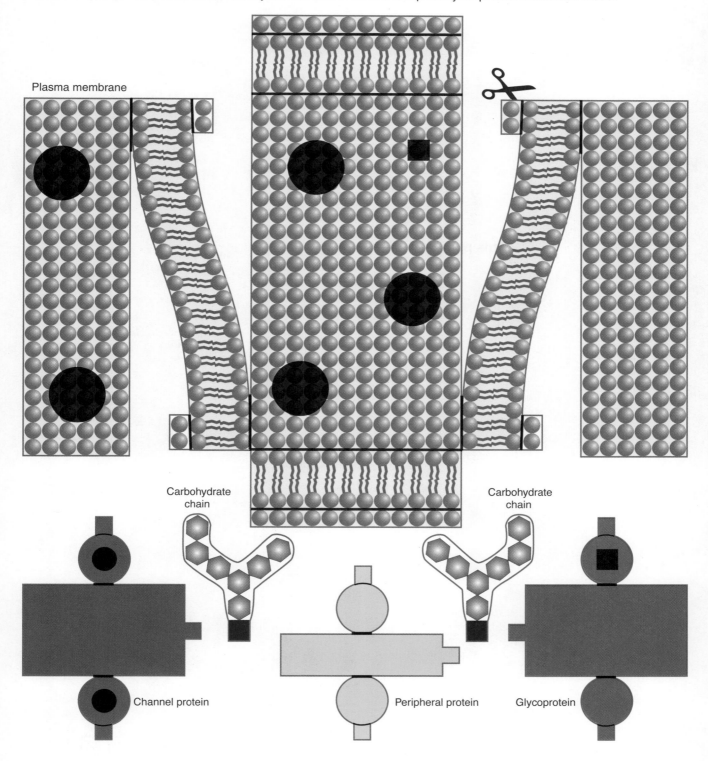

Plasma membrane

Carbohydrate chain

Carbohydrate chain

Channel protein

Peripheral protein

Glycoprotein

This page is deliberately left blank

38 Diffusion in Cells

Key Question: What is diffusion, and what are the factors that affect the rate of diffusion of a particle from one point to another?

What is diffusion?

▶ Diffusion is the movement of particles from regions of high concentration to regions of low concentration. Diffusion is a passive process, meaning it needs no input of energy to occur. During diffusion, molecules move randomly about, becoming evenly dispersed.

▶ Most diffusion in biological systems occurs across membranes. Simple diffusion occurs directly across a membrane, whereas facilitated diffusion involves helper **proteins.** Neither requires the cell to expend energy.

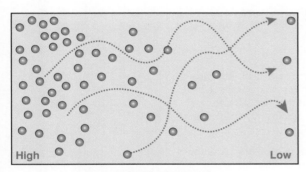

Concentration gradient

If molecules can move freely, they move from high to low concentration (down a concentration gradient) until evenly dispersed. Net movement then stops.

Factors affecting the rate of diffusion

Concentration gradient	Diffusion rate is higher when there is a greater concentration difference between two regions.
The distance moved	Diffusion occurs at a greater rate over shorter distances than over larger distances.
The surface area involved	The larger the area across which diffusion occurs, the greater the rate of diffusion.
Barriers to diffusion	Rate of diffusion is slower across thick barriers than across thin barriers.
Temperature	Rate of diffusion increases with temperature.

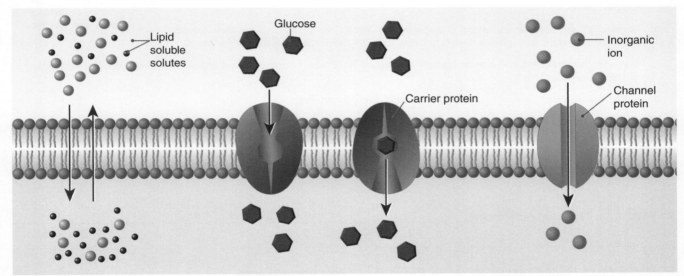

Simple diffusion

Molecules move directly through the membrane without assistance and without any energy expenditure. Example: O_2 diffuses into the blood and CO_2 diffuses out.

Facilitated diffusion by carriers

Carrier proteins allow large lipid-insoluble molecules that cannot cross the membrane by simple diffusion to be transported into the cell. Example: the transport of glucose into red blood cells.

Facilitated diffusion by channels

Channel proteins (hydrophilic pores) in the membrane allow inorganic ions to pass through the membrane. Example: K^+ ions leaving nerve cells to restore membrane resting potential.

1. What is diffusion? _____

_____

2. (a) How is facilitated diffusion different from simple diffusion? _____

_____

(b) How is it the same? _____

©2022 **BIOZONE** International
ISBN: 978-1-98-856692-4
Photocopying Prohibited

3. Why would a thin flat cell have a greater rate of diffusion to and from its center than a thick spherical cell?

_____

_____

_____

Observing diffusion

▶ Diffusion through a partially permeable membrane can be modelled using dialysis tubing. The pores of the dialysis tubing determine the size of the molecules that can pass through. In the experiment described below, you will investigate how glucose will diffuse down its concentration gradient from a high glucose concentration to a low glucose concentration and demonstrate, via the model, the selective permeability of the plasma membrane.

Investigation 2.2 Simple diffusion across a membrane

See appendix for equipment list.

1. Add 200 mL of distilled water to a clean 200 mL beaker. Remove a 1 mL sample and place in a clean test tube. Use a glucose dipstick to test for the presence and concentration of glucose in the 1 mL sample. If glucose is present, the indicator window will change color. The color change can be compared against a reference to determine the concentration of glucose present.

2. Now add a few drops of Lugol's indicator to test for the presence of starch. Lugol's indicator contains iodine, and turns blue/black in the presence of starch.

3. Obtain a short section of dialysis tubing, approximately 10 cm long. Use thread or nylon line to tie off one end (or tie a knot in the tubing if long enough).

4. You may need to rinse the tubing under water to make it pliable enough to open.

5. Fill the dialysis tubing with 5 mL each of a 1% starch solution and a 10% glucose solution.

6. Remove a 1 mL sample and place in a clean test tube. Tie off the top of the dialysis tubing, rinse well with distilled water, then place in the beaker of distilled water.

7. Test for the presence and concentration of glucose and then starch in the sample from the dialysis tubing as in steps 1 and 2.

8. Leave the dialysis tubing in the distilled water for 30 minutes.

9. Remove 1 mL of water from the beaker and place in a clean test tube. Use a glucose dipstick to test for the presence and concentration of glucose. Test for the presence of the starch using Lugol's indicator.

10. Remove a 1 mL sample from the dialysis tubing and place in a clean test tube. Use a glucose dipstick to test for the presence and concentration of glucose. Test for the presence of the starch using Lugol's indicator.

Distilled water

Solution containing starch and glucose

4. What is the aim of the experiment? _____

_____

5. What part of a cell does the dialysis tubing represent? _____

6. In the spaces provided (below) draw the distribution of starch and glucose at the start and at the end of the experiment. Use the colored symbols shown under the table to represent starch and glucose:

Dialysis tubing start	Beaker start		Dialysis tubing end	Beaker end

● Starch

● Glucose

7. Explain your results: _____

_____

_____

©2022 **BIOZONE** International
ISBN: 978-1-98-856692-4
Photocopying Prohibited

39 Osmosis in Cells

Key Question: How is the movement of water affected by separating solutions containing high and low solute concentrations with a partially permeable membrane?

Osmosis

▸ Osmosis is the diffusion of water molecules from regions of lower solute concentration (higher free water concentration) to regions of higher solute concentration (lower free water concentration) across a partially permeable membrane.

▸ A partially permeable membrane lets some, but not all, molecules pass through. The plasma membrane of a cell is an example of a partially permeable membrane.

▸ Osmosis is a passive process (it requires no energy to occur).

Osmotic potential

The presence of solutes (dissolved substances) in a solution increases the tendency of water to move into that solution. This tendency is called the osmotic potential or osmotic pressure. The greater a solution's concentration, i.e. the more total dissolved solutes it contains, the greater the osmotic potential.

Dialysis tubing ready for use

Demonstrating osmosis

Osmosis can be demonstrated using the simple experiment described below.

A glucose solution (high solute concentration) is placed into dialysis tubing, and the tubing is placed into a beaker of water (low solute concentration). The difference in concentration of glucose (solute) between the two solutions creates an osmotic gradient. Water moves by osmosis into the glucose solution and the volume of the glucose solution inside the dialysis tubing increases.

The dialysis tubing acts as a partially permeable membrane, allowing water to pass freely, while keeping the glucose inside the dialysis tubing.

DID YOU KNOW?

A solvent dissolves other substances. A solute is the substance dissolved by a solvent to form a solution.

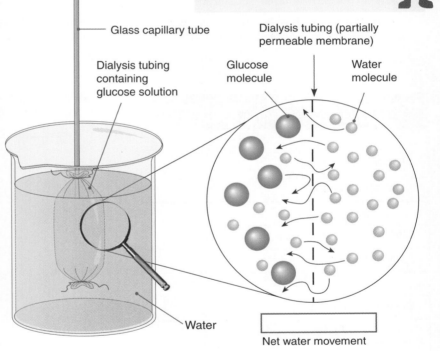

Glass capillary tube

Dialysis tubing containing glucose solution

Dialysis tubing (partially permeable membrane)

Glucose molecule

Water molecule

Water

Net water movement

1. What is osmosis? _____

_____

_____

2. (a) In the blue box on the diagram above, draw an arrow to show the direction of net water movement.

   (b) Why did water move in this direction? _____

_____

_____

_____

©2022 **BIOZONE** International
ISBN: 978-1-98-856692-4
Photocopying Prohibited

LS1.A

Estimating the total osmolarity of potato cells

▶ The osmolarity (a measure of solute concentration) of a cell or tissue can be estimated by placing part of the cell or tissue into a series of solutions of known concentration and observing if the tissue loses (hypertonic solution) or gains (hypotonic solution) water.

▶ The solution in which the tissue remains unchanged indicates the osmolarity of the tissue.

Investigation 2.3   Estimating osmolarity

See appendix for equipment list.

1. Prepare 6 beakers of sucrose ($C_{12}H_{22}O_{11}$, table sugar) solution with the concentrations of 0.0 (distilled water), 0.1, 0.2, 0.3, 0.4, and 0.5 mol/L of sucrose (0, 34.2 g , 68.5 g, 102.6 g, 136.9 g, and 171.1 g per litre). Label the beakers so that they can be easily identified at the end of the experiment.

2. Peel a potato and cut it into 18 identical cubes 1 cm^3 (1 cm x 1 cm x 1 cm) or use a cork borer to produce 18 identical cylinders of potato. Pat the potato cubes dry with a paper towel.

3. Weigh three cubes together, record their mass in the table below under initial mass. Place the cubes in the beaker of distilled water.

4. Repeat step 3 with the other 15 potato cubes and concentrations. Make sure you identify each beaker so the cubes can be weighed at the end of the experiment.

5. Leave the potato cubes in the solutions for at least 40 minutes (or up to 24 hours).

6. Remove the potato cubes from the distilled water and pat dry with a paper towel. Weigh all three together and record their mass in the table below under final mass.

7. Repeat for all the other concentrations of sucrose.

8. Calculate the change in mass (if any) for all the concentrations. Then calculate the % change (+ or –) (this removes any error based on the masses of the potato cubes not being identical).

9. Plot the % change vs sucrose concentration on the grid provided.

Sucrose concentration (mol/L)	Initial mass (I) (g)	Final mass (F) (g)
0.00		
Change (C) = (F-I) g		
% Change (C/I x 100)		
0.1		
Change (C) = (F-I) g		
% Change (C/I x 100)		
0.2		
Change (C) = (F-I) g		
% Change (C/I x 100)		
0.3		
Change (C) = (F-I) g		
% Change (C/I x 100)		
0.4		
Change (C) = (F-I) g		
% Change (C/I x 100)		
0.5		
Change (C) = (F-I) g		
% Change (C/I x 100)		

1. Use the grid below to draw a line graph of the sucrose concentration vs total % change in mass:

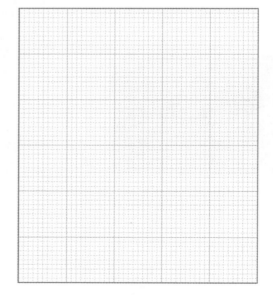

2. Use the graph to estimate the osmolarity of the potato (the point where there is no change in mass):

_____

3. Which of the solutions are hypotonic? Which are hypertonic?

_____

_____

40 Diffusion and Cell Size

Key Question: How does the surface area to volume ratio of a cell affect how substances diffuse into the cell's center?

Single-celled organisms

Single-celled organisms, e.g. *Amoeba*, are small and have a large surface area relative to the cell's volume. The cell's requirements can be met by the diffusion or active transport of materials directly into and out of the cell (below).

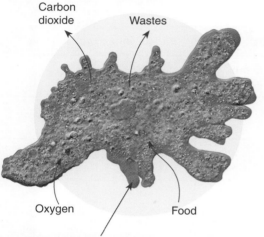

Carbon dioxide

Wastes

Oxygen

Food

The **plasma membrane**, which surrounds every cell, regulates the movement of substances into and out of the cell. For each square micrometer of membrane, only so much of a particular substance can cross per second.

Multicellular organisms

Multicellular organisms, e.g. plants and animals, are often large, and large organisms have a relatively small surface area compared to their volume. Diffusion alone is not sufficient to supply their cells with everything they need, so multicellular organisms need specialized systems to transport materials to and from their cells.

In a multicellular organism such as an elephant, the body's need for respiratory gases cannot be met by diffusion through the skin.

A specialized gas exchange surface (lungs) and circulatory (blood) system are required to supply the body's cells with oxygen and remove carbon dioxide.

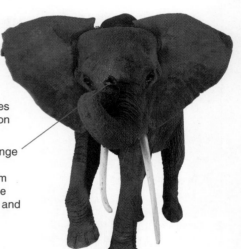

Smaller is better for diffusion

One large cube

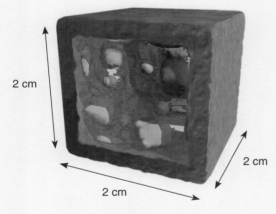

2 cm

2 cm

2 cm

Volume = 8 cm³

Surface area = 24 cm²

Eight small cubes

1 cm

1 cm

1 cm

Volume = 8 cm³ for 8 cubes

Surface area = 6 cm² for 1 cube

= 48 cm² for 8 cubes

The eight cells and the single large cell have the same total volume, but their surface areas are different. The small cells together have twice the total surface area of the cell, because there are more exposed (inner) surfaces. Real organisms have complex shapes, but the same principles apply.

The surface-area to volume relationship has important implications for the processes involving transport into and out of the cells across membranes. For activities such as gas exchange, the surface area available for diffusion is a major factor limiting the rate which the oxygen can be supplied to tissues.

©2022 **BIOZONE** International
ISBN: 978-1-98-856692-4
Photocopying Prohibited

SF LS1.A

1. Calculate the volume, surface area, and the ratio of surface area to volume for each of the four cubes below (the first has been done for you). Show your calculations as you complete the table below.

| 2 cm cube | 3 cm cube | 4 cm cube | 5 cm cube |

Cube size	Surface area (cm²)	Volume (cm³)	Surface area to volume ratio
2 cm cube	2 x 2 x 6 = 24 cm² (2 cm x 2 cm x 6 sides)	2 x 2 x 2 = 8 cm³ (height x width x depth)	24 to 8 = 3:1
3 cm cube			
4 cm cube			
5 cm cube			

2. Create a graph of the surface area against the volume of each cube, on the grid on the right. Draw a line connecting the points and label axes and units.

NEED HELP? See Activity 17

3. Which increases the fastest with increasing size: the volume or the surface area?

_____

_____

4. Explain what happens to the ratio of surface area to volume with increasing size:

_____

_____

_____

_____

_____

5. (a) Diffusion of substances into and out of a cell occurs across the cell surface. Describe how increasing the size of a cell affects the ability of diffusion to transport materials into and out of a cell:

_____

_____

(b) Describe how this places constraints on cell size, and explain how multicellular organisms have overcome this:

_____

_____

_____

©2022 **BIOZONE** International
 ISBN: 978-1-98-856692-4
 Photocopying Prohibited

41 Observing Diffusion in Cells

Key Question: How does the shape of a cell affect its surface area to volume ratio, and how does this affect how substances diffuse into it?

  Investigation 2.4 How cell shapes affect diffusion

See appendix for equipment list.

⚠ **Caution is required when using scalpels or razors. NaOH can burn skin and clothes, and irritates the eyes. You should wear protective eyewear.**

1. The diffusion of substances into cells of varying shapes and sizes can be modeled by soaking agar cubes infused with phenolphthalein indicator in sodium hydroxide (NaOH). Phenolphthalein turns pink in the presence of a base (NaOH). As the NaOH diffuses into the agar, the phenolphthalein changes to pink and indicates how far the NaOH has diffused into the agar.

2. Cut an agar block infused with phenolphthalein into three model cells of 1 x 1 x 1 cm, 2 x 2 x 2 cm, and 4 x 4 x 4 cm. Cut a fourth block into a shape 4 x 1 x 1 cm and a fifth shape 3 x 0.5 x 3 cm.

3. Fill a 200 ml beaker with 0.1 mol/L NaOH. Using tongs, place the model cells into the NaOH and leave to soak for 5 minutes.

4. Use tongs to remove the cells. Allow excess NaOH to drip off into the beaker before placing on folded paper towels. Dry the agar block carefully,

5. Cut each block in half and observe the extent to which the infused phenolphthalein has turned pink.

1. In the space below, draw scale cross sections of the model cells, showing where the indicator has changed their color:

2. Compare the three cubes. Comment on the effect of increasing the size of the cell in terms of diffusion:

_____

_____

_____

3. Which of the five shapes appears to be the best, in terms of nutrients being able to diffuse easily into the cells's center?

_____

_____

_____

©2022 **BIOZONE** International
ISBN: 978-1-98-856692-4

LS1.A

42 Factors Affecting Membrane Permeability

Key Question: What is the effect of increasing temperature on the permeability of cellular membranes?

Beetroot cubes

▸ Membrane permeability can be disrupted if membranes are subjected to high temperatures. At high temperatures, the membrane proteins become denatured, i.e. lose their structure. They no longer function properly and the membrane loses its selective permeability and becomes leaky.

▸ Plant cells often contain a large central vacuole surrounded by a membrane called a tonoplast. In beetroots, the cell vacuoles contain a water-soluble red pigment called betacyanin, which gives beetroot its color. If the tonoplast is damaged, the red pigment leaks out into the surrounding environment. The amount of leaked pigment relates to the amount of damage to the tonoplast.

Investigation 2.5 The effect of temperature on membrane permeability

See appendix for equipment list.

1. Use a cork borer with an internal diameter of 4 mm to produce 15 cylinders of beetroot 20 mm long. Place them in a beaker of distilled water.

2. Set up five sets of three test tubes of 5 mL of distilled of water at the following temperatures using water baths: 0°C (ice bath), 20°C, 40°C, 60°C, 90°C. Leave for a few minutes to equalise the distilled water temperatures with the water baths.

3. Remove the beetroot from the distilled water and pat dry with a paper towel. Place one cylinder of beetroot into each test tube. Leave them for 30 minutes.

4. Remove the beetroot from the test tubes. Observe each group of test tubes and record the color of the water in the table below.

5. Zero a colorimeter set to 530 nm with distilled water then use it to measure the absorbance of each beetroot sample and record the absorbance in the table below.

6. Calculate the mean absorbance for each temperature.

NEED HELP? See Activity 23

Temperature (°C)	Observation	Absorbance of beetroot samples at varying temperatures			Mean
		Absorbance at 530 nm			
		Sample 1	Sample 2	Sample 3	
0					
20					
40					
60					
90					

1. Why is it important to wash the beetroot cubes in distilled water prior to carrying out the experiment? _____

_____

2. (a) Based on the results in the table above, describe the effect of temperature on membrane permeability: _____

_____

_____

(b) Explain why this effect occurs: _____

_____

_____

    LS1.A

©2022 **BIOZONE** International
ISBN: 978-1-98-856692-4
Photocopying Prohibited

43 Active Transport

Key Question: What is active transport, and how does it transport molecules and ions across a cellular membrane?

- Active transport is the movement of molecules (or ions) from regions of low concentration to regions of high concentration across a plasma membrane.

- Active transport needs energy to proceed because molecules are being moved against their concentration gradient.

- The energy for active transport comes from ATP (adenosine triphosphate). Energy is released when ATP is hydrolyzed (water is added) forming ADP (adenosine diphosphate) and inorganic phosphate (P_i).

- Transport (carrier) **proteins** in the plasma membrane use energy to transport molecules across a membrane (below).

- Active transport can be used to move molecules into and out of a cell.

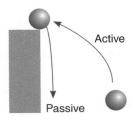

A ball falling is a passive process; it requires no energy input. Replacing the ball requires active energy input.

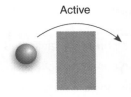

It requires energy to actively move an object across a physical barrier.

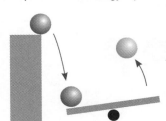

Sometimes, the energy of a passively moving object can be used to actively move another. For example, a falling ball can be used to catapult another (left).

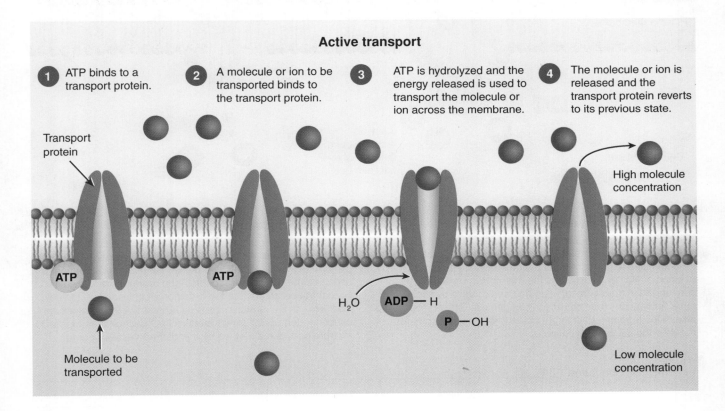

Active transport

1 ATP binds to a transport protein.

2 A molecule or ion to be transported binds to the transport protein.

3 ATP is hydrolyzed and the energy released is used to transport the molecule or ion across the membrane.

4 The molecule or ion is released and the transport protein reverts to its previous state.

Transport protein

High molecule concentration

ATP

ATP

H_2O ADP — H

P — OH

Molecule to be transported

Low molecule concentration

1. What is active transport? _____

_____

2. (a) Why does active transport require energy? _____

_____

(b) Where does the energy for active transport come from? _____

_____

©2022 **BIOZONE** International
ISBN: 978-1-98-856692-4
Photocopying Prohibited

LS1.A

44 What is an Ion Pump?

Key Question: How do ion pumps transport ions and molecules across cellular membranes?

▸ **Proteins** play an important role in the movement of molecules into and out of cells. When molecules are moved against their concentration gradient, energy is needed.

▸ Some lipid-soluble molecules can readily cross the cell membrane. Water soluble or charged molecules cross the membrane by facilitated diffusion or by active transport involving special membrane proteins called ion pumps.

▸ Ion pumps directly or indirectly use energy to transport ions across the membrane against a concentration gradient.

▸ The sodium-potassium pump (below, left) is found in almost all animal cells and is also common in plant cells. The concentration gradient created by ion pumps is often coupled to the transport of other molecules, such as glucose or sucrose, across the membrane (below right).

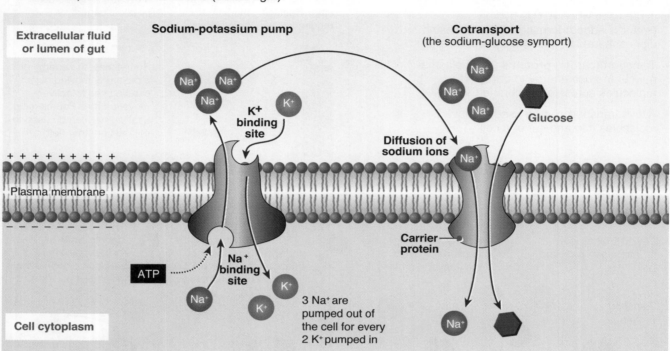

Sodium-potassium (Na+/K+) pump

The Na+/K+ pump is a protein in the membrane that uses energy in the form of ATP to exchange sodium ions (Na+) for potassium ions (K+) across the membrane. The unequal balance of Na+ and K+ across the membrane creates large concentration gradients that can be used to drive transport of other substances, e.g. cotransport of glucose. The Na+/K+ pump also helps to maintain the right balance of ions and so helps regulate the cell's water balance.

Cotransport (coupled transport)

A specific carrier protein controls the entry of glucose into the intestinal epithelial cells from the gut where digestion is taking place. The energy for this is provided indirectly by a gradient in sodium ions. The carrier "couples" the return of Na+ down its concentration gradient to the transport of glucose into the cell. The process is therefore called cotransport. A low intracellular concentration of Na+ (and therefore the concentration gradient for transport) is maintained by a sodium-potassium pump.

1. What is an ion pump? _____

_____

_____

2. (a) Explain what is meant by cotransport: _____

_____

(b) How is cotransport used to move glucose into the intestinal epithelial cells? _____

_____

_____

  LS1.A   SF

©2022 **BIOZONE** International
ISBN: 978-1-98-856692-4
Photocopying Prohibited

45 Specialization in Plant Cells

Key Question: How does cell specialization allow plant cells to carry out specialist functions?

Cell specialization

▸ A **specialized cell** is a cell with the specific features needed to perform a particular function in the organism.

▸ Cell specialization occurs during development when specific **genes** are switched on or off.

▸ Multicellular organisms have many types of specialized cells. These work together to carry out the essential functions of life.

▸ The size and shape of a cell allows it to perform its function. The number and type of **organelles** in a cell is also related to the cell's role in the organism.

Cells in the leaves of plants are often green because they contain the pigment chlorophyll, which is needed for photosynthesis.

Specialized cells in vascular tissue are needed to transport water and sugar around the plant.

Some cells are strengthened to provide support for the plant, allowing it to keep its form and structure.

Plants have root-hair cells so they can get water and nutrients (mineral ions) from the soil.

Many plant cells have a regular shape because of their semi-rigid cell wall. Specialized cells form different types of tissues. Simple tissues, like this onion epidermis, have only one cell type.

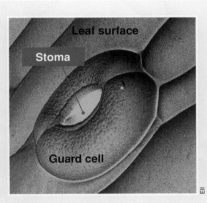

Specialized guard cells surround the stomata (pores) on plant leaves. The guard cells control the opening and closing of stomata and prevent too much water being lost from the plant.

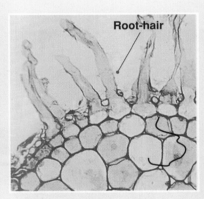

A plant root-hair is a tube-like outgrowth of a plant root cell. Their long, thin shape greatly increases their surface area. This allows the plant to absorb water and minerals efficiently.

1. What is a specialized cell? _____

_____

2. (a) Name the specialized cell that helps to prevent water loss in plants: _____

   (b) How does this cell prevent water being lost? _____

   _____

   _____

3. How do specialized root hairs help plants to absorb more water and minerals from the soil? _____

_____

_____

©2022 **BIOZONE** International
ISBN: 978-1-98-856692-4
Photocopying Prohibited

46 Specialization in Animal Cells

Key Question: How does cell modification allow animal cells to carry out specialist functions?

Specialization in animal cells

▸ There are over 200 different types of cell in the human body.

▸ Animal cells lack a cell wall, so they can take on many different shapes. Therefore, there are many more types of animal cells than there are plant cells.

▸ **Specialized cells** often have modifications or exaggerations to a normal cell feature to help them do their job. For example, nerve cells have long, thin extensions to carry nerve impulses over long distances in the body.

▸ Specialization improves efficiency because each cell type is highly specialized to perform a particular task.

Fat cell

Thin, flat epithelial cells line the walls of blood vessels (arrow). Large fat cells store lipid.

Some nerve cells are over 1 m long.

SEM: White blood cell

Louisa Howard, Katherine Connolly Dartmouth College

TEM: Cellular projections of intestinal cell

RBC

SEM: Egg and sperm

Some animal cells can move or change shape. A sperm cell must be able to swim so that it can fertilize an egg. A white blood cell changes its shape to engulf and destroy foreign materials, e.g. bacteria.

Cells that line the intestine have extended cell membranes. This increases their surface area so that more nutrients (food) can be absorbed. Red blood cells (RBCs) have no nucleus so they have more room inside to carry oxygen around the body.

The egg (ovum) is the largest human cell. It is about 0.1 mm in diameter and can be seen with the naked eye. The smallest human cells are sperm cells and red blood cells.

1. What is the advantage of cell specialization in a multicellular organism? _____

_____

_____

_____

2. For each of the following specialized animal cells, name a feature that helps it carry out its function:

   (a) White blood cell: _____

   (b) Sperm cell: _____

   (c) Nerve cell: _____

   (d) Red blood cell: _____

 LS1.A SF

©2022 **BIOZONE** International
ISBN: 978-1-98-856692-4

47 What is DNA?

Key Question: What does DNA do in the cell, and what does it look like when extracted from the cell?

About DNA

▸ **DNA** stands for deoxyribonucleic acid.

▸ DNA is called the blueprint for life because it contains all of the information an organism needs to develop, function, and reproduce.

▸ DNA stores and transmits genetic information.

▸ DNA is found in every **cell** of all living organisms.

▸ DNA has a double-helix structure (left). If the DNA in a single human cell was unwound, it would be more than two meters long! The long DNA molecules are tightly packed in an organized way so that they can fit into the nucleus.

DNA is found in every cell of all living organisms: animals, plants, fungi, protists, and bacteria.

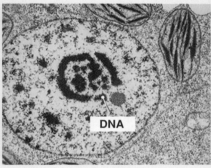

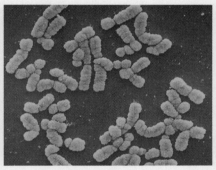

DNA contains the instructions an organism needs to develop, survive, and reproduce. Small differences in DNA cause differences in appearance.

In eukaryotes, most of the cell's DNA is located in the nucleus (above). A very small amount is located in mitochondria and in the chloroplasts of plants.

The DNA in eukaryotes is packaged into chromosomes (above). Each chromosome is made up of a DNA molecule and associated **proteins**. The proteins help to package the DNA into the nucleus.

1. (a) What does DNA stand for? _____

   (b) Where is most of the DNA found in eukaryotes? _____

   (c) What does DNA do? _____

   _____

   _____

2. (a) How is DNA packaged in eukaryotes? _____

   _____

   (b) Why does DNA have to be tightly packaged? _____

   _____

   _____

©2022 **BIOZONE** International
ISBN: 978-1-98-856692-4
Photocopying Prohibited

LS1.A

Extracting DNA from cells

▶ In a lab, scientists usually use extraction kits to separate **DNA** from cells. These will contain all the parts needed to accurately remove the DNA, but in general they are all based on the same technique. This includes breaking open the cells, precipitating the DNA, and removing contaminants.

▶ In a classroom, DNA is easily extracted by precipitating it out of solution using ice cold ethanol. It is good to use strawberries for this method because they are octaploid (have 8 sets of chromosomes) and their color makes it easy to see the precipitating DNA. However, other fruits and vegetables, such as kiwifruit, bananas, and broccoli can also be used.

Investigation 2.6   Extracting DNA

See appendix for equipment list.

Work in pairs for this activity.

1. Take 5-6 strawberries and place them in a large zip-lock bag. Squash the strawberries into a smooth paste. This mechanically breaks up the cells, but does not release the DNA.

2. To release the DNA, add 100 mL of water, 5 mL of detergent, and a pinch of salt to the paste. Reseal the bag and mix the contents by squashing and crumpling the bag. The detergent breaks down the cellular membranes and deactivates DNases, which would chop up the DNA. The salt helps to remove the proteins bound to the DNA and keeps them in solution. Positive ions in the salt also neutralise the negative charge of the DNA.

3. Place a piece of filter paper in a funnel and position the funnel so the excess fluid can drain into a beaker. Pour the contents of the bag into the filter funnel and allow it to drain. It should produce a clear reddish solution (right).

4. Gently add the ethanol on top of the strawberry solution by placing a clean glass rod into the beaker and carefully pouring the ethanol down the rod. Add ethanol until there are equal volumes of strawberry solution and ethanol.

5. Ethanol removes the water from around the DNA so it precipitates where the ethanol and the solution meet, forming whitish glue-like strands. Low temperatures speed up the precipitation and limit DNase activity.

6. The DNA strands can be centrifuged with ethanol to isolate the DNA as a pellet.

Yaminchhipa10 CC 4.0

3. In the extraction and isolation of DNA:

(a) Why is it necessary to disrupt the cellular membranes? _____

(b) Why does the DNA precipitate out in ethanol? _____

(c) For a DNA extraction, why is it helpful that strawberries are octaploid? _____

(d) Why is salt added? _____

_____

(e) What is the purpose of the filter funnel? _____

_____

4. In a DNA extraction, student A obtained DNA in long threads, whereas student B obtained DNA that appeared fluffy. Account for the differences in these two results and suggest what student B might have done incorrectly?

_____

_____

_____

_____

©2022 **BIOZONE** International
ISBN: 978-1-98-856692-4
Photocopying Prohibited

48 Nucleotides

Key Question: What is the structure and function of nucleotides and what are the three components that make them up?

The structure of a nucleotide

Nucleotides are the building blocks of nucleic acids (**DNA** and **RNA**). Nucleotides have three parts to their structure (see diagrams below):

▶ A nitrogen containing base

▶ A five carbon sugar

▶ A phosphate group

Symbolic form of a nucleotide
(showing positions of the 5 C atoms on the sugar)

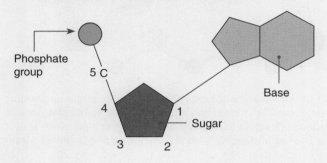

Chemical structure of a nucleotide

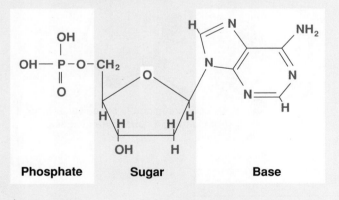

| Phosphate | Sugar | Base |

Nucleotide bases

Five different kinds of nitrogen bases are found in nucleotides. These are:

Adenine (A)

Guanine (G)

Cytosine (C)

Thymine (T)

Uracil (U)

DNA contains adenine, guanine, cytosine, and thymine.

RNA also contains adenine, guanine, and cytosine, but uracil (U) is present instead of thymine.

Purines:
(two-ring bases)

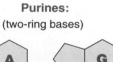

Adenine   Guanine

Pyrimidines:
(single-ring bases)

Cytosine   Thymine   Uracil
              (DNA only)   *(RNA only)*

Sugars

Nucleotides contain one of two different sorts of sugars. **Deoxyribose** sugar is only found in DNA. **Ribose** sugar is found in RNA.

Deoxyribose sugar   Ribose sugar
(found in DNA)   (found in RNA)

1. What are the three components of a nucleotide? _____

_____

2. List the nucleotide bases present:

(a) In DNA: _____

(b) In RNA: _____

3. Name the sugar present:

(a) In DNA: _____   (b) In RNA: _____

SF   LS1.A

49 DNA and RNA

Key Question: What is the difference between DNA and RNA, and what are their functions in the cell?

The structure of DNA

▸ **Nucleotides** join together to form nucleic acids.

▸ Deoxyribonucleic acid (**DNA**) is a nucleic acid.

▸ DNA consists of two strands of nucleotides linked together to form a double helix. A double helix is like a ladder twisted into a corkscrew shape. The rungs of the ladder are the two nitrogen bases joined by hydrogen bonds. The double helix is "unwound" to show its structure in the diagram (right).

Who discovered the DNA double helix?

Two scientists, James Watson and Francis Crick, are credited with discovering the structure of DNA. However, they used X-ray pictures of DNA from another scientist, Rosalind Franklin, to confirm their hypothesis.

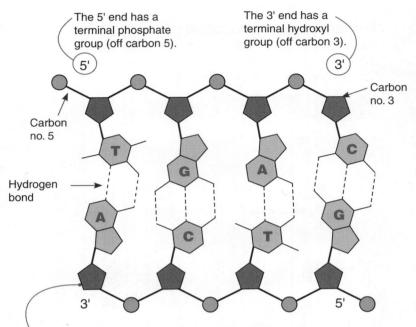

The 5' end has a terminal phosphate group (off carbon 5).

The 3' end has a terminal hydroxyl group (off carbon 3).

Carbon no. 3

Carbon no. 5

Hydrogen bond

The DNA backbone is made up of alternating phosphate and sugar molecules. Each DNA strand has a direction. The single strands run in the opposite direction to each other (they are anti-parallel). The ends of a DNA strand are labeled 5' (five prime) and 3' (three prime).

The structure of RNA

Ribonucleic acid (RNA) is a type of nucleic acid. Like DNA, the nucleotides are linked together through a condensation reaction. RNA is single stranded, and has many functions including **protein** synthesis, and cell regulation. There are 3 types of RNA:

▸ Messenger RNA (mRNA)

▸ Transfer RNA (tRNA)

▸ Ribosomal RNA (rRNA)

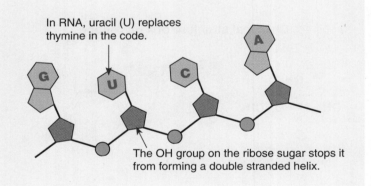

In RNA, uracil (U) replaces thymine in the code.

The OH group on the ribose sugar stops it from forming a double stranded helix.

1. The diagram on the right shows a double-stranded DNA molecule. Fill in the boxes (a-e) using the following terms:

   Sugar group   Purine bases
   Phosphate group   Pyrimidine bases
   Hydrogen bonds

2. If you wanted to use a radioactive or fluorescent tag to label only the RNA in a cell and not the DNA, what molecule(s) would you label?

3. If you wanted to use a radioactive or fluorescent tag to label only the DNA in a cell and not the RNA, what molecule(s) would you label?

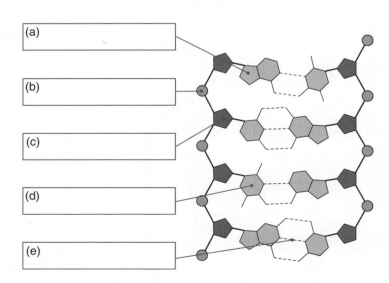

(a)

(b)

(c)

(d)

(e)

LS1.A   SF

©2022 **BIOZONE** International
ISBN: 978-1-98-856692-4
Photocopying Prohibited

50 Modeling the Structure of DNA

Key Question: How does the base pairing rule determine the way nucleotides join together to form DNA?

The exercise on the following pages is designed to help you understand the structure of **DNA**, and learn the base pairing rule for DNA.

The way the nucleotide bases pair up between strands is very specific. The chemistry and shape of each base means they can only bond with one other DNA nucleotide. Use the information in the table below if you need help remembering the base pairing rule while you are constructing your DNA molecules.

DID YOU KNOW?

Chargaff's rules

Before Watson and Crick described the structure of DNA, an Austrian chemist called Chargaff analyzed the base composition of DNA from a number of organisms. He found that the base composition varies between species but that within a species the percentage of A and T bases are equal and the percentage of G and C bases are equal. Validation of Chargaff's rules was the basis of Watson and Crick's base pairs in the DNA double helix model.

DNA base pairing rule				
Adenine	always pairs with	**Thymine**	A	⟷ T
Thymine	always pairs with	**Adenine**	T	⟷ A
Cytosine	always pairs with	**Guanine**	C	⟷ G
Guanine	always pairs with	**Cytosine**	G	⟷ C

1. Cut out each of the nucleotides on page 71 by cutting along the columns and rows (see arrows indicating two such cutting points). Although drawn as geometric shapes, these symbols represent chemical structures.

2. Place one of each of the four kinds of nucleotide on their correct spaces below:

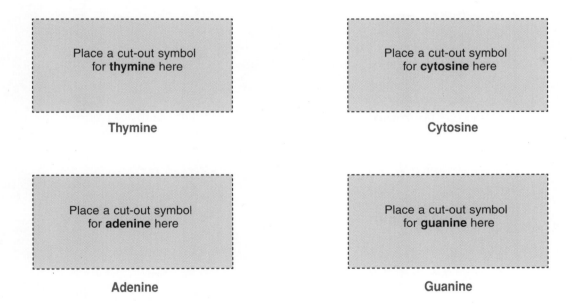

> Place a cut-out symbol for **thymine** here
>
> **Thymine**

> Place a cut-out symbol for **cytosine** here
>
> **Cytosine**

> Place a cut-out symbol for **adenine** here
>
> **Adenine**

> Place a cut-out symbol for **guanine** here
>
> **Guanine**

3. Identify and label each of the following features on the adenine nucleotide immediately above: **phosphate, sugar, base, hydrogen bonds**.

4. Create one strand of the DNA molecule by placing the 9 correct "cut out" nucleotides in the labeled spaces on the following page (DNA molecule). Make sure these are the right way up (with the P on the left) and are aligned with the left hand edge of each box. Begin with thymine and end with guanine.

5. Create the complementary strand of DNA by using the base pairing rule above. Note that the nucleotides have to be arranged upside down.

6. Once you have checked that the arrangement is correct, glue, paste, or tape these nucleotides in place.

©2022 **BIOZONE** International
ISBN: 978-1-98-856692-4

SF SSM LS1.A

DNA molecule

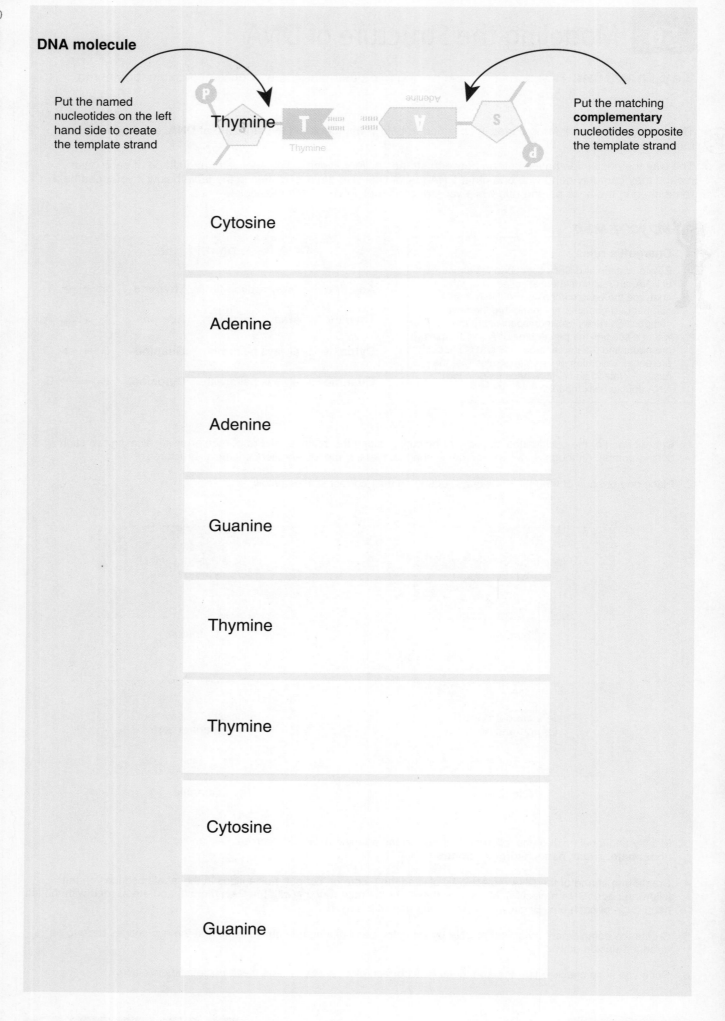

Put the named nucleotides on the left hand side to create the template strand

Thymine

Cytosine

Adenine

Adenine

Guanine

Thymine

Thymine

Cytosine

Guanine

Put the matching **complementary** nucleotides opposite the template strand

©2022 **BIOZONE** International
ISBN: 978-1-98-856692-4
Photocopying Prohibited

Nucleotides

Cut out this page and separate each of the 24 nucleotides
by cutting along the columns and rows (see arrows indicating the cutting points).

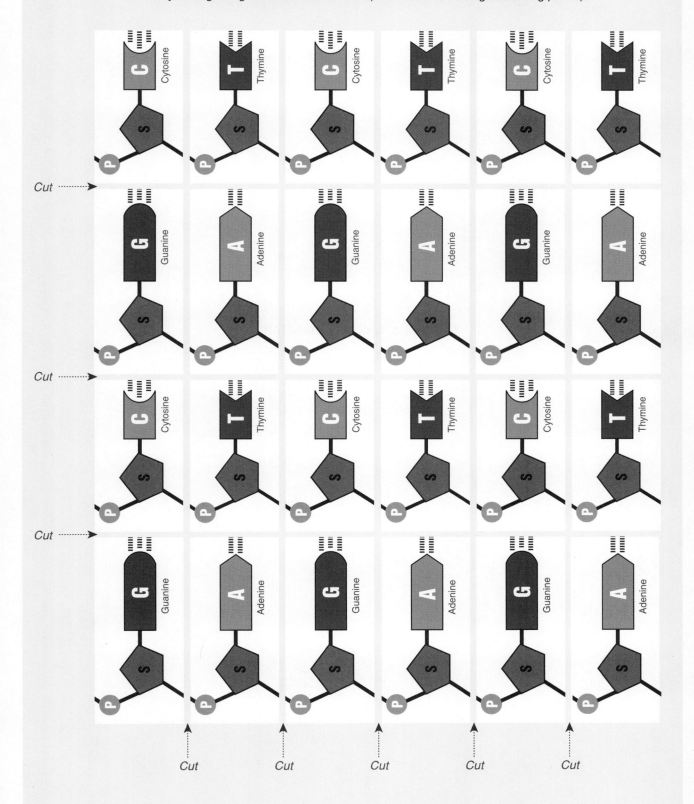

This page is deliberately left blank

51 Genes Code for Proteins

Key Question: What are genes? What is the relationship between genes and proteins?

▸ A **gene** is a section of **DNA** that codes for a **protein**. Gene expression is the process of rewriting a gene into a protein. It involves two stages: transcription of the DNA and translation of the mRNA into protein.

▸ A gene is bounded by a start (promoter) region, upstream of the gene, and a terminator region, downstream of the gene. These regions control transcription by telling RNA polymerase where to start and stop.

▸ RNA polymerase binds to the promoter region to begin transcription of the gene.

▸ The one gene-one protein model is helpful for visualizing the processes involved in gene expression, although it is overly simplistic for eukaryotes. The information flow for gene to protein is shown below.

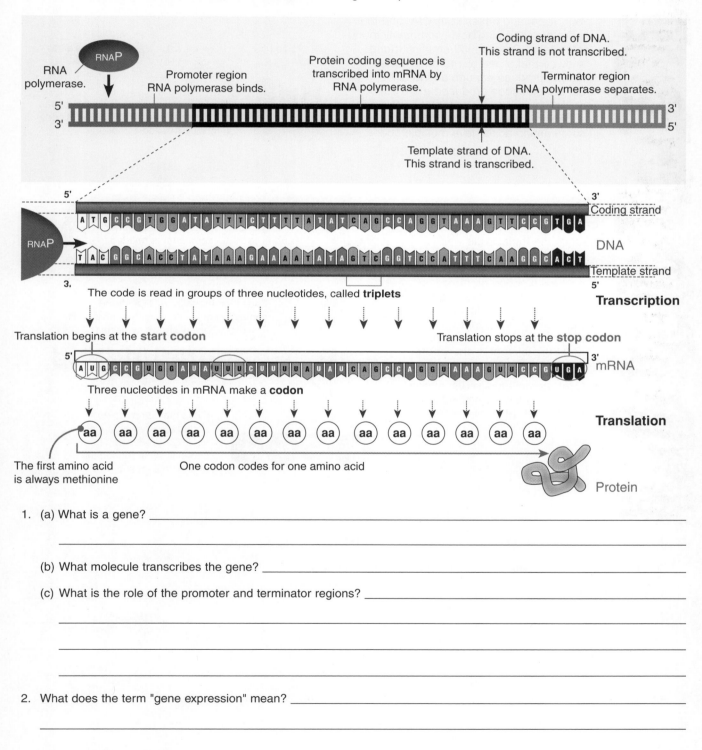

1. (a) What is a gene? _____

    _____

    (b) What molecule transcribes the gene? _____

    (c) What is the role of the promoter and terminator regions? _____

    _____

    _____

    _____

2. What does the term "gene expression" mean? _____

    _____

    _____

©2022 **BIOZONE** International
ISBN: 978-1-98-856692-4

52 Cracking the Genetic Code

Key Question: How did scientists discover which three letter triplets coded for which animo acids found in proteins?

The genetic code

▶ Once it was discovered that **DNA** carries the genetic code needed to produce **proteins**, the race was on to "crack the code" and find out how it worked.

▶ The first step was to find out how many nucleotide bases code for an amino acid. Scientists knew that there were four nucleotide bases in mRNA, and that there are 20 amino acids commonly found in proteins. Simple mathematics (right) showed that a one or two base code did not produce enough amino acids, but a triplet code produced more amino acids than existed. The triplet code was accepted once scientists confirmed that some amino acids have multiple codes.

Number of bases in the code	Working	Number of amino acids produced
Single (4^1)	4	4 amino acids
Double (4^2)	4 x 4	16 amino acids
Triple (4^3)	4 x 4 x 4	64 amino acids

A triplet (three nucleotide bases) codes for a single amino acid. The triplet code on mRNA is called a codon.

How was the genetic code cracked?

▶ Once the triplet code was discovered, the next step was to find out which amino acid each codon produced. In 1961 two scientists, Marshall Nirenberg and Heinrich Matthaei, developed an experiment to crack the code. Their experiment is shown on the right.

▶ Over the next few years, similar experiments were carried out using different combinations of nucleotides until all of the codes were known.

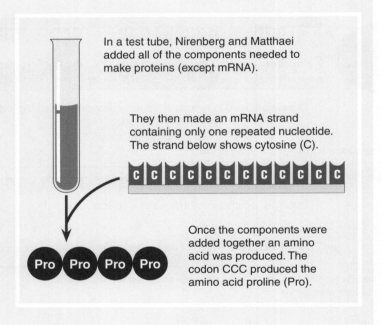

In a test tube, Nirenberg and Matthaei added all of the components needed to make proteins (except mRNA).

They then made an mRNA strand containing only one repeated nucleotide. The strand below shows cytosine (C).

Once the components were added together an amino acid was produced. The codon CCC produced the amino acid proline (Pro).

1. (a) How many types of nucleotide bases are there in mRNA? _____

(b) How many different amino acids are commonly found in proteins? _____

(c) Why did scientists reject a one or two base code when trying to work out the genetic code? _____

_____

_____

_____

2. A triplet code could potentially produce 64 amino acids. Why are only 20 amino acids produced? _____

_____

_____

   LS1.A SF

©2022 **BIOZONE** International
ISBN: 978-1-98-856692-4
Photocopying Prohibited

53 Amino Acids Make Up Proteins

Key Question: How does the sequence of amino acids in a protein determine a protein's shape and function?

Proteins are made up of amino acids

▶ **Proteins** are large molecules made up of many smaller units called amino acids joined together. The amino acids are joined together by peptide bonds. The sequence of amino acids in a protein is determined by the order of **nucleotides** in **DNA**.

▶ All amino acids have a common structure (right) consisting of an amine group, a carboxyl group, a hydrogen atom, and an 'R' group. Each type of amino acid has a different 'R' group (side chain). Each 'R' group has a different chemical property.

▶ The chemical properties of the amino acids are important because the chemical interactions between amino acids cause a protein to fold into a specific, three-dimensional shape. The protein's shape helps it carry out its specialized role.

The general structure of an amino acid

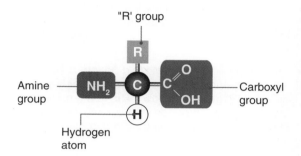

A polypeptide chain

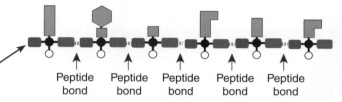

The order of amino acids in a protein is directed by the order of nucleotides in DNA (and therefore mRNA).

Peptide bond   Peptide bond   Peptide bond   Peptide bond   Peptide bond

The shape of a protein determines its role

▶ The sequence of amino acids determines how a protein will fold up, i.e. the shape it will form.

▶ The shape of a protein determines its role. Proteins generally fall into two groups, globular and fibrous (right).

▶ The shape of a protein is so important to its function, that if the structure of a protein is **denatured** (destroyed) it can no longer carry out its function.

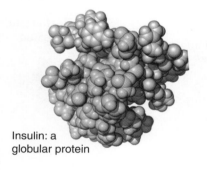

Insulin: a globular protein

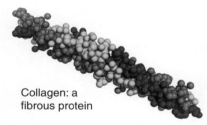

Collagen: a fibrous protein

Globular proteins

▶ Globular proteins are round and water soluble. Their functions include:
- Catalytic, e.g. enzymes.
- Regulation, e.g. hormones.
- Transport, e.g. hemoglobin.
- Protective, e.g. antibodies.

Fibrous proteins

▶ Fibrous proteins are long and strong. Their functions include:
- Support and structure, e.g. connective tissue.
- Contractile, e.g. myosin, actin.

1. (a) Name the four components of an amino acid: _____

_____

(b) What makes each type of amino acid unique? _____

2. Why are proteins important in organisms? _____

_____

3. (a) Why is the shape of a protein important? _____

_____

(b) What happens to a protein if it loses its shape? _____

_____

SF   LS1.A

54 The Functional Structure of Proteins

Key Question: How does modeling help us understand the structure of a protein?

Proteins fold up into a functional structure

▸ The amino acid sequence of a protein is only the first step in making a functional protein. A protein must fold into a functional structure in order to carry out its biological role. This is where the 'R' groups become important.

▸ The amino acid chain will fold up into a specific shape depending on the interactions between the different 'R' groups (below). These interactions include hydrogen bonds, disulfide (S – S) bonds, and hydrophobic (water-hating) and hydrophilic (water-loving) interactions.

▸ First, the amino acid chain folds into coils (helices) and sheets to create a secondary structure. These shapes are created and maintained by hydrogen bonds between CO and NH groups.

▸ More distant parts of the folded chain can then interact to create a highly organized, tertiary structure. Disulfide bridges are important in maintaining the folded tertiary structure.

▸ Some functional proteins, such as haemoglobin (below) consist of two or more polypeptide chains. When multiple polypeptides come together, the protein has a quaternary structure. The functional structure of hemoglobin also includes four iron atoms. These enable hemoglobin to bind oxygen.

Primary structure
Amino acid chain

Secondary structure
Coiled helix

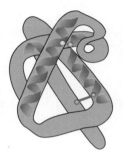

Tertiary structure
Folded helices

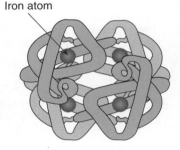

Iron atom

Quaternary structure
Multi-unit protein

Investigation 2.7 Modeling protein structure

See appendix for equipment list.

Work in pairs for this activity.

1. You will need pipe cleaners with four colors. We have used 2 white, 2 pink, 2 purple, and 4 blue but you can swap out for the colors you have. Each color represents a different amino acid.

2. Twist a loop in the center of each pipe cleaner (figure 1). The twist represents the amino acid's functional group.

3. Join the amino acids together (figure 2) by twisting their arms together in the following sequence:
   1) white 2) pink 3) blue 4) purple 5) blue 6) pink 7) blue 8) white 9) blue 10) purple.

What level of protein organization does the structure in figure 2 represent? _____

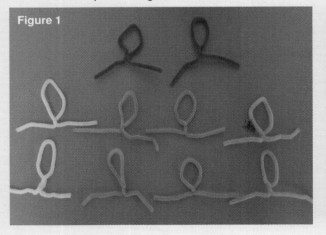

Figure 1

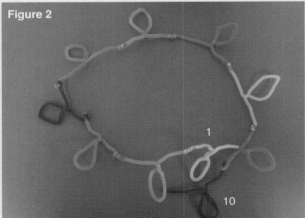

Figure 2

1

10

    LS1.A

©2022 **BIOZONE** International
ISBN: 978-1-98-856692-4

4. Attach sticky tape to the loops of the purple pipe cleaners and to one arm of each of the blue pipe cleaners. These represent places where hydrogen bonding can occur.

5. Join the sticky tape together between amino acids 3 and 5 and also between amino acids 7 and 9 (figure 3).

   Describe what happens to the shape of the model when you do this: _____

   _____

   _____

   What level of protein structure does this represent? _____

6. Attach binder clips or paper clips to the loops of the two pink amino acids and then use the clips to join the two pink amino acids together. The clips represent a disulfide bond.

7. Join the sticky tape together on the two purple amino acids (figure 4). Your protein has now formed its fully functional structure.

   What level of protein structure does this represent? _____

Figure 3   H = hydrogen bond

Figure 4

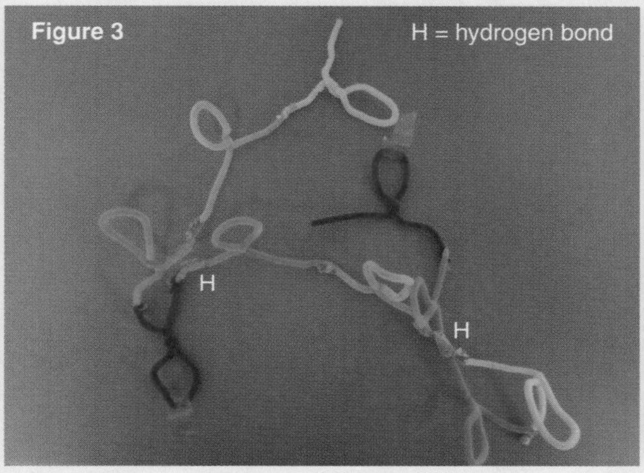

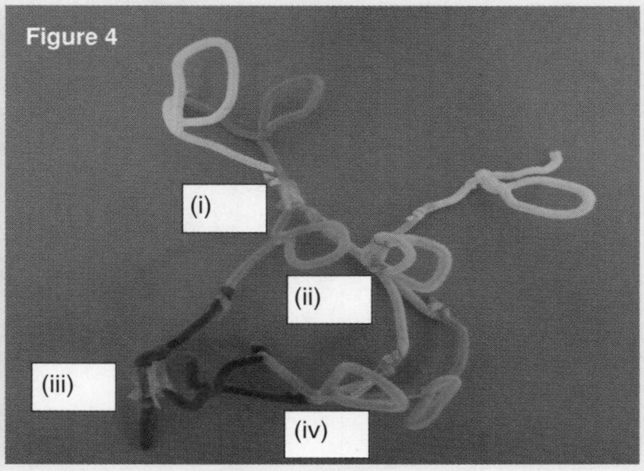

8. (a) Label figure 4 to show the location of all of the hydrogen bonds (H) and the disulfide bond (S).

   (b) Based on the properties of your model and its components, which of these bonds is likely the strongest?

   _____

9. Break the hydrogen bonds between amino acids 3 and 5 and also between 7 and 9 in your molecule.

   (a) What happens to the shape of the protein? _____

   _____

   (b) What process does breaking these bonds represent? _____

   _____

   (c) What effect will this process have on the protein's ability to carry out its job? _____

   _____

   _____

   _____

1. How could you adapt your model to demonstrate quaternary structure? _____

   _____

   _____

   _____

   _____

55 Proteins Have Many Roles in Cells

Key Question: What kinds of proteins are found in the body and what are their numerous roles?

▸ In **eukaryotic cells**, most of the genetic information (**DNA**) is found in the nucleus. DNA provides instructions that code for the formation of proteins. Proteins carry out most of a cell's work. A cell produces many different types of proteins and each carries out a specific task in the cell.

▸ Proteins are involved in the structure, function, and regulation of the body's cells, tissues, and organs. Without functioning proteins, a cell cannot carry out its specialized role and the organism may die.

The nucleus is the control center of a cell

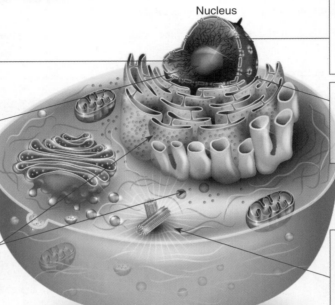

Nucleus

The DNA within the nucleus carries the instructions for the cell's structure and function. This involves producing proteins.

Ribosomes are made in the nucleolus (a dense region within the nucleus).

Proteins are made outside the nucleus by ribosomes. These may be free in the cytoplasm or associated with the rough endoplasmic reticulum (rER).

Genes (sections of DNA) code for specific proteins. A cell can control the type of protein it makes by only expressing the genes for the proteins it needs.

The double-layered nuclear membrane has pores to allow materials to move between the nucleus and the cytoplasm.

Microtubules made of protein form the cell's internal skeleton. This includes the centrioles and spindle fibers involved in cell division.

A generalized animal cell

▸ While a generalized cell produces a range of proteins, some body cells are highly specialized to produce large amounts of a specific protein. This specialization defines their functional role. Three examples are pictured below.

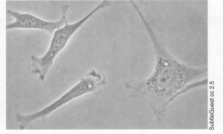
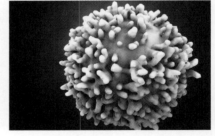

Cells within specialized regions of the pancreas produce and release a protein hormone called insulin. Insulin (red in photo) helps regulate blood glucose.

Fibroblasts are specialized cells that continuously produce and secrete the materials that form connective tissue, including the protein called collagen.

B lymphocytes (B cells) are white blood cells that are specialized to produce and secrete proteins called antibodies which protect the body against diseases.

1. Suggest what might happen to a protein's functionality if it was incorrectly encoded by the DNA. Explain your answer:

    _____

    _____

    _____

    _____

2. The opposite page shows six pictograms of proteins in action, six protein functions, six protein examples, and six photographs. These are not in any matched order. Cut out the 24 boxes and paste or tape them into the grid on the next page so that each pictogram is matched with its correct function, example, and illustrative photograph.

LS1.A

©2022 **BIOZONE** International
ISBN: 978-1-98-856692-4
Photocopying Prohibited

PICTOGRAM	FUNCTION	EXAMPLE	PHOTO EXAMPLE

Internal defense

Antibodies (also called immunoglobulins) are "Y" shaped proteins that protect the body by identifying and killing disease-causing organisms such as bacteria and viruses.

Immunoglobulin A

IgA is found in the gut and airways. It destroys disease-causing organisms growing in these areas.

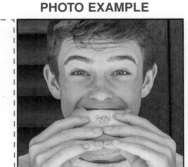

Catalytic

Thousands of different chemical reactions take place in an organism every minute. Each chemical reaction is catalyzed by enzymes. The word-ending "ase" indicates an enzyme.

Actin & myosin

Two proteins that work together to bring about contraction (movement) in all the muscles of the body, including those that work without your awareness.

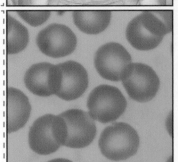

Regulation

Regulatory proteins such as hormones act as signal molecules to control biological processes and coordinate responses in cells, tissues, and organs.

Hemoglobin

A protein found in red blood cells. It binds oxygen and carries it through the blood, delivering it to cells.

Movement

Contractile proteins are involved in movement of muscles and form the internal supporting structures of cells.

Collagen

Found in the skin and connective tissues, including bones, tendons, and ligaments. It is the most abundant protein in the body.

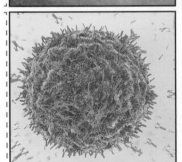

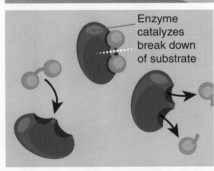

Transport

Proteins can carry substances around the body or across membranes. In the blood, they transport and store oxygen. In cell membranes, they help molecules move into and out of cells.

Estrogen

A hormone that is critical for reproduction in females. Estrogen levels increase during pregnancy to maintain a healthy pregnancy.

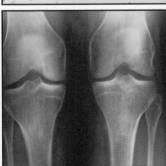

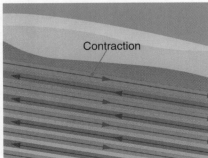

Structural

Structural proteins provide physical support or protection. They are strong, fibrous (thread like) and stringy.

Amylase

An enzyme that breaks down starch into sugars in the first stage of digestion.

This page is deliberately left blank

PICTOGRAM	FUNCTION	EXAMPLE	PHOTO EXAMPLE

56 Reactions in Cells

Key Question: How do anabolic and catabolic reactions build or break down molecules in the body?

▸ Metabolism refers to all of the chemical reactions carried out within a living organism to maintain life. All metabolic reactions are controlled by **enzymes**.

▸ There are two categories of metabolic reactions: anabolic and catabolic.

Anabolic reactions

▸ Anabolic reactions are reactions that result in the production (synthesis) of a more complex molecule from smaller components or smaller molecules. During anabolic reactions, simple molecules are joined to form a larger, more complex molecule.

▸ Anabolic reactions need a net input of energy to proceed. They are called endergonic reactions.

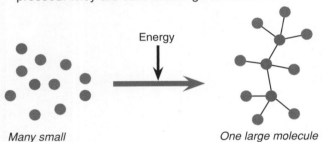

Many small molecules   *One large molecule*

Catabolic reactions

▸ Catabolic reactions are reactions that break down large molecules into smaller components.

▸ Catabolic reactions involve a net release of energy. They are called exergonic reactions.

▸ Catabolic reactions are the opposite of anabolic reactions.

▸ The energy released from catabolic reactions can be used to drive other metabolic processes.

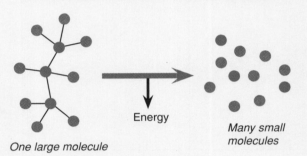

One large molecule   *Many small molecules*

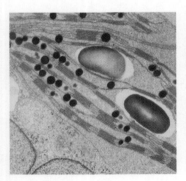

Plants carry out photosynthesis in **organelles** called chloroplasts (left). Photosynthesis is an anabolic process because it converts carbon dioxide and water into glucose. Energy from the sun is required to drive photosynthesis.

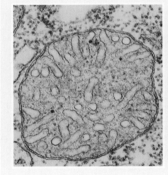

Cellular respiration is an example of a catabolic reaction. Glucose is broken down in a series of reactions to release carbon dioxide, water, and ATP (energy). The energy is used to fuel other activities in the cell. Some stages of cellular respiration take place in the mitochondria (left).

1. What is an anabolic reaction? _____
   _____

2. (a) What is a catabolic reaction? _____
   _____

   (b) Why are catabolic reactions considered to be the opposite to anabolic reactions? _____
   _____
   _____

3. Identify the following reactions as either catabolic or anabolic:

   (a) Protein synthesis: _____   (c) Digestion: _____

   (b) ATP conversion to ADP: _____   (d) DNA synthesis: _____

©2022 **BIOZONE** International
ISBN: 978-1-98-856692-4
Photocopying Prohibited

LS1.A

57 Enzymes Catalyze Reactions in Cells

Key Question: What are enzymes and what role do they play in biological reactions?

What are enzymes?

▶ **Enzymes** are **proteins**. They control all the metabolic reactions that take place in a cell.

▶ Enzymes are called biological **catalysts** because they speed up biochemical reactions.

▶ During the reaction, the enzyme itself remains unchanged, and is not used up.

▶ Each enzyme controls a very specific metabolic reaction, or series of related metabolic reactions.

▶ Enzymes may break down a single substrate molecule (catabolism), or join two or more substrate molecules together (anabolism).

▶ Extremes of temperature or pH can alter the enzyme's active site where catalysis occurs. This can lead to loss of function and is called denaturation.

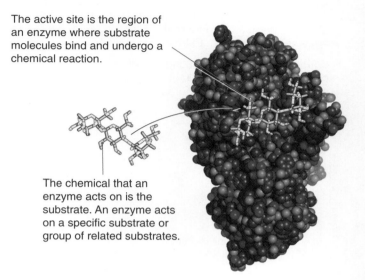

The active site is the region of an enzyme where substrate molecules bind and undergo a chemical reaction.

The chemical that an enzyme acts on is the substrate. An enzyme acts on a specific substrate or group of related substrates.

How enzymes work

▶ An early model to explain enzyme activity described the enzyme and its substrate as a lock and key, where the substrate fitted neatly into the active site of the enzyme. Evidence showed this model to be flawed and it has since been modified to recognize the flexible nature of enzymes (the induced fit model).

▶ The induced fit model for enzyme action is shown below. The shape of the enzyme changes when the substrate fits into the active site. The substrate becomes bound to the enzyme by weak chemical bonds. This weakens bonds within the substrate itself, allowing the reaction to proceed more readily.

Induced fit model of enzyme activity

1 Substrate molecules — **Enzyme** — Active site

2 **Enzyme** — Enzyme changes shape

2b **Enzyme**

3 **Enzyme** — End product released

Two substrate molecules are drawn into the active site of the enzyme.

The binding of the substrate causes the enzyme to change shape and the catalytic parts of the enzyme contact the substrate. The reaction can then occur.

The end product is released and the enzyme returns to its previous shape.

1. (a) Why are enzymes so important in metabolism? _____

_____

_____

(b) Why do you think an organism needs so many different enzymes? _____

_____

_____

2. Describe the induced fit model of enzyme activity: _____

_____

_____

_____

©2022 **BIOZONE** International
ISBN: 978-1-98-856692-4
Photocopying Prohibited

SF LS1.A

58 Enzymes Have Optimal Conditions to Work

Key Question: What conditions are optimal for enzymes, and what happens to their structure and function outside of these conditions?

▸ **Enzymes** usually have a set of conditions (e.g. pH and temperature) where their activity is greatest. This is called their optimum. At low temperatures, the activity of most enzymes is very slow, or does not proceed at all. Enzyme activity increases with increasing temperature, but falls off after the optimum temperature is exceeded and the enzyme is denatured. Extremes in pH can also cause denaturation.

▸ Within their normal operating conditions, enzyme reaction rates are influenced by enzyme and substrate concentration in a predictable way (below). In the graphs below, the rate of reaction or degree of enzyme activity is plotted against each of four factors that affect enzyme function.

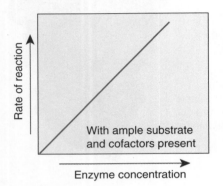

1. **Enzyme concentration**

   (a) Describe the change in the rate of reaction when the enzyme concentration is increased (assuming there is plenty of the substrate present):

   _____

   _____

   (b) Suggest how a cell may vary the rate of an enzyme controlled reaction:

   _____

   _____

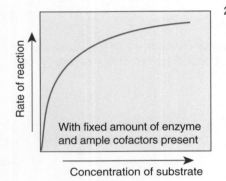

2. **Substrate concentration**

   (a) Describe the change in the rate of reaction when the substrate concentration is increased (assuming a fixed amount of enzyme):

   _____

   _____

   (b) Explain why the rate changes the way it does: _____

   _____

   _____

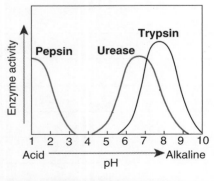

3. **Temperature**

   Higher temperatures speed up all reactions, but few enzymes can tolerate temperatures higher than 50–60°C. The rate at which enzymes are denatured (change their shape and become inactive) increases with higher temperatures.

   (a) Describe what is meant by an optimum temperature for enzyme activity:

   _____

   _____

   (b) Explain why most enzymes perform poorly at low temperatures:

   _____

   _____

4. **Acidity and alkalinity (pH)**

   Each enzyme has a optimum pH range for activity and will be denatured by conditions where the pH is outside this range, e.g. very acid or alkaline. Most enzymes in humans have pH optima that are close to the pH of the body's cells (near neutral pH). However, some enzymes work outside of this range in order to perform their specific function.

   State the optimum pH for each of the enzymes:

   Pepsin: _____   Trypsin: _____   Urease: _____

©2022 **BIOZONE** International
ISBN: 978-1-98-856692-4
Photocopying Prohibited

Investigating enzyme function

▶ **Enzymes** are a group of proteins that act as **catalysts**. This means they cause or speed up (catalyze) chemical reactions, including all those occurring in our bodies. If it wasn't for enzymes, these reactions would happen very slowly, too slowly to keep us alive! There are thousands of different types of enzymes. Each enzyme catalyzes only one type of chemical reaction.

▶ All enzymes have an ideal "operating temperature". This is called the optimum temperature and it is where the enzyme's catalytic activity is at its highest.

▶ In the experiment below, you will find out how temperature affects the activity of an enzyme called salivary amylase. Salivary amylase is a digestive enzyme found in saliva. It breaks down the starch in food into smaller sugar molecules. It works best at pH 7.0 (the pH of the mouth). The reaction is described below:

$$\text{Starch (starting material/substrate)} \xrightarrow{\text{Amylase}} \text{Sugars (product)}$$

▶ We can track the activity of amylase using the iodine test. Iodine solution is a yellow/brown solution and shows if starch is present.

▶ If the iodine indicator (usually potassium iodide) stays yellow/brown after the addition of a sample it means there is no starch present in the sample. If the sample turns blue/black, it means that starch is present (image right).

 Investigation 2.8 Effect of temperature on enzyme activity

See appendix for equipment list.

Work in pairs for this activity.

1. Each pair or group should set up a reaction plate (photo, right) by adding a single drop of 0.1 M iodine solution (I_2KI) into each well.

2. Add 2 mL of 1% amylase solution and 1 mL of a buffer solution at pH 7.0 into a clean test tube.

3. Place the test tube in a water bath set at 10°C. Let the tube sit for 5 minutes until it is at the correct temperature.

4. Add 1 mL of 1% starch solution into the test tube. Mix well and immediately start a timer.

5. After one minute, use a plastic pipette to take one or two drops from the test tube.

6. Add a drop of the sample into the first well of the reaction plate. Record below if starch is present (Y) or not present (N).

7. Continue to take samples every minute until no starch is detected (the color stays brown after the sample is added).

8. Repeat the procedure at temperatures of 20, 30, 40, 50, and 60°C, using a clean spotting plate each time.

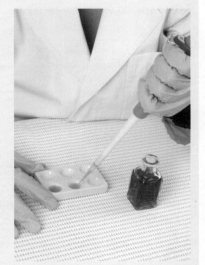

Temperature (°C)	Time (minutes)														
	1	2	3	4	5	6	7	8	9	10	11	12	13	14	15
10															
20															
30															
40															
50															
60															

©2022 **BIOZONE** International
ISBN: 978-1-98-856692-4

5. Use enzyme vocabulary, e.g. substrate, product, to describe how the iodine test is used to follow the progress of the reaction of amylase on starch:

_____

_____

_____

_____

_____

6. Write a suitable aim for this investigation: _____

_____

7. No controls have been included in this investigation. What would a suitable control have been? _____

_____

8. (a) Was there any temperature(s) where denaturation occurred (there was no enzyme activity)? If so, identify it here:

   _____

   _____

   (b) How did you know denaturation had occurred?

   _____

   _____

   _____

   _____

9. On the grid, plot the time taken for all the starch to be digested against temperature. If enzyme denaturation occurred, do not plot that result.

   NEED HELP? See Activities 17, 18

10. Describe how temperature affects the activity of amylase:

_____

_____

_____

_____

_____

11. Based on your experiment, what is the optimum temperature for amylase? _____

_____

12. Predict amylase activity below 10°C and give a reason for your prediction: _____

_____

_____

_____

59 Investigating Catalase Activity

Key Question: How does the germination stage of mung beans affect the activity of catalase, as measured by the breakdown of hydrogen peroxide?

▶ Enzyme activity can be measured easily in simple experiments. This activity describes an experiment in which germinating seeds of different ages were tested for their level of catalase activity. Hydrogen peroxide solution is used as the substrate and simple apparatus is used to measure oxygen production (see backgound).

The aim and hypothesis

To investigate the effect of germination age on the level of catalase activity in mung beans. The students hypothesized that if metabolic activity increased with germination age, catalase activity would also increase with germination age.

Background

Germinating seeds are metabolically very active. The metabolism produces reactive oxygen species, including hydrogen peroxide (H_2O_2). H_2O_2 helps germination by breaking dormancy, but it is also toxic.

To counter the toxic effects of H_2O_2 and prevent cellular damage, germinating seeds produce catalase, an **enzyme** that breaks down H_2O_2 to water and oxygen.

A class was divided into six groups, with each group testing the seedlings of each age. Each group's set of results (for 0.5, 2, 4, 6, and 10 days) represents one trial.

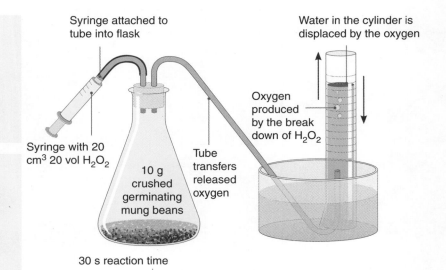

Syringe attached to tube into flask

Water in the cylinder is displaced by the oxygen

Syringe with 20 cm^3 20 vol H_2O_2

10 g crushed germinating mung beans

Tube transfers released oxygen

Oxygen produced by the break down of H_2O_2

30 s reaction time

The apparatus and method

Ten grams of germinating mung bean seeds (0.5, 2, 4, 6, or 10 days old) were crushed and placed in a conical flask. There were six trials at each of the five seedling ages. With each trial, 20 cm^3 of 20 vol H_2O_2 was added to the flask at time 0 and the reaction was run for 30 seconds.

The oxygen released was collected via a tube into an inverted measuring cylinder. The volume of oxygen produced is measured by the amount of water displaced from the cylinder. The results are presented in the table below:

NEED HELP? See Activities 6, 23, 24

Stage of germination (days)	Group (trial) # Volume of oxygen collected after 30 s (cm^3)								Mean rate (cm^3/s/g)
	1	2	3	4	5	6	Mean	Standard deviation	
0.5	9.5	10	10.7	9.5	10.2	10.5			
2	36.2	30	31.5	37.5	34	40			
4	59	66	69	60.5	66.5	72			
6	39	31.5	32.5	41	40.3	36			
10	20	18.6	24.3	23.2	23.5	25.5			

1. Write the equation for the catalase reaction with hydrogen peroxide: _____

_____

2. Complete the table above to summarize the data from the six trials:

   (a) Calculate the mean volume of oxygen for each stage of germination and enter the values in the table.

   (b) Calculate the standard deviation for each mean and enter the values in the table (you may use a spreadsheet).

   (c) Calculate the mean rate of oxygen production in cm^3 per second per gram. For the purposes of this exercise, assume that the weight of germinating seed in every case was 10.0 g.

©2022 **BIOZONE** International
ISBN: 978-1-98-856692-4
Photocopying Prohibited

LS1.A

3. (a) What sort of graph would you use to plot the results of this experiment? _____

   (b) Explain your choice: _____

   _____

   (c) Use the tabulated data to plot the results on the grid provided below. Include the standard deviation as error bars above and below each mean.

4. (a) Describe the trend in the data: _____

   _____

   _____

   (b) Explain the relationship between stage of germination and catalase activity shown in the data: _____

   _____

   _____

   (c) Do the results support the students' hypothesis? _____

   _____

5. Describe any potential sources of errors in the apparatus or the procedure: _____

   _____

   _____

   _____

6. Describe two things that might affect the validity of findings in this experimental design: _____

   _____

   _____

   _____

60 Organ Systems Work Together

Key Question: How do the muscular and skeletal systems work together to make the body move?

- An **organ system** is a group of organs that work together to perform a certain group of tasks.

- Although each system has a specific job, e.g. digestion, reproduction, internal transport, or gas exchange, the organ systems must interact to maintain the functioning of the organism.

- There are 11 organ (body) systems in humans. Plants have fewer organ systems.

- The example on this page shows how the muscular and skeletal organ systems in humans work together to achieve movement in the arm.

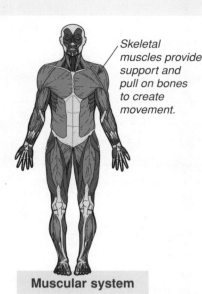

Skeletal muscles provide support and pull on bones to create movement.

Muscular system

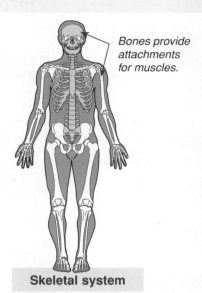

Bones provide attachments for muscles.

Skeletal system

Flex your muscles

Bones act with muscles to form levers that enable movement

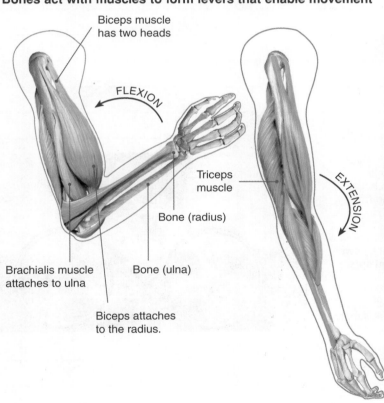

Biceps muscle has two heads

FLEXION

Triceps muscle

Bone (radius)

Brachialis muscle attaches to ulna

Bone (ulna)

Biceps attaches to the radius.

EXTENSION

1. What is an organ system? _____

_____

2. (a) What is the role of the skeletal system in movement? _____

_____

(b) What is the role of the muscular system in movement? _____

_____

_____

©2022 **BIOZONE** International
ISBN: 978-1-98-856692-4
Photocopying Prohibited

Investigation 2.9 Forearm movements

1. What makes different parts of your body move? Try this quick investigation to find out.

2. Hang the hand of your right arm down by your side, thumb forward. Put your left hand around the middle of your upper right arm so that you can feel the muscle underneath. Make a fist, and bend your right arm at the elbow towards you (forearm flexion).

3. Describe how the shape of the muscle on your upper arm changes as you do this:

   _____

   _____

4. What do you think is happening to the muscle? _____

   _____

5. Now move your left hand around a bit more so that you can feel the muscle at the back of your upper right arm while it is in the flexed (bent) position. Make a fist, and then straighten your arm from the elbow (forearm extension).

6. What do you feel now? _____

   _____

3. Use the diagram on the previous page and your own investigation to explain how the interaction of body systems can generate movement of body parts:

   _____

   _____

   _____

   _____

   _____

   _____

4. Cycling is a sport that requires strong leg muscles. Cyclists often use specialized shoes that clip into the cycle's pedals to firmly attach their feet to the pedal. This helps with stability and stops the feet from slipping when rapidly accelerating the cycle.

   Clipping the feet to the pedals also helps with leverage and, with the right timing, effectively doubles the cyclist's power and ability to accelerate the cycle from any current speed.

   Thinking about how muscles work in pairs, how might clipping the feet to the cycle's pedals help a cyclist accelerate a cycle much faster than if the feet were simply resting on the pedal as with typical bicycle pedals?

   _____

   _____

   _____

   _____

   _____

   _____

   _____

   _____

©2022 **BIOZONE** International
ISBN: 978-1-98-856692-4
Photocopying Prohibited

61 Circulation and Gas Exchange Interactions

Key Question: How do the circulatory and respiratory systems interact to provide the body's tissues with oxygen and remove carbon dioxide?

Circulatory system

Function

Delivers oxygen (O_2) and nutrients to all cells and tissues. Removes carbon dioxide (CO_2) and other waste products of metabolism. CO_2 is transported to the lungs.

Components

▸ Heart
▸ Blood vessels:
  • Arteries
  • Veins
  • Capillaries
▸ Blood

Interaction between systems

In vertebrates, the respiratory system and cardiovascular system interact to supply oxygen and remove carbon dioxide from the body.

Respiratory system

Function

Provides surface for gas exchange. Moves fresh air into and stale air out of the body.

Components

▸ Airways:
  • Pharynx
  • Larynx
  • Trachea
▸ Lungs:
  • Bronchi
  • Bronchioles
  • Alveoli
▸ Diaphragm

Head and upper body

Oxygen (O_2) from inhaled air moves from the lungs into the circulatory system and is transported within red blood cells to the heart. The heart pumps the blood to the body where O_2 is released and carbon dioxide (CO_2) is picked up. The blood returns to the heart and is pumped to the lungs where CO_2 is released into the lungs to be breathed out.

Lung

Heart

Lung

Lower body

Bronchiole

Capillaries

The airways of the lungs end at the **alveoli**, which are microscopic air sacs that enable gas exchange.

From the heart to the lungs.

Red blood cells are replenished with oxygen from the alveolus and carbon dioxide is released from the blood into the alveolus.

CO_2

CO_2
O_2

O_2

The carbon dioxide released from the blood exits the body during exhalation. Inhalation brings in fresh air, containing oxygen.

The respiratory system and the circulatory system come together at the alveoli (singular alveolus). Oxygen and carbon dioxide diffuse across the thin walls of capillaries and alveoli.

Capillary

From the lungs to the heart.

Red blood cell

Responses to exercise

▶ During exercise, your body needs more oxygen to meet the extra demands placed on the muscles, heart, and lungs. At the same time, more carbon dioxide must be expelled. To meet these increased demands, blood flow must increase. This is achieved by increasing the rate of heart beat. As the heart beats faster, blood is circulated around the body more quickly, and exchanges between the blood and tissues increase.

▶ The arteries and veins must be able to resist the extra pressure of higher blood flow and must expand (dilate) to accommodate the higher blood volume. If they didn't, they could rupture (break). During exercise, the muscular, cardiovascular, and nervous systems interact to maintain the body's systems in spite of increased demands (right).

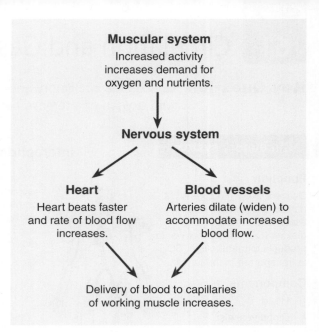

Muscular system
Increased activity increases demand for oxygen and nutrients.

Nervous system

Heart
Heart beats faster and rate of blood flow increases.

Blood vessels
Arteries dilate (widen) to accommodate increased blood flow.

Delivery of blood to capillaries of working muscle increases.

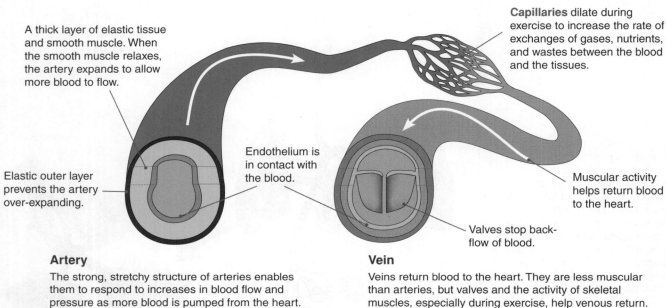

A thick layer of elastic tissue and smooth muscle. When the smooth muscle relaxes, the artery expands to allow more blood to flow.

Capillaries dilate during exercise to increase the rate of exchanges of gases, nutrients, and wastes between the blood and the tissues.

Elastic outer layer prevents the artery over-expanding.

Endothelium is in contact with the blood.

Muscular activity helps return blood to the heart.

Valves stop back-flow of blood.

Artery
The strong, stretchy structure of arteries enables them to respond to increases in blood flow and pressure as more blood is pumped from the heart.

Vein
Veins return blood to the heart. They are less muscular than arteries, but valves and the activity of skeletal muscles, especially during exercise, help venous return.

1. In your own words, describe how the circulatory system and respiratory system work together to provide the body with oxygen and remove carbon dioxide:

_____

_____

_____

_____

_____

_____

2. (a) What happens to blood flow during exercise? _____

   (b) How do body systems interact to accommodate the extra blood flow needed when a person exercises?

_____

_____

_____

_____

©2022 **BIOZONE** International
ISBN: 978-1-98-856692-4
Photocopying Prohibited

62 Circulation and Digestive Interactions

Key Question: How do the circulatory and digestive systems interact to provide the body's tissues with nutrients?

Circulatory system

Function

Delivers oxygen (O_2) and nutrients to all cells and tissues. Removes carbon dioxide (CO_2) and other waste products of metabolism.

Components

- Heart
- Blood vessels:
  - Arteries
  - Veins
  - Capillaries
- Blood

Food is digested in the stomach and small intestine and is then absorbed and passed to the circulatory system. The capillaries around the stomach and intestines collect nutrients and then drain to the hepatic portal vein, which carries the blood directly to the liver. The liver then processes this nutrient rich blood, e.g. glucose is stored as glycogen. The hepatic vein then transports nutrients from the liver to supply the other tissues of the body.

Interaction between systems

In mammals, the **digestive system** and **cardiovascular system** interact to supply nutrients to the body.

Digestive system

Function

Digest food and absorb useful molecules from it, and eliminate undigested material.

Components

- Mouth and pharynx
- Esophagus
- Stomach
- Liver and gall bladder (accessory organs)
- Pancreas (accessory organ)
- Small intestine
- Large intestine

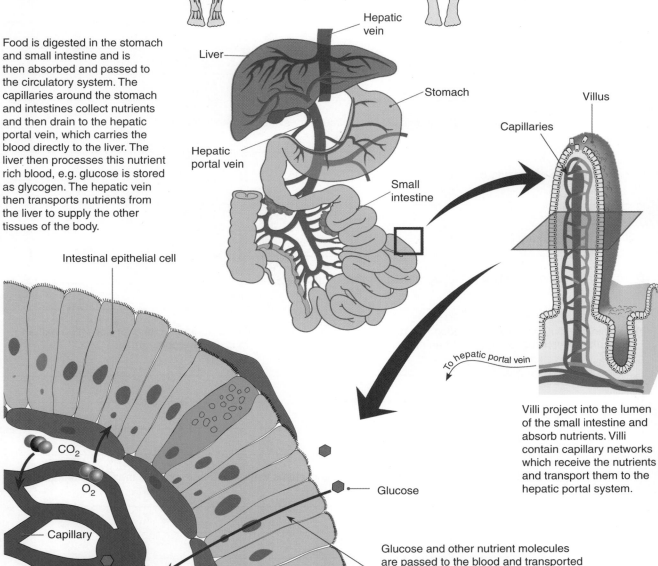

Villi project into the lumen of the small intestine and absorb nutrients. Villi contain capillary networks which receive the nutrients and transport them to the hepatic portal system.

Glucose and other nutrient molecules are passed to the blood and transported to other parts of the body. Oxygen passes to the intestinal cells, while carbon dioxide passes into the blood.

©2022 **BIOZONE** International
ISBN: 978-1-98-856692-4
Photocopying Prohibited

 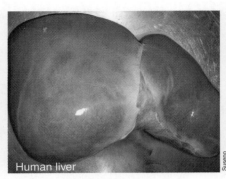 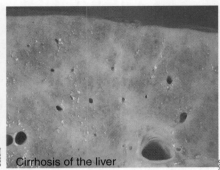

Human liver

Suseno

Cirrhosis of the liver

CDC

Blood flow to the digestive tract increases steadily after a meal and remains elevated for about 2.5 hours, reaching a maximum after about 30 minutes. During exercise, blood flow in the digestive tract is reduced as it is redirected to the muscles.

Nutrients, e.g. minerals, sugars, and amino acids, are transported in the blood plasma to the liver. The liver receives nutrient-rich deoxygenated blood from the digestive system via the hepatic portal vein, and oxygen rich blood from the hepatic artery.

Scarring of the liver tissue, or cirrhosis, can result in portal hypertension (high blood pressure). The scarred tissue obstructs blood flow in the liver. This causes pressure to build up in upstream blood vessels, resulting in swelling and possible hemorrhage.

1. How are nutrients transported in the blood? _____

_____

2. Explain how liver cirrhosis affects the circulatory system: _____

_____

_____

_____

_____

3. (a) At which two points in the body do the digestive and circulatory systems directly interact? _____

_____

   (b) Explain what is happening at these points: _____

_____

_____

_____

4. (a) What happens to blood flow to the digestive tract after a meal? _____

_____

   (b) Explain why it is often recommended that a person should exercise within 2.5 hours of eating, or eat within half an hour after exercising to gain most benefit from the exercise (in terms of muscle development):

_____

_____

_____

_____

5. In your own words, describe how the circulatory and digestive systems work together to provide the body with nutrients:

_____

_____

_____

_____

_____

©2022 **BIOZONE** International
ISBN: 978-1-98-856692-4
Photocopying Prohibited

63 Plant Organ Systems

Key Question: What are the different parts of the plant organ system?

Plants have fewer **organ systems** than animals because they are simpler and have lower energy demands. The two primary plant organ systems are the shoot system and the root system.

Shoot system

The above-ground parts of the plant: including organs such as leaves, buds, stems, and the flowers and fruit (or cones) if present. All parts of the shoot system produce hormones.

The shoot and root systems of plants are connected by transport tissues (xylem and phloem) that are continuous throughout the plant.

Leaves

▶ Manufacture food via photosynthesis.
▶ Exchange gases with the environment.
▶ Store food and water.

Stems

▶ Transport water and nutrients between roots and leaves.
▶ Support and hold up the leaves, flowers and fruit.
▶ Produce new tissue for photosynthesis and support.
▶ Store food and water.

Root system

The below-ground parts of the plant, including the roots and root hairs.

Roots

▶ Anchor the plant in the soil.
▶ Absorb and transport minerals and water.
▶ Store food.
▶ Produce hormones.
▶ Produce new tissue for anchorage and absorption.

Structures for sexual reproduction

▶ Reproductive structures are concerned with passing on **genes** to the next generation.
▶ Flowers or cones are the reproductive structures of seed plants (angiosperms and gymnosperms).
▶ Fruits provide flowering plants with a way to disperse the seeds.

1. Describe how each of the following systems provides for the essential functions of life for the plant:

   (a) Root system: _____

   _____

   (b) Shoot system: _____

   _____

2. In the boxes below, place the following list of plant functions: *Photosynthesis, transport, absorption, anchorage, storage, sexual reproduction, hormone production, growth*

Functions shared by the root and shoot system	Functions unique to the shoot system	Functions unique to the root system

©2022 **BIOZONE** International
ISBN: 978-1-98-856692-4
Photocopying Prohibited

64 Interacting Systems in Plants

Key Question: How do the shoot and root systems of plants interact to balance water uptake and loss, so that the plant can maintain the essential functions of life?

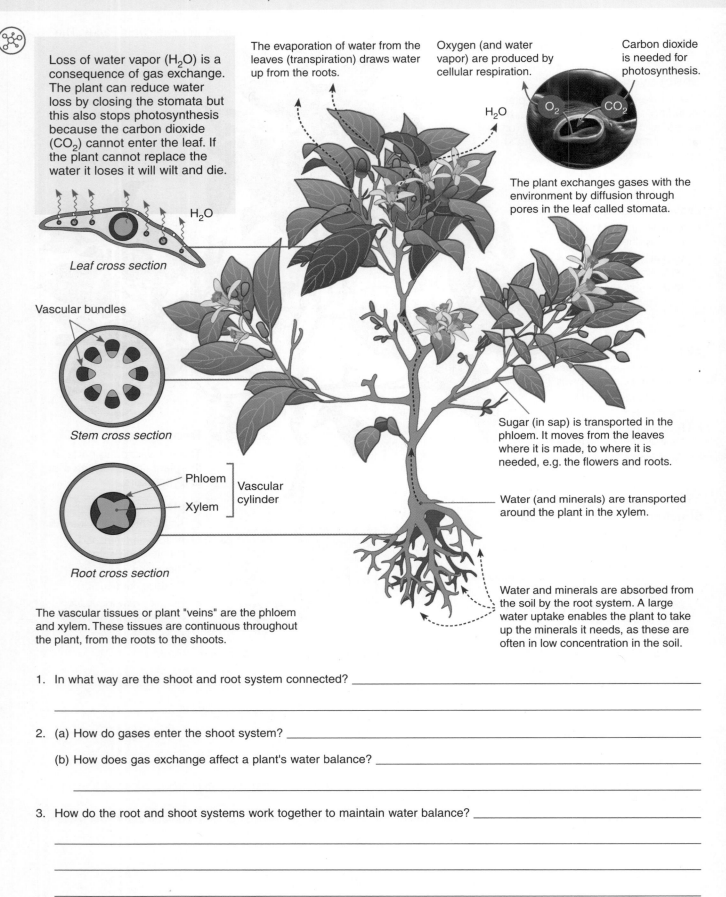

Loss of water vapor (H_2O) is a consequence of gas exchange. The plant can reduce water loss by closing the stomata but this also stops photosynthesis because the carbon dioxide (CO_2) cannot enter the leaf. If the plant cannot replace the water it loses it will wilt and die.

H_2O

Leaf cross section

The evaporation of water from the leaves (transpiration) draws water up from the roots.

H_2O

Oxygen (and water vapor) are produced by cellular respiration.

O_2 CO_2

Carbon dioxide is needed for photosynthesis.

The plant exchanges gases with the environment by diffusion through pores in the leaf called stomata.

Vascular bundles

Stem cross section

Phloem   Vascular
Xylem   cylinder

Root cross section

The vascular tissues or plant "veins" are the phloem and xylem. These tissues are continuous throughout the plant, from the roots to the shoots.

Sugar (in sap) is transported in the phloem. It moves from the leaves where it is made, to where it is needed, e.g. the flowers and roots.

Water (and minerals) are transported around the plant in the xylem.

Water and minerals are absorbed from the soil by the root system. A large water uptake enables the plant to take up the minerals it needs, as these are often in low concentration in the soil.

1. In what way are the shoot and root system connected? _____
_____

2. (a) How do gases enter the shoot system? _____

   (b) How does gas exchange affect a plant's water balance? _____
_____

3. How do the root and shoot systems work together to maintain water balance? _____
_____
_____
_____

©2022 **BIOZONE** International
ISBN: 978-1-98-856692-4
Photocopying Prohibited

65 Review Your Understanding

Key Question: How does the wood frog survive freezing in winter?

▸ At the beginning of the chapter you were introduced to the wood frog, a multicellular vertebrate that is able to survive freezing temperatures down to -6°C. The mechanisms the frog used to survive freezing included:

- Increasing the amount of glucose in the blood around the organs by up to 100 times the normal amount.

- Increasing urea in the extracellular spaces (spaces outside the cells).

- Having proteins that operate at lower temperatures than similar proteins in other frogs.

- Proteins in the cells plasma membranes allow glucose and urea to enter the cell as freezing proceeds.

- Both breathing and the heartbeat stop.

- Antifreeze proteins bind to ice crystals.

▸ You should now be able to explain how these mechanisms stop the frog from freezing to death.

1. Annotate the diagram below to help explain how the points above help the frog survive freezing:

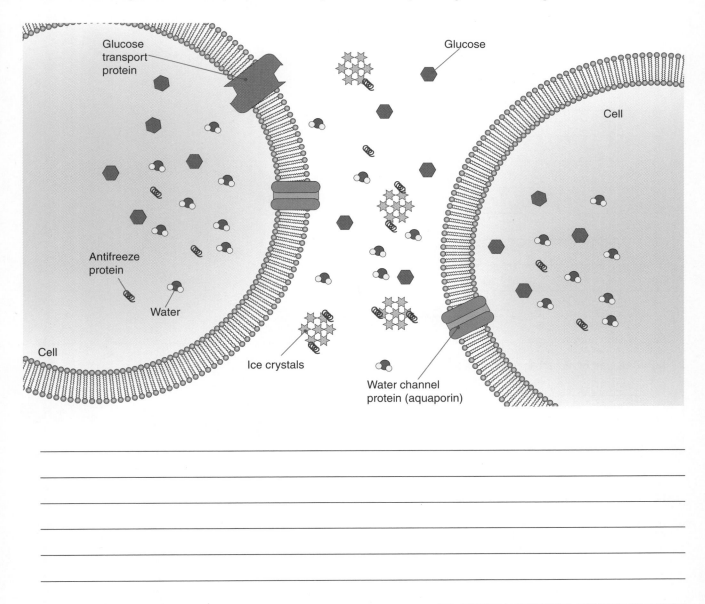

_____

_____

_____

_____

_____

_____

66 Summing Up

1. The diagrams below show a nerve cell and a muscle cell (fiber). Each cell comes from the same person and has the same DNA in it. Use diagrams to help explain how it is possible that a multicellular organism is able to develop all the different cells it needs to build its body from a single fertilized cell:

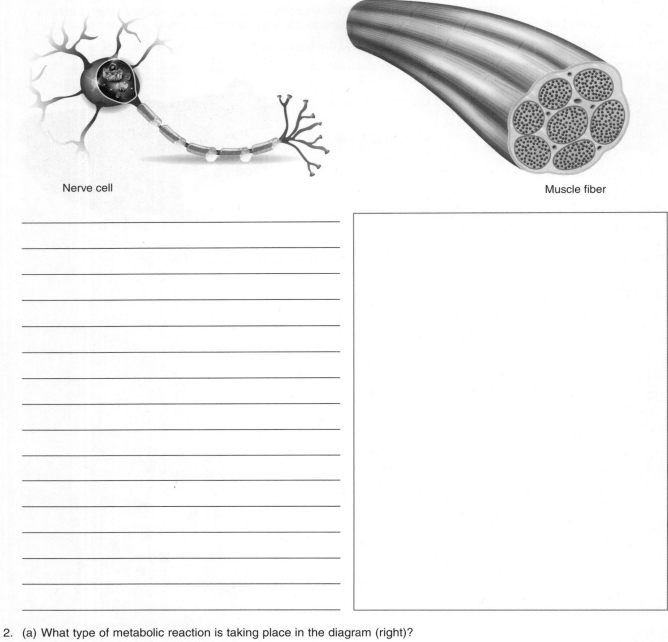

Nerve cell

Muscle fiber

2. (a) What type of metabolic reaction is taking place in the diagram (right)?

_____

(b) What is occurring during this reaction? _____

_____

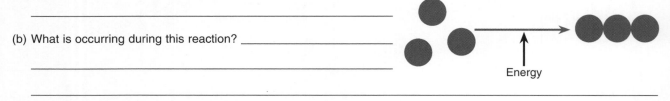

Energy

(c) Give an example of this type of metabolic reaction: _____

3. A grasshopper has the following percentages of nucleotides in its DNA: A = 29.3, G = 20.5, C = 20.7, T = 29.3, %GC = 41.2, %AT = 58.6. For a rat, the percentages are A = 28.6, G = 21.4, C = 20.5, T = 28.4, %GC = 42.9, %AT = 57.0. How do these numbers demonstrate Chargaff's rules?

_____

_____

©2022 **BIOZONE** International
ISBN: 978-1-98-856692-4
Photocopying Prohibited

4. The body of large multicellular organisms has a hierarchical structure. Organ systems, made out of different cells and tissues, work together to carry out tasks essential for the functioning of the body. Two of those organ systems are the muscular system and the nervous system. In the spaces below, draw labeled diagrams of these systems to show their hierarchical structure, including the cells and tissues that make them up. You may need to do some research to help identify the cells and tissues involved.

5. In the space below, create a simple diagram(s) that shows how the nervous system and muscular system interact to produce the movement for catching a ball.

©2022 **BIOZONE** International
ISBN: 978-1-98-856692-4
Photocopying Prohibited

CHAPTER 3

Feedback Mechanisms

Anchoring Phenomenon

Hot Dog: How do mammals manage to maintain stable conditions in their bodies, despite facing a wide range of external conditions?　　67　82

How do feedback mechanisms maintain the internal environment?

☐ 1 Define homeostasis. Use a model to identify the mechanisms that help the body maintain a constant internal environment, even when external conditions are changing.　　68

☐ 2 Discuss how multiple body systems work together to maintain homeostasis.　　69 83

☐ 3 Understand how negative feedback discourages changes inside the body and stabilizes the system.　　70 79 83

☐ 4 Understand how positive feedback amplifies a response and destabilizes a system. Identify features of positive feedback in the process of childbirth.　　71

How are body temperature and blood glucose regulated?

☐ 5 Distinguish between endotherms and ectotherms. Understand how ectotherms utilize heat energy from an external source to raise their internal temperature, in comparison to heat generated metabolically in endotherms. Discuss advantages and disadvantages of ectothermy as a means of maintaining a stable body temperature.　　72

☐ 6 Define thermoregulation as a means of stabilizing a thermal body system. Define poikilotherms. Interpret data to analyze the difference in body temperature in both endotherms and poikilotherms when environmental temperature changes. Investigate a range of insulating materials for their effectiveness in heat conductivity.　　73

☐ 7 Use a model to understand the human thermoregulation negative feedback system. Discuss how phenomena such as fever can impact negative feedback, and how the body can stabilize temperature, to compensate.　　74

☐ 8 Interpret data on body shape and heat loss to explain adaptations to hot and cold environments to stabilize body temperature. Investigate how body shape can help regulate body temperature. Analyze data to form a valid conclusion from a heat loss investigation.　　75

☐ 9 Use both a mathematical and graphical model to investigate blood glucose homeostasis in humans. Explain how type 2 diabetes disrupts blood glucose feedback mechanisms.　　76 77

How can we investigate feedback mechanisms?

☐ 10 Use data to calculate blood flow during exercise, in the context of circulatory and respiratory system homeostasis. Investigate the effect of exercise on heart rate. Plan and conduct an investigation on the effect of exercise on breathing rate.　　78 79

☐ 11 Use a graphical model of plant transpiration to discuss water balance homeostasis mechanisms. Explain how negative feedback controls stomata opening and closing to maintain water balance stability. Investigate how external environmental changes affect plant transpiration rates. Process data to analyze the relationship between the effect of the external environment on plant water loss due to transpiration rate.　　80 81

67 Hot Dog

Key Question: How do mammals manage to maintain stable conditions in their bodies, despite facing a wide range of external conditions?

▸ Mammals have body temperatures that remain constant despite changes in the environmental temperature. As a group they have evolved a wide variety or adaptations that allow them to maintain constant body temperatures even in extreme environments.

▸ Some of these adaptations are commonly seen in dogs:

▸ A hot dog pants. Air is rapidly drawn rapidly over the tongue and inner mouth, which evaporates moisture from their surface and helps cool the body.

▸ A cold dog curls itself up into a tight ball, reducing its surface area exposed to the cold air and conserving its body heat.

1. (a) How do you cool down when you are hot? Discuss this in groups of 3-4:

    _____

    _____

    _____

    (b) Identify the ways of keeping cool in (a) that are shared with other mammals: _____

    _____

    _____

2. (a) How do you warm up when you are cold? Discuss this in groups of 3-4:

    _____

    _____

    _____

    (b) Identify the ways of keeping cool in (a) that are shared with other mammals: _____

    _____

    _____

3. Suggest why it is important for mammals to maintain a stable body temperature: _____

    _____

4. What advantage might maintaining a stable body temperature have?_____

    _____

68 Homeostasis

Key Question: How do organisms maintain a constant internal environment despite changes in their external environment?

What is homeostasis?

Homeostasis literally means "constant state". Organisms maintain homeostasis, i.e. a relatively constant internal environment, even when the external environmental is changing. This requires energy.

For example, when you exercise, your body must keep your temperature constant, at about 37.0 °C, despite the increased heat generated by activity. Similarly, you must regulate blood glucose (sugar) levels and blood pH, water and electrolyte balance, and blood pressure. Your body's organ systems carry out these tasks.

To maintain homeostasis, the body's receptors must detect changes in the environment, process this sensory information, and respond to it appropriately. The response provides new feedback to the receptor. These three components are illustrated below.

The analogy of a temperature setting on a heat pump is a good way to explain how homeostasis is maintained. A heat pump has sensors (a receptor) to monitor room temperature. It also has a control center to receive and process the data from the sensors. Depending on the data it receives, the control center activates the effector (heating/cooling unit), switching either on or off.

When the room is too cold, the heating unit switches on, and the cooling unit is off. When it is too hot, the heating unit switches off and the cooling unit is switched on. This system maintains a constant temperature, similar to homeostasis in the body.

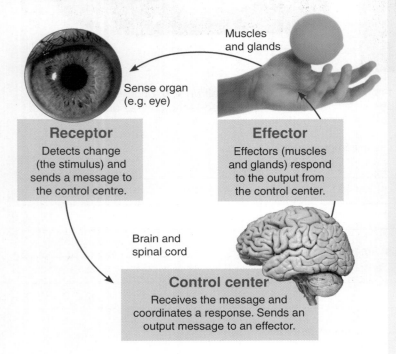

Receptor
Detects change (the stimulus) and sends a message to the control centre.

Sense organ (e.g. eye)

Muscles and glands

Effector
Effectors (muscles and glands) respond to the output from the control center.

Brain and spinal cord

Control center
Receives the message and coordinates a response. Sends an output message to an effector.

The analogy of staying upright on a mountain bike, using body weight, arms, pedals, brakes, and steering, demonstrates that many homeostasis systems have multiple mechanisms to maintain a steady state.

1. What is homeostasis? _____

_____

_____

2. What are the roles of the following components in maintaining homeostasis:

(a) Receptor: _____

_____

(b) Control center: _____

_____

(c) Effector: _____

_____

  LS1.A SC

©2022 **BIOZONE** International
ISBN: 978-1-98-856692-4
Photocopying Prohibited

69 Keeping in Balance

Why is homeostasis important?

▶ An organism must constantly regulate its internal environment in order to carry out essential life processes, such as growing and responding to the environment. Changes outside of normal levels for too long can stop the body systems from working properly, and can result in illness or death.

▶ **Homeostasis** relies on monitoring all the information received from the internal and external environments and coordinating appropriate responses. This often involves many different organ systems working together.

▶ Most of the time, an organism's body systems are responding to changes at the subconscious level, but sometimes homeostasis is achieved by changing a behavior, e.g. finding shade if the temperature is too high.

Some examples of how the body keeps in balance

Regulating respiratory gases

Cellular respiration, which requires glucose, takes place in all cells. It requires a constant supply of oxygen and produces carbon dioxide as a waste product. The changing demands of the body for supply of oxygen and removal of CO_2, e.g. when exercising, are met by the gas exchange and circulatory systems of the body.

Maintaining nutrient supply

Food and drink provide the energy and nutrients that the body needs to carry out the essential processes of life. Factors that change the demand for nutrients include activity level and environmental challenges, e.g. the body requires more energy when active or in extremes of temperature.

Coordinating responses

The body is constantly bombarded by stimuli from the environment. The brain must prioritize its responses and decide which stimuli are important and require a response, and which ones do not.

Repairing injuries

Damage to the body's tissues triggers responses to repair it and return it to a normal state as quickly as possible. For example, blood clotting stops too much blood from leaving the body from an open wound. If too much blood is lost, a person will go into fatal shock.

Maintaining fluid and electrolyte balance

Fluid and salts are taken in with food and drink but imbalances quickly lead to cellular disruption and death, if not treated. Water balance is maintained by ensuring that the amount of fluid consumed and generated by metabolism equals the amount of water lost in urine and feces, sweat, and breathing. Maintaining the body's levels of fluid and salts is the job of the kidneys.

Protecting the body against disease

Like all organisms, we are under constant attack from pathogens (disease-causing organisms) which can cause damage to the body's systems. The body's skin and immune system produce chemicals and cells that act to prevent the entry of pathogens and limit the damage they cause if they do enter. The cardiovascular system circulates these components through the body.

SC   LS1.A

The body's organ systems work together to maintain homeostasis

Organ systems work together to maintain the environment necessary for the functioning of the body's cells. A constant internal environment allows an organism to be somewhat independent of its external environment, so that it can move about, even as its environment changes. The simplified example below illustrates how organ systems interact with each other to maintain a constant internal environment.

Once food has been digested (broken down) in the digestive system, it is absorbed (taken up) into the blood of the circulatory system.

The **digestive system** (gut and associated digestive glands) is responsible for the breakdown and absorption of food. Ultimately, it provides the energy and nutrients required by all the body's systems.

Unabsorbed digestive matter is expelled from the digestive system as feces.

CO_2

O_2

Food

Heart

Urine is produced by the kidneys of the urinary system. It contains the waste products of metabolism, particularly nitrogen-containing wastes and excess ions.

The **gas exchange system**, i.e. the lungs and airways, brings in a supply of oxygen for the body's cells and tissues and expels waste carbon dioxide through breathing.

The **urinary system** (the kidneys and associated ducts) has several roles including disposing of nitrogen-containing and other waste products from the body, regulating ion balance, and controlling the volume and pressure of the blood.

Once nutrients have been absorbed, they are carried in the blood and delivered to cells all around the body (solid arrows). Wastes (dashed arrow) move from the cells back into the blood and are transported and removed.

The **circulatory system** (the heart, blood vessels and blood) distributes respiratory gases, nutrients, and other substances, e.g. hormones, to the cells and tissues of the body.

1. Why is it important that body systems are kept in balance? _____

_____

_____

_____

_____

2. Why is it important that the brain prioritizes the importance of incoming stimuli? _____

_____

_____

_____

3. Using an example, briefly explain why homeostasis often involves more than one body system: _____

_____

_____

_____

©2022 **BIOZONE** International
ISBN: 978-1-98-856692-4
Photocopying Prohibited

70 Negative Feedback Mechanisms

Negative feedback is a control system which maintains the body's internal environment at a steady state. Negative feedback has a stabilizing effect and discourages variations from a set point. It works by returning internal conditions back to a steady state when variations are detected (right).

Most body systems achieve **homeostasis** through negative feedback. Body temperature, blood glucose levels, and blood pressure are all controlled by negative feedback mechanisms.

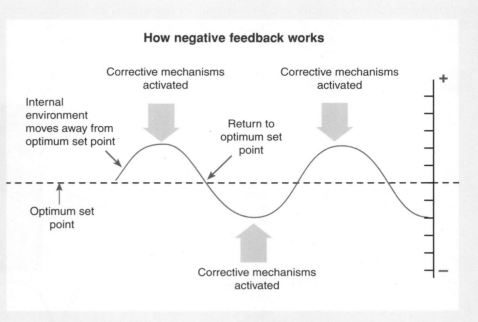

How negative feedback works

Stomach emptying: an example of negative feedback

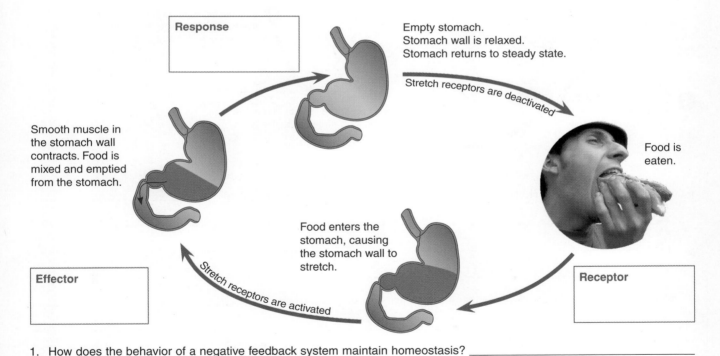

1. How does the behavior of a negative feedback system maintain homeostasis? _____

_____

_____

_____

2. (a) On the diagram of a stomach emptying, name the receptor, effector, and response in the spaces provided.

   (b) What is the steady state for this example? _____

©2022 **BIOZONE** International
ISBN: 978-1-98-856692-4
Photocopying Prohibited

71 Positive Feedback Mechanisms

Key Question: How do positive feedback mechanisms work to amplify a physiological response in order to achieve a particular outcome?

Positive feedback mechanisms amplify (increase) or speed up a physiological response, usually to achieve a particular outcome. Examples of positive feedback include fruit ripening, fever, blood clotting, labor (childbirth), and lactation (production of milk). A positive feedback mechanism stops when the end result is achieved, e.g. the baby is born, a pathogen is destroyed by a fever, or ripe fruit falls off a tree.

Positive feedback is less common than **negative feedback** because it creates an escalation in response, which is unstable. This response can be dangerous or even cause death if it is prolonged.

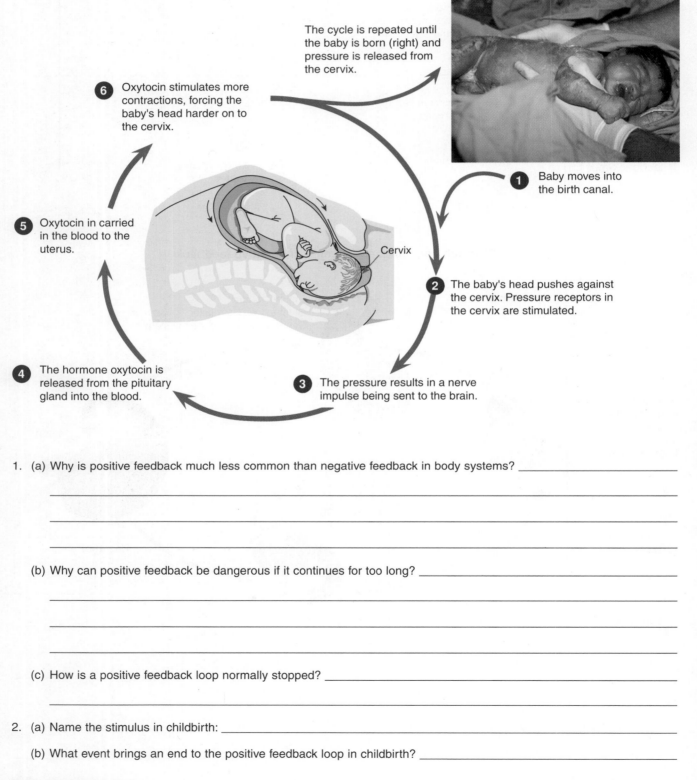

The cycle is repeated until the baby is born (right) and pressure is released from the cervix.

6 Oxytocin stimulates more contractions, forcing the baby's head harder on to the cervix.

1 Baby moves into the birth canal.

5 Oxytocin in carried in the blood to the uterus.

Cervix

2 The baby's head pushes against the cervix. Pressure receptors in the cervix are stimulated.

4 The hormone oxytocin is released from the pituitary gland into the blood.

3 The pressure results in a nerve impulse being sent to the brain.

1. (a) Why is positive feedback much less common than negative feedback in body systems? _____

_____

_____

_____

(b) Why can positive feedback be dangerous if it continues for too long? _____

_____

_____

_____

(c) How is a positive feedback loop normally stopped? _____

_____

2. (a) Name the stimulus in childbirth: _____

(b) What event brings an end to the positive feedback loop in childbirth? _____

72 Sources of Body Heat

Key Question: What are the two major ways that organisms obtain heat to enable their body metabolism to function effectively?

Why is body heat important?

The essential processes of life are regulated by enzymes and require a certain temperature in order to operate most efficiently. This optimal temperature varies, depending on the organism. A bacterium living in a geothermal hot pool will have a very different optimal temperature from a soil bacterium in a woodland! Below the optimal temperature, metabolic reactions proceed very slowly. Above the optimal temperature, the enzymes may be damaged and the reaction canot proceed.

The average body temperature of mammals is ~38°C. For birds, it is ~40°C. Most snakes and lizards operate best in the range 24-35°C, although some operate in the mammalian range.

Reptiles are ectotherms. These captive turtles are basking under a lamp to raise their body temperature for activity. In the wild, a low body temperature reduces their ability to forage and escape predators.

Where do animals get their body heat from?

Heat can come from external sources, such as the Sun, or it can come from metabolic activity. These two strategies divide animals into two groups.

▶ **Ectotherms** depend on external sources of heat from the environment for their heat energy, e.g. heat from the Sun.

▶ **Endotherms** generate most of their body heat from internal metabolic processes.

In reality, animals often fall somewhere between these two extremes. For example, some large insects use muscular activity to raise the temperature of their thorax prior to flight and some snakes use metabolic heat to incubate eggs.

All birds are endothermic. Even in cold Antarctic temperatures, the metabolic activity of these emperor penguins provides the warmth to sustain life processes. Endothermy requires a lot of energy, so endotherms cannot go without food for long.

The ectotherm-endotherm continuum

Most, but not all, fish are fully ectothermic and rely solely on the environment for their body heat.

Snakes use heat energy from the environment to increase their body temperature for activity.

Some large insects, such as bumblebees, may raise their temperature for short periods through muscular activity.

Mammals and birds achieve high body temperatures through metabolic activity and reduction of heat loss.

Increasingly endothermic →

1. Distinguish between ectotherms and endotherms in terms of their sources of body heat: _____
   _____
   _____

2. (a) Why are the movements of many ectotherms slow in the early morning? _____
   _____
   _____

   (b) Why could this be a disadvantage?_____
   _____

   (c) What might be an advantage of ectothermy? _____
   _____

©2022 **BIOZONE** International
ISBN: 978-1-98-856692-4
Photocopying Prohibited

73 Thermoregulation

Thermoregulation refers to the regulation of body temperature in the face of changes in the temperature of the environment. Animals show two extremes of body temperature tolerance:

▶ **Homeotherms** (all birds and mammals) maintain a constant body temperature, independent of environmental variation, which represents a large energetic cost.

▶ **Poikilotherms** allow their body temperature to vary with the temperature of the environment. Most fish and all amphibians cannot regulate body temperature at all, but most reptiles use behavior both to warm up and to avoid overheating (below).

We have seen that animals are classed as ectotherms or endotherms, depending on their sources of heat energy. Most endotherms are also strict homeotherms (always thermoregulate) and most, but not all, ectotherms allow their body temperatures to vary with environmental fluctuations. Thermoregulation relies on physical, physiological, and behavioral mechanisms.

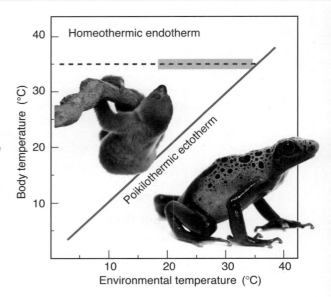

Mechanisms of thermoregulation

Homeothermic endotherm (mammal)

Always thermoregulate, largely by physiological mechanisms. This requires a large amount of energy, especially when outside the range of their normal body temperature.

Wool, hair, or fur traps air next to the skin, providing an insulating layer to reduce heat loss and slow heat gain.

Panting and sweating cool through evaporation. Mammals usually sweat or pant but not both.

Heat can be generated by shivering.

In cold weather, many mammals cluster together to retain body heat.

Poikilothermic ectotherm (reptile)

Thermoregulate at the extremes of their temperature range. The energetic costs of this are much lower than for homeotherms because the environment provides heat energy for warming.

Increasing blood flow to the surface can help remove heat quickly.

Basking in the sun is common in lizards and snakes. The sun warms the body. They seek shade to cool down.

Some lizards reduce points of contact with hot ground, e.g. standing on two legs instead of four, reducing heat uptake via conduction.

1. What is thermoregulation? _____

_____

2. (a) The graph (top right) shows body temperature variations in a mammal and amphibian with change in environmental temperature. Compare and explain the different responses:

_____

_____

(b) The blue shaded area on the graph marks the region where the energy cost of thermoregulation is lowest for the sloth. Why is the sloth using less energy to thermoregulate in this temperature range?

_____

_____

©2022 **BIOZONE** International
ISBN: 978-1-98-856692-4
Photocopying Prohibited

Liolaemus

Core body temperature falls during sleep

The Peruvian mountain lizard (*Liolaemus*) emerges in the morning when the air temperature is below freezing. It exposes itself to the sun, rapidly heating up to a body temperature that enables it to be fully active (below). Once warm, it maintains its preferred body temperature of around 35°C by changing posture and orientation to the sun and thereby controlling the amount of heat absorbed.

Although human core body temperature is relatively stable around 36-37°C, it does vary slightly over a 24 hour period (below). These variations occur in response to the body's internal rhythm and not the environment. Assuming a person sleeps at night and is awake during the day, the lowest body temperatures occur in the early morning and are highest in the afternoon.

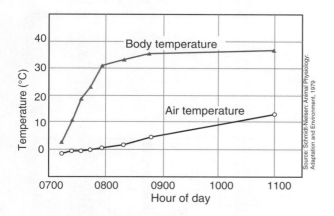

Source: Schmidt-Nielsen: Animal Physiology: Adaptation and Environment, 1979

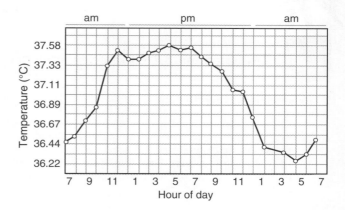

3. Most thermoregulation in homeotherms occurs through physiological processes, e.g. sweating. Describe another physiological mechanism that mammals use to thermoregulate, and describe its effect:

_____

_____

4. Thermoregulation can be aided by both physical features (structural) and behavior. Describe an example of each:

   (a) Behavior: _____

   _____

   (b) Physical features: _____

   _____

5. Describe the main difference between the thermoregulatory mechanisms of a mammal and a lizard such as *Liolaemus*:

_____

_____

6. Both humans and the Peruvian mountain lizard show a daily variation in core body temperature. Describe two main differences between these daily variations:

_____

_____

_____

_____

7. Gather information from a range of difference sources, including print and digital, to compare endotherm mammal and ectotherm reptile or amphibian thermoregulation methods, which may be physiological, structural, and/or behavioral. Present your findings in a suitable format, including a reference list and, in your discussion, identify the strengths and weaknesses of information types in the context of this activity.

Feathered dinosaurs and insulation

▶ Scientists have now found sufficient evidence, mostly from fossilized remains, to understand that a significant number of dinosaur species had feathers on part or all of their bodies. The *Deinonychus* (left) was a species of dromaeosaurid theropods, the group made famous by the Jurassic Park velociraptor. It is likely that all species in this group were covered in feathers.

▶ Although some avian dinosaurs, the ancestors of birds, used feathers for flight, scientists concluded that a key function of feathers was most likely insulation to retain body heat, because:

• Most feathers found on the meat-eating dinosaur fossils have a different structure from those allowing flight.

• Fossils of feathered dinosaurs, who survived many cold, dark months each year, have been found in what was then polar regions.

• Evidence suggests many feathered species were warmblooded (endothermic) with a high metabolic rate. The ability for smaller dinosaur species to retain body heat would be a distinct advantage.

Investigation 3.1   Exploring insulation

See appendix for equipment list.

1. You will work in small groups or pairs. Your teacher may ask you to test all of the materials listed below, or only some. You can compare your results with the other groups. Four insulating materials will be studied: fat (lard), feathers, wool, and cotton balls.

2. Predict the best insulator: _____

   Predict the worst insulator: _____

3. Set up the control by placing a 100 mL beaker directly inside a 250 mL beaker (no insulation).

4. Set up your test by packing your chosen insulating material into a 250 mL beaker. Leave space to insert a 100 mL beaker.

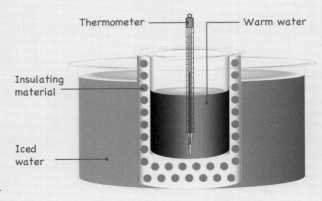

5. Pour warm water (approx. 45°C) into both 100 mL beakers then place each set-up into separate containers of iced water (above). Place a thermometer into each of the 100 mL beakers. You may need to tape or weigh the beakers down to stop them floating and tipping over. Start a stop watch and record the temperature every two minutes for 20 minutes in the table below.

| Minutes | Temperature (°C) | | | | |
	Control	Fat	Feathers	Wool	Cotton balls
2					
4					
6					
8					
10					
12					
14					
16					
18					
20					

©2022 **BIOZONE** International
ISBN: 978-1-98-856692-4
Photocopying Prohibited

74 Thermoregulation in Humans

Key Question: What role does the hypothalamus play in regulating body temperature in humans?

The hypothalamus regulates temperature

▶ In humans, the temperature regulation center is a region of the brain called the hypothalamus. It has thermoreceptors that monitor core body temperature and has a "set-point" temperature of 36.7°C (98.6°F).

▶ The hypothalamus acts like a thermostat. It registers changes in the core body temperature and receives information about temperature change from thermoreceptors in the skin. It then coordinates the appropriate nervous and hormonal responses to counteract any changes and restores normal body temperature, as shown in the diagram below.

▶ When normal temperature is restored, the corrective mechanisms are switched off. This is an example of a **negative feedback** regulation.

▶ Infection can reset the set-point of the hypothalamus to a higher temperature. Homeostatic mechanisms then act to raise the body temperature to the new set point, resulting in a fever (right).

Fever

Infection can reset normal temperature control so that body temperature increases above the normal range, resulting in a fever. Fever is an important defence against infection.

Regulating body temperature

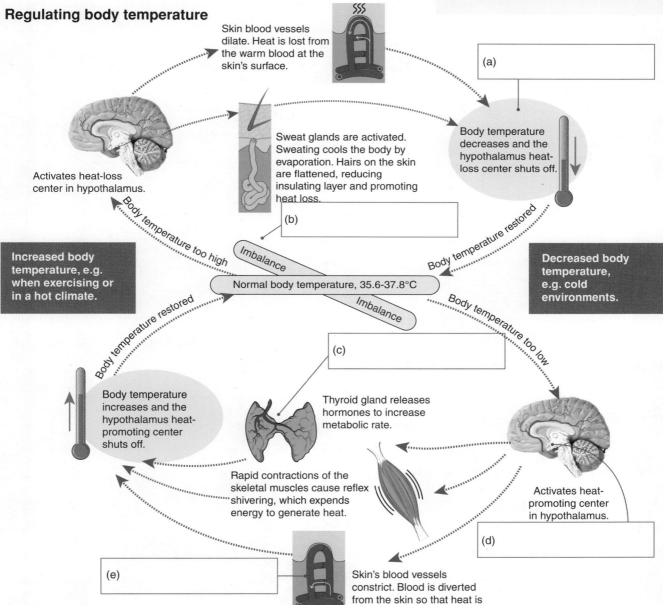

Skin blood vessels dilate. Heat is lost from the warm blood at the skin's surface.

(a)

Body temperature decreases and the hypothalamus heat-loss center shuts off.

Activates heat-loss center in hypothalamus.

Sweat glands are activated. Sweating cools the body by evaporation. Hairs on the skin are flattened, reducing insulating layer and promoting heat loss.

(b)

Body temperature too high

Body temperature restored

Increased body temperature, e.g. when exercising or in a hot climate.

Imbalance

Normal body temperature, 35.6-37.8°C

Imbalance

Decreased body temperature, e.g. cold environments.

Body temperature restored

Body temperature too low

(c)

Thyroid gland releases hormones to increase metabolic rate.

Body temperature increases and the hypothalamus heat-promoting center shuts off.

Rapid contractions of the skeletal muscles cause reflex shivering, which expends energy to generate heat.

Activates heat-promoting center in hypothalamus.

(d)

(e)

Skin's blood vessels constrict. Blood is diverted from the skin so that heat is

1. Label the diagram above by appropriately adding the labels: stimulus, receptors, control centre, and effectors.

©2022 **BIOZONE** International
ISBN: 978-1-98-856692-4
Photocopying Prohibited

SC | LS1.A |

Thermoregulation in newborns

▶ Newborn babies cannot fully thermoregulate until six months of age. They can become too cold or too hot very quickly.

▶ Newborns minimize heat loss by reducing the blood supply to the periphery (skin, hands, and feet). This helps to maintain the core body temperature. Increased brown fat activity and general metabolic activity generates heat. Newborns are often dressed in a hat to reduce heat loss from the head, and tightly wrapped to trap heat next to their bodies.

▶ Newborns lower their temperature by increasing peripheral blood flow. This allows heat to be lost, cooling the core temperature. Newborns can also reduce their body temperature by sweating, although their sweat glands are not fully functional until four weeks after birth.

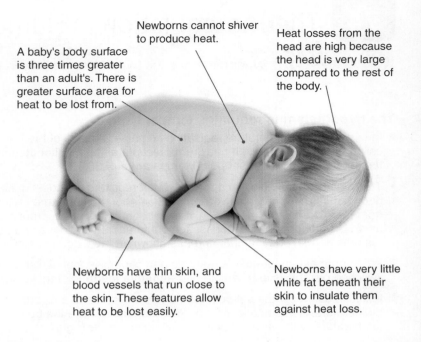

A baby's body surface is three times greater than an adult's. There is greater surface area for heat to be lost from.

Newborns cannot shiver to produce heat.

Heat losses from the head are high because the head is very large compared to the rest of the body.

Newborns have thin skin, and blood vessels that run close to the skin. These features allow heat to be lost easily.

Newborns have very little white fat beneath their skin to insulate them against heat loss.

2. (a) Where is the temperature regulation center in humans located? _____

   (b) How does it carry out this role? _____
   _____

3. Describe the role of the following in maintaining a constant body temperature in humans:

   (a) The skin: _____
   _____

   (b) The muscles: _____
   _____

   (c) The thyroid gland: _____
   _____

4. How is negative feedback involved in keeping body temperature within narrow limits? _____
   _____
   _____
   _____

5. (a) Why does infection result in an elevated core body temperature? _____
   _____

   (b) What is the purpose of this? _____
   _____

   (c) Explain why a prolonged fever can be fatal: _____
   _____

6. (a) What features of a newborn cause it to lose heat quickly? _____
   _____

   (b) What mechanisms do newborns have to control body temperature? _____
   _____
   _____

©2022 **BIOZONE** International
ISBN: 978-1-98-856692-4
Photocopying Prohibited

75 Body Shape and Heat Loss

Key Question: How does body shape influence how quickly heat is lost from the body's surface?

Body shape influences heat loss

▸ Body shape influences how heat is retained or lost. Animals with a lower surface area to volume ratio will lose less body heat per unit of mass than animals with a high surface area to volume ratio.

▸ In humans there is a negative relationship between surface area and latitude. Indigenous people living near the equator tend to have a higher surface area to volume ratio so they can lose heat quickly. In contrast, people living in higher latitudes near the poles have a lower surface area to volume ratio so that they can conserve heat.

▸ People from low latitudes (equatorial regions) have a taller, more slender body and proportionately longer limbs. Those at high latitudes are stockier, with shorter limbs.

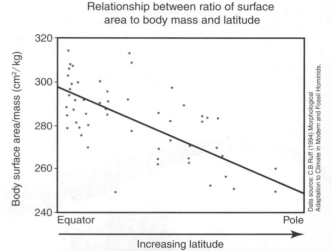

Relationship between ratio of surface area to body mass and latitude

Data source: C.B Ruff (1994) Morphological Adaptation to Climate in Modern and Fossil Hominids.

The Inuit people of Arctic regions are stocky, with relatively short extremities. This body shape is well suited to reducing the surface area over which heat can be lost from the body.

The indigenous peoples of equatorial Africa, such as these young Kenyan men, are tall and slender, with long limbs. This body shape increases the surface area over which heat can be lost.

The relationship holds true for fossil hominids. Neanderthals, which inhabited Eurasia during the last glacial, had robust, stocky bodies relative to modern humans, as this reconstruction shows.

1. Why would having a reduced surface area to volume ratio be an advantage in a cold climate? _____

_____

_____

2. Hypothermia is a condition that occurs when the body cannot generate enough heat and the core body temperature drops below 35°C. Prolonged hypothermia is fatal. Exposure to cold water results in hypothermia more quickly than exposure to the same temperature of air because water is much more effective than air at conducting heat away from the body.
In the graph (right), hypothermia resulting in death is highly likely in region 1 and highly unlikely in region 2.

(a) Which body shape has best survival at 15°C?

_____

(b) Explain your choice: _____

_____

_____

_____

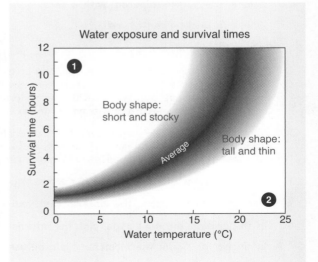

Water exposure and survival times

How does body shape help regulate temperature?

▶ It is often debated whether the dinosaurs were endothermic or ectothermic, and to what degree they maintained a stable body temperature. Some dinosaurs had large, sail-like structures on their backs or heads. One hypothesis suggests these structures might have been used in **thermoregulation**. Other hypotheses suggest they may have served a signalling purpose.

▶ Many dinosaurs did not have sail-like structures. The sheer size of some dinosaurs, e.g. large sauropods like *Brachiosaurus* or *Alamosaurus,* would have protected them from external temperature variations so that maintaining a stable body temperature would have been relatively simple.

▶ Many other dinosaur groups had feathers or feather-like body coverings, suggesting their thermoregulation was somewhat independent of body shape and that they may have been endothermic.

The enormous bulk of some dinosaurs, e.g. *Alamosaurus*, may have protected their internal temperature against changes in environmental temperature.

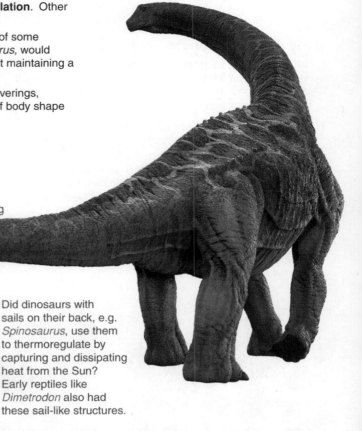

30 m long

16 m long

Did dinosaurs with sails on their back, e.g. *Spinosaurus*, use them to thermoregulate by capturing and dissipating heat from the Sun? Early reptiles like *Dimetrodon* also had these sail-like structures.

Investigation 3.2   Investigating body shape and temperature regulation

See appendix for equipment list.

1. Divide the class into six groups. Each group should choose a shape/orientation from the descriptions below. You will pool class results to complete the table on the next page.

    Shape 1: Foil sphere 3 cm diameter

    Shape 2: Foil sphere 6 cm diameter

    Shape 3: Fusiform (tapered), long side to heat

    Shape 4: Fusiform (tapered), end on to heat

    Shape 5: Fusiform (tapered) with a sail

    Shape 6: Snake-like (thin roll)

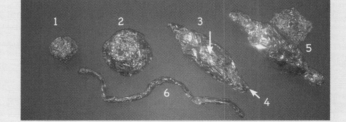

2. Record the shape your group is using: _____

3. Create your shape using aluminum foil. Use scissors or a compass to make a small entry hole and push a thermometer* into the center of the shape. Switch the heat lamp on and place the shape 30 cm from it. If you are using shape 3 or 4, you will need to orientate the shape long side (3) or end on (4) to the heat lamp. In the table below, record the temperature every minute for a total of 6 minutes.

4. a. Switch the lamp OFF and record the temperature after 2 minutes in the table on the next page.
    b. Switch it back ON and record the temperature after 2 minutes in the table on the next page
    c. Repeat this ON/OFF sequence twice more, recording the temperature in the table on the next page.

5. Switch the lamp OFF and record the temperature every minute for 6 minutes in the table on the next page.

6. Share the results among the class so that you can complete the table for all six aluminum shapes.

7. On the grid on the next page, plot the internal temperature over time for your shape.

* A datalogger or digital thermometer could be used for more accurate temperature recording.

©2022 **BIOZONE** International
ISBN: 978-1-98-856692-4
Photocopying Prohibited

Shape	Temperature change																
	Lamp ON 6 minutes						OFF	ON	OFF	ON	OFF	ON	Lamp OFF 6 minutes				
1																	
2																	
3																	
4																	
5																	
6																	

NEED HELP?
See Activities
17, 18

3. Which shape heated up and cooled down the fastest? _____

4. Which shape heated up and cooled down the slowest? _____

5. (a) Which shape had the most stable temperature during the phase of switching the lamp on and off?

_____

(b) Why do you think this was? _____

_____

6. Does body shape or size have any effect on temperature regulation? Explain: _____

_____

_____

_____

_____

7. It is now thought that most *Coelurosauria* (a group of theropod dinosaurs that included *Tyrannosaurus rex*) had feathers of some kind. What advantage might the feathers have given these dinosaurs over non-feather dinosaurs?

_____

_____

©2022 **BIOZONE** International
ISBN: 978-1-98-856692-4
Photocopying Prohibited

76 Controlling Blood Glucose Levels

Key Question: How is a constant blood glucose level maintained in the body?

The importance of blood glucose

▶ **Glucose** is the body's main energy source. It is chemically broken down during cellular respiration to generate ATP, which is used to power metabolism. Glucose is the main sugar circulating in blood, so it is often called blood sugar. Blood glucose levels are regulated by **negative feedback** involving two hormones, insulin and glucagon.

▶ Blood glucose levels are tightly controlled because cells must receive an adequate and regular supply of fuel. Prolonged high or low blood glucose causes serious physiological problems and even death. Normal activities, such as eating and exercise, alter blood glucose levels, but the body's control mechanisms regulate levels so that fluctuations are minimized and generally occur within a physiologically acceptable range. For humans, this is 60-110 mg/dL, indicated by the shaded area in the graph below.

Insulin enables cells to take up glucose

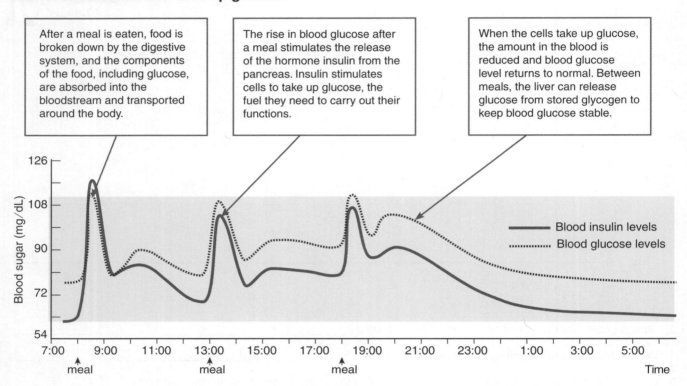

After a meal is eaten, food is broken down by the digestive system, and the components of the food, including glucose, are absorbed into the bloodstream and transported around the body.

The rise in blood glucose after a meal stimulates the release of the hormone insulin from the pancreas. Insulin stimulates cells to take up glucose, the fuel they need to carry out their functions.

When the cells take up glucose, the amount in the blood is reduced and blood glucose level returns to normal. Between meals, the liver can release glucose from stored glycogen to keep blood glucose stable.

Blood insulin levels

Blood glucose levels

What happens if your body does not produce insulin?

▶ In some people, the insulin-producing cells of the pancreas are damaged ,e.g. by infection, and the body cannot produce insulin. This life threatening disorder, which commonly affects children and teenagers, is called **type 1 diabetes mellitus**.

▶ The cells of type 1 diabetes cannot take up glucose from the blood, so glucose remains in the blood and blood glucose levels are elevated. The kidneys try to rid the body of the apparently "excess" glucose so sufferers produce large volumes of "sweet" urine, as glucose is excreted in the urine. They feel tired and weak and are constantly hungry and thirsty. Fats are metabolized for fuel.

▶ The only current treatment is regular injection with human insulin, together with careful dietary management to control blood glucose levels.

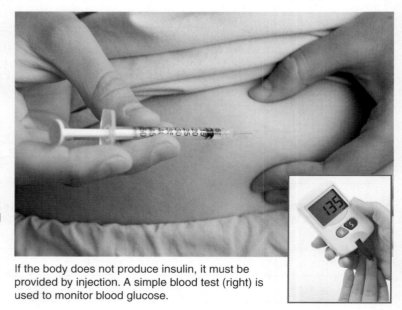

If the body does not produce insulin, it must be provided by injection. A simple blood test (right) is used to monitor blood glucose.

LS1.A   SC

©2022 **BIOZONE** International
ISBN: 978-1-98-856692-4
Photocopying Prohibited

Controlling blood glucose levels

Blood glucose (BG) is controlled by two hormones produced by special endocrine cells in the pancreas. The hormones work antagonistically (oppose each other) and levels are tightly controlled by **negative feedback**.

▸ Insulin lowers blood glucose by promoting glucose uptake by cells and glycogen storage in the liver.

▸ Glucagon increases blood glucose by promoting release of glucose from the breakdown of glycogen in the liver.

▸ When normal blood glucose levels are restored, negative feedback stops hormone secretion.

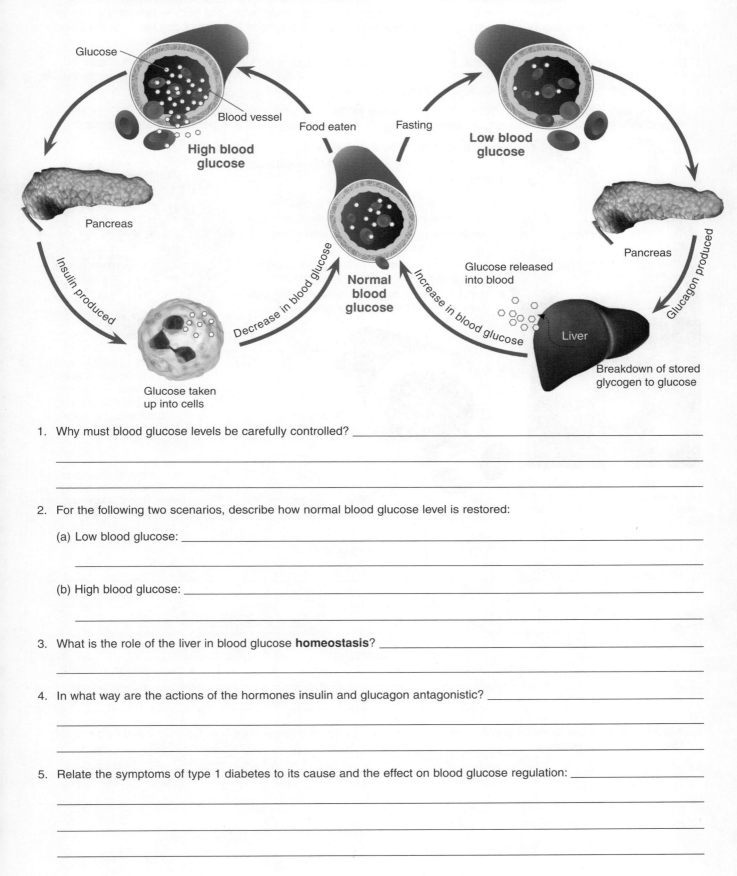

1. Why must blood glucose levels be carefully controlled? _____

_____

_____

2. For the following two scenarios, describe how normal blood glucose level is restored:

   (a) Low blood glucose: _____

   _____

   (b) High blood glucose: _____

   _____

3. What is the role of the liver in blood glucose **homeostasis**? _____

_____

4. In what way are the actions of the hormones insulin and glucagon antagonistic? _____

_____

_____

5. Relate the symptoms of type 1 diabetes to its cause and the effect on blood glucose regulation: _____

_____

_____

_____

77 Type 2 Diabetes

Key Question: What is type 2 diabetes, and how does it differ from type 1 diabetes?

▶ As we have seen, high blood glucose levels are the result of a condition called **diabetes mellitus**. In type 1 diabetes, the body cannot produce insulin and the body's cells cannot take up glucose at all, resulting in blood glucose levels that are too high. However, a second form of diabetes occurs when the body is resistant to insulin's effects. The pancreas produces insulin, but the body's cells stop responding to it, and glucose levels in the blood remain high.

▶ Symptoms are similar to type 1 diabetes and are mild at first. The body's cells do not respond appropriately to the insulin present and blood glucose levels become elevated. Type 2 diabetes typically affects older people but is increasingly common in young adults and overweight children.

▶ Type 2 diabetes is a disease that becomes progressively worse over time, eventually damaging blood vessels and leading to heart disease and stroke. Treatment is through dietary management and prescribed anti-diabetic drugs.

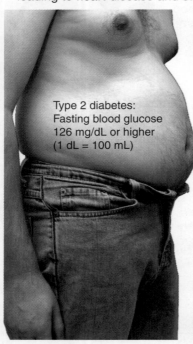

Type 2 diabetes:
Fasting blood glucose
126 mg/dL or higher
(1 dL = 100 mL)

Cells are starved of fuel, leading to increased appetite and overeating. This may contribute to an existing obesity problem.

Risk factors for type 2 diabetes

These factors interfere with the body's normal blood glucose regulation:

▶ Obesity and sedentary lifestyle: inactivity increases risk through its effects on body weight.

▶ Family history: there is a strong genetic link for type 2 diabetes.

▶ High blood pressure: up to 60% of people with undiagnosed diabetes have high blood pressure.

▶ High blood lipids: more than 40% of people with diabetes have abnormally high blood lipids.

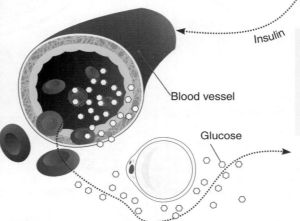

Insulin

Blood vessel

Glucose

The beta cells of the pancreatic islets (above) produce insulin, the hormone responsible for the cellular uptake of glucose. In type 2 diabetes, the body's cells do not use the insulin properly.

Cellular uptake of glucose is impaired and glucose remains in the blood. Type 2 diabetes is also called insulin resistance.

1. How does type 2 diabetes differ from type 1 diabetes with respect to:

(a) Cause: _____

_____

(b) Symptoms: _____

_____

(c) Treatment: _____

_____

2. Antidiabetic drugs can work in several ways. For each of the statements below, state how the drug would help restore blood glucose homeostasis:

(a) Drug increases sensitivity of cells to insulin: _____

_____

(b) Drug increases pancreatic secretion of insulin: _____

_____

LS1.A SC

©2022 **BIOZONE** International
ISBN: 978-1-98-856692-4
Photocopying Prohibited

78 Homeostasis During Exercise

Key Question: How do the circulatory and respiratory systems function to maintain homeostasis during exercise?

▸ During exercise, greater metabolic demands are placed on the body and it must work harder to maintain **homeostasis.**

▸ Maintaining homeostasis during exercise is principally the job of the circulatory and respiratory systems, although the skin, kidneys, and liver are also important.

Working muscles need more ATP than muscles at rest.

Increased body temperature

During exercise, the extra heat produced by muscle contraction must be dispersed to prevent overheating. Thermoregulatory mechanisms, such as sweating and increased blood flow to the skin, release excess heat into the surrounding environment and help cool the body.

Increased heart rate

An increased heart rate circulates blood around the body more quickly. This increases the rate at which exchanges can be made between the blood and the working tissues. Oxygen and glucose are delivered and metabolic wastes, e.g. carbon dioxide, are removed.

Increased glucose production

During exercise, working muscles quickly take up and use the freely available blood glucose. Glucose is mobilized from glycogen stores in the liver and supplies the body with fuel to maintain ATP production.

Increased breathing rate

Exercise increases the body's demand for energy (ATP). Oxygen is required for cellular respiration and ATP production. Increasing the rate of breathing delivers more oxygen to working tissues and enables them to make the ATP they need to keep working. An increased breathing rate also increases the rate at which carbon dioxide is expelled from the body.

1. The graph (right) compares the change in cardiac output (a measure of total blood flow in L) during rest and during exercise. The color of the bars indicates the proportion of blood flow in skeletal muscle relative to other body parts.

   ▇ Blood flow to muscle
   ▢ Blood flow to other body parts

   (a) What percentage of the blood goes to the muscles at rest?

   _____

   _____

   (b) What percentage of the blood goes to the muscles during exercise?

   _____

   _____

   Cardiac output (L)

   Total 5.5 L: muscle 0.9 L

   Resting

   Total 22.5 L: muscle 17 L

   Heavy exercise

2. (a) What happens to the total blood flow during heavy exercise compared to at rest? _____

   _____

   _____

   (b) Why does this occur? _____

   _____

   (c) What would be happening to breathing rate during this time? _____

79 Effect of Exercise on Breathing and Heart Rate

Key Question: What effect does exercise have on breathing and heart rate?

Investigation 3.3 Investigating effect of exercise on heart rate.

See appendix for equipment list.

In this practical, you will work in groups of three to see how exercise affects heart rates. The body's response to exercise can be measured by monitoring changes in heart rate before and after a controlled physical effort. Choose one person to carry out the exercise and one person each to record heart rate.

Heart rate (beats per minute) is obtained by measuring the pulse (right) for 15 seconds and multiplying by four.

CAUTION: The person exercising should have no known pre-existing heart or respiratory conditions.

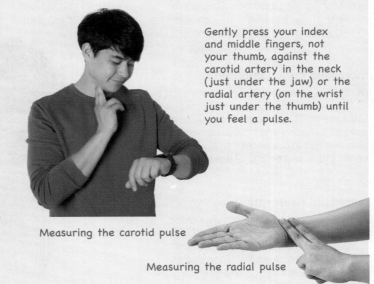

Gently press your index and middle fingers, not your thumb, against the carotid artery in the neck (just under the jaw) or the radial artery (on the wrist just under the thumb) until you feel a pulse.

Measuring the carotid pulse

Measuring the radial pulse

1. Resting measurements: The person carrying out the exercise should sit down on a chair for 5 minutes and try not to move. After 5 minutes of sitting, measure their heart rate. Record the resting data in the table (below).

2. Exercising measurements: Choose an exercise to perform. Some examples include: step ups onto a chair, skipping rope, jumping jacks, or running in place.

3. Begin the exercise, and take measurements after 1, 2, 3, and 4 minutes. The person exercising should stop just long enough for the measurements to be taken. Record the results in the table.

4. Post-exercise measurements: After the exercise period has finished, the exerciser should sit down on a chair. Take pulse measurements 1 and 5 minutes after finishing the exercise. Record the results in the table, below.

Activity	Resting before exercise	At 1 minute during exercise	At 2 minutes during exercise	At 3 minutes during exercise	At 4 minutes during exercise	1 minute after exercise	5 minutes after exercise
Pulse Rate							

1. (a) Graph your results on separate piece of paper. You will use the vertical axis to plot heart rate. When you have finished answering the questions below, attach it to this page.

NEED HELP?
See Activities 17, 18

   (b) Analyze your graph and describe what happened to heart rate during exercise:

   _____

2. (a) Describe what happened to heart rate after exercise:

   _____

   _____

   (b) Why did this change occur? Link your answer to the idea that **feedback mechanisms** maintain **homeostasis**.

   _____

   _____

   _____

   _____

LS1.A   SC

©2022 **BIOZONE** International
ISBN: 978-1-98-856692-4
Photocopying Prohibited

Investigation 3.4   Investigating effect of exercise on breathing rate.

See appendix for equipment list.

You could work in both small groups and individually to see how exercise affects breathing rates.

1. Complete questions (3), (4), and (5) below, to plan your investigation.

2. Conduct your investigation following your planned method and collect data.

3. Complete question (6) after the data collection, data processing, and investigation conclusion have been completed.

3. How is this phenomenon linked to feedback mechanisms that maintain homeostasis? _____

_____

_____

_____

_____

4. Identifying Evidence: Develop an investigation plan and consider the following:

(a) How will the change in the external environment be determined? _____

_____

(b) How will the response of the living system be identified? _____

_____

_____

(c) How will this data contribute to investigating the phenomenon in question (1)? _____

_____

_____

5. Planning for the investigation: Construct an investigation plan. Complete the chart to develop the structure:

Steps to consider	Your investigation
(a) How will the change in the external environment be measured, i.e type, length, units, quantity of exercise?	
(b) How will the response of the living system (breathing) be measured, i.e. units, length, timing?	
(c) How are the internal conditions (homeostasis of oxygen demand) linked to what is being measured?	

6. Use the following space to construct a draft method. Develop a final investigation procedure (on separate paper), including a data table in which to record your results, incorporating the information provided previously. Ensure control of other variables. You may wish to include a labeled diagram.

7. Considering your collected data: Attach your written procedure and data table to this page. Collect your data from your developed investigation procedure, then use your findings to answer the following questions:

(a) Explain how your conclusion was able to answer "How does exercise affect breathing rates?"

_____

_____

_____

(b) How accurate and precise was your data and what steps did you take to ensure accuracy? _____

_____

_____

(c) How generalizable was your data and how did you ensure reliability? _____

_____

_____

(d) What limitations did your group experience in collecting the data? _____

_____

(e) How you could you refine the design of the investigation to increase accuracy and generalizability of your data?

_____

_____

_____

_____

_____

_____

80 Homeostasis in Plants

Key Question: How does the process of transpiration help maintain water homeostasis in plants?

Maintaining water balance

▶ Like animals, plants need water for life processes. Water gives cells turgor, transports dissolved substances, and is a medium in which metabolic reactions can take place. Maintaining water balance is an important homeostatic function in plants. In plants, evaporative water loss from stomata drives a transpiration stream that ensures plants have a constant supply of water to support essential life processes.

▶ Vascular plants obtain water from the soil. Water enters the plant via the roots, and is transported throughout the plant by a specialized tissue called xylem. Water is lost from the plant by evaporation. This evaporative water loss is called **transpiration**.

▶ Transpiration has several important functions:

• Provides a constant supply of water needed for essential life processes such as photosynthesis.
• Cools the plant by evaporative water loss.
• Helps the plant take up minerals from the soil.

▶ However, if too much water is lost by transpiration, a plant will become dehydrated and may die.

The role of stomata

▶ Water loss occurs mainly through stomata (pores in the leaf). The rate of water loss can be regulated by specialized guard cells either side of the stoma, which open or close the pore.

• Stomata open: transpiration rate increases.
• Stomata closed: transpiration rates decrease.

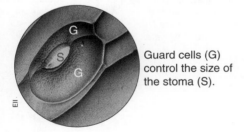

Guard cells (G) control the size of the stoma (S).

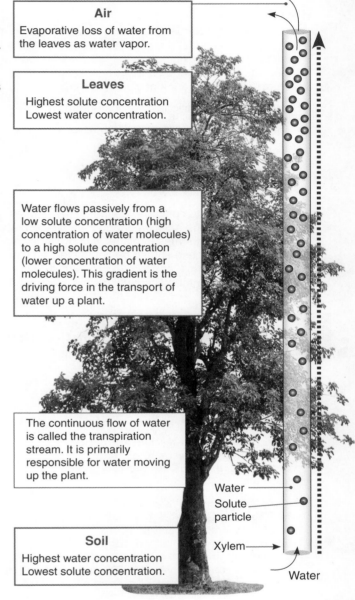

Air
Evaporative loss of water from the leaves as water vapor.

Leaves
Highest solute concentration
Lowest water concentration.

Water flows passively from a low solute concentration (high concentration of water molecules) to a high solute concentration (lower concentration of water molecules). This gradient is the driving force in the transport of water up a plant.

The continuous flow of water is called the transpiration stream. It is primarily responsible for water moving up the plant.

Water

Solute particle

Xylem

Water

Soil
Highest water concentration
Lowest solute concentration.

1. (a) What is transpiration? _____

   (b) How does transpiration provide water for essential life processes in plants? _____
   _____
   _____

2. How do plants regulate the amount of water lost from the leaves? _____
   _____

3. (a) What would happen if too much water was lost by transpiration? _____
   _____

   (b) When might this happen? _____
   _____

©2022 **BIOZONE** International
ISBN: 978-1-98-856692-4
Photocopying Prohibited

81 Measuring Transpiration in Plants

Key Question: What effects do physical factors in the environment, such as humidity, temperature, light level, and air movement, have on transpiration rate in plants?

The potometer

▶ A potometer is a simple instrument for investigating **transpiration** rate (water loss per unit time). The equipment is simple to use and easy to obtain. A basic potometer, such as the one shown right, can easily be moved around so that transpiration rate can be measured under different environmental conditions.

▶ Some physical conditions investigated are:

- Humidity or vapor pressure (high or low)

- Temperature (high or low)

- Air movement (still or windy)

- Light level (high or low)

▶ It is also possible to compare the transpiration rates of plants with different adaptations e.g. comparing transpiration rates in plants with rolled leaves to rates in plants with broad leaves. If possible, experiments like these should be conducted simultaneously using replicate equipment. If conducted sequentially, care should be taken to keep the environmental conditions the same for all plants used.

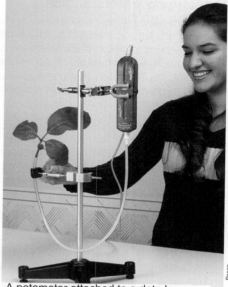

A potometer attached to a data logger

Investigation 3.5 Investigating plant transpiration

1. Four different conditions that influence transpiration will be tested: room conditions (ambient), wind, bright light, and high humidity.

2. Before starting, your teacher will decide if your group is to test one of these conditions (and which one) and pool class data for all four.

3. Set up the potometer and plant as in the diagram. It is best if the plant leaves used are large and few (4-6 leaves) rather than small and many. Alternatively, the plant can be placed in a 250 mL conical flask with 200 mL of water and a thin layer of cooking oil floated on top. This is weighed before the experiment and then every 3 minutes (or as the experiment requires). The difference in mass in grams is equal to the volume of water transpired in mL. After setting up the potometer, let the apparatus equilibrate for 10 minutes, and then record the position of the air bubble in the pipette (or the mass of the equipment for the alternative method). This is time 0 and position 0.

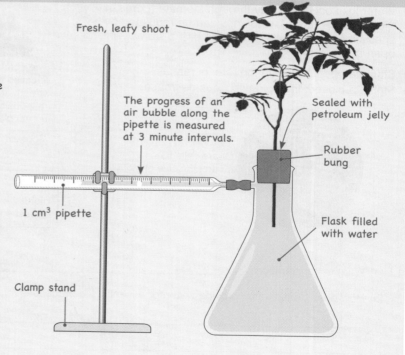

Fresh, leafy shoot

The progress of an air bubble along the pipette is measured at 3 minute intervals.

Sealed with petroleum jelly

Rubber bung

1 cm^3 pipette

Flask filled with water

Clamp stand

4. The plant can now be exposed to one of the four conditions. Record results in Table 1.

5. For the <u>ambient environment</u> the equipment can be placed on the bench away from bright light or wind. Record the net movement of the bubble every 3 minutes for 30 minutes.

6. For the <u>high wind environment</u> the equipment can be placed on the bench in front of a fan set on a moderate speed (away from bright light). Record the net movement of the bubble every 3 minutes for 30 minutes.

      LS1.A SC

©2022 **BIOZONE** International
ISBN: 978-1-98-856692-4
Photocopying Prohibited

7. For the bright <u>light environment,</u> the equipment can be placed on the bench in front of a bright light (about 40 cm away). Record the net movement of the bubble every 3 minutes for 30 minutes.

8. For the <u>high humidity environment</u> the equipment can be placed on a bench away from bright light, in a sealed plastic bag with 2–3 sprays of water from a spray bottle. Record the net movement of the bubble every 3 minutes for 30 minutes.

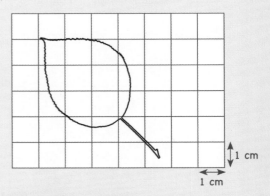

9. It is important that for fair comparison of transpiration the area of leaf used in each environment (or by different groups) should be calculated and the volume of water lost per square centimeter compared (mL/cm^2).

10. Leaf area can be measured by tracing the leaves onto graph paper and counting the squares, or by tracing or photocopying the leaves onto a paper of a known mass per area, then cutting out the shapes and weighing them. For both methods, multiply by 2 for both leaf surfaces.

11. Once the area of the leaf is calculated, the transpiration (water lost) in mL/cm^2 can be calculated for each time recording and recorded in Table 2.

Table 1. Potometer readings (in mL water loss)

Time (min) / Treatment	0	3	6	9	12	15	18	21	24	27	30
Ambient											
Wind											
High humidity											
Bright light											

Table 2. Potometer readings in mL per cm^2

Time (min) / Treatment	0	3	6	9	12	15	18	21	24	27	30
Ambient											
Wind											
High humidity											
Bright light											

1. Measure the area of the leaves you used: _____

2. Why is comparing water loss per square cm over time more important than just comparing the water loss over time?

_____

_____

_____

©2022 **BIOZONE** International
ISBN: 978-1-98-856692-4
Photocopying Prohibited

3. Plot the data in Table 2 on the grid provided:

NEED HELP?
See Activities
17, 18

4. Identify the independent variable: _____

5. (a) Identify the control: _____

   (b) Explain the purpose of including an experimental control in an experiment:

   _____

   _____

   _____

6. (a) Which factors increased water loss? _____

   (b) How does each environmental factor influence water loss? _____

   _____

   _____

   _____

   _____

   _____

7. From your results predict how each of the following conditions might influence transpiration:

   (a) Low humidity, e.g. dry desert: _____

   (b) Low light levels, e.g. overcast day: _____

   (c) Hot dry winds: _____

8. How might different types of plants affect the results? _____

   _____

   _____

©2022 BIOZONE International
ISBN: 978-1-98-856692-4
Photocopying Prohibited

82 Review Your Understanding

Key Question: How do mammals manage to maintain stable conditions in their bodies, despite facing a wide range of external conditions?

▶ At the beginning of this chapter you were asked how mammals were able to maintain a stable internal environment, even when the external conditions changed, or were extreme. You should now be able to answer those questions.

1. Mammals are endotherms, whereas reptiles are ectotherms. Compare and contrast these two heat regulation methods:

   _____

   _____

   _____

   _____

2. The fennec fox (right) is a small canine mammal found in the North African Sahara desert. The horned desert viper is a snake also found in the same environment. Compare how the two species produce and maintain body heat:

   (a) Fennec fox: _____

   _____

   (b) Horned desert viper: _____

   _____

3. The fox and snake eat a similar range of food: small vertebrates and invertebrates. The Sahara desert can drop from as much as 38°C to -4°C (below freezing) at night. Both species actively hunt at night, but the snake becomes slower as the temperature drops. How might the constantly high activity level of the fox, linked to warm-bloodedness, be both an advantage and disadvantage compared to the snake?

   _____

   _____

   _____

   _____

4. The Arctic fox belongs to the same genus as the fennec fox, *Vulpes*, and can be found in very cold northern hemisphere tundra, including parts of Alaska. Many other mammals also live in these habitats. Only one reptile species, the common lizard, lives on the edges of this cold environment. Explain why mammals have been able to successfully spread and survive in the colder extreme environments on Earth, yet reptiles have not, linking to their method of temperature regulation:

   _____

   _____

   _____

   _____

   _____

5. Summarize how the process of homeostasis allows mammals, including humans, to survive in an extreme environment:

   _____

   _____

   _____

©2022 **BIOZONE** International
ISBN: 978-1-98-856692-4

83 Summing Up

1. The body is constantly interacting with the environment, with daily fluctuations in temperature, food and fluid intake, and physical exertion. The maintenance of a steady state is the job of all the body's systems working together. Complete the diagram below to summarize how the steady state (homeostasis) is achieved in each situation.

a) Walking to work in cold winter conditions:

b) Exercising for 30 minutes mid-morning:

c) Feeling hungry before lunch:

d) Straight after eating a large lunch:

©2022 **BIOZONE** International
ISBN: 978-1-98-856692-4

CHAPTER 4

Growth and Development

Activity number

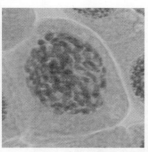

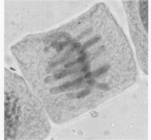

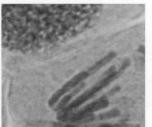

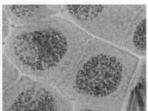

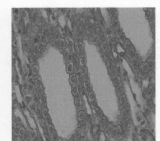

Anchoring Phenomenon

The power to rebuild: How can the axolotl have such superpowers of regeneration, with the ability to regrow amputated limbs, damaged spinal cords, and even parts of its brain?

84 97

How do organisms grow and develop through mitosis?

☐ 1 Explain how mitosis enables a multicellular organism to develop from a single cell (a fertilized egg) called a zygote.

85

☐ 2 Explain why DNA replication must take place before a cell can divide. Link DNA replication to the ability to produce two identical copies of DNA, where a copy goes to each new cell produced during mitosis. Explain the importance of semi-conservative replication. From text and models, paraphrase the process of DNA replication.

86 87

☐ 3 Develop models to demonstrate semi-conservative DNA replication. Compare and evaluate your models to the experimental models of Meselson and Stahl.

88

☐ 4 Discuss the three primary functions of mitosis.

89

☐ 5 Briefly outline key steps of the cell cycle, using information from text, images, and models.

90

☐ 6 Describe the general process of mitosis (cell division), using information from text, images, and models.

91

☐ 7 Develop a more detailed explanation of mitosis and cytokinesis, using information from text, images, and models. Summarize key features of each phase of mitosis. Compare the process of cytokinesis in plant and animal cells.

92

☐ 8 Investigate the process of mitosis by developing a physical model.

93

Why is cell differentiation needed to carry out specialized roles?

☐ 9 Define stem cells, using information provided from text, images, and models. Explain how the properties of stem cells allow a multicellular organism to be made up of many different types of specialized cells.

94 95

☐ 10 Understand that body tissues are composed of related cell types that work together to carry out specific functions, and come together to form organs that help the organism to function, as a whole. Describe the function of the main types of tissue in the body.

96

☐ 11 Calculate the approximate number of cells in a human body. Compare your estimate to actual data, and discuss why they might be different. Model hypothetical scenarios showing the process of mitosis.

98

84 The Power to Rebuild

Key Question: How can the axolotl have such superpowers of regeneration, with the ability to regrow amputated limbs, damaged spinal cords, and even parts of its brain?

Humans and most other vertebrates are able to repair damaged tissue when injured, but that's about the limit of our regenerative powers. Many invertebrates, such as octopuses and crabs, can grow new limbs if they are damaged or amputated. However, the axolotl has advanced regenerative powers. Young axolotls often attack and bite each other, inflicting severe wounds. Its amazing powers of healing allow it to recover, without scarring, from intense injury.

▶ If a limb is damaged, or amputated it grows back with no scarring - over and over again, if necessary.

▶ Axolotls can also grow a new tail, a new jaw, replace a damaged spinal cord and even some parts of their brain. Organs can be transplanted from one axolotl to another without any rejection issues and the axlotl is more resistant to cancer than mammals.

1. Think about last time you scraped your knees or accidentally cut yourself. How did the wound heal - from the center outwards or from the edges to the center?

_____

2. What is special about the axolotl's ability to heal itself, compared to humans? _____

_____

_____

_____

3. Do you know of an organ in the human body that has the power to regenerate if a large piece of it has to be removed due to damage or disease:

_____

_____

_____

4. What stages do you think cells go through, in order to divide and replicate correctly? _____

_____

_____

_____

_____

5. How do you think a cell "knows" that is is to develop into skin, bone, brain tissue, blood etc?

_____

_____

_____

_____

©2022 **BIOZONE** International
ISBN: 978-1-98-856692-4
Photocopying Prohibited

85 Growth and Development of Organisms

Key Question: How do multicellular organisms develop from a single cell to produce genetically identical copies?

Organisms grow and develop by a process called **mitosis**.

▸ Multicellular organisms begin as a single cell (a fertilized egg) and develop into complex organisms made up of many cells. This is achieved by the process of mitosis (mitotic **cell division**).

▸ In multicellular organisms, mitosis is responsible for growth and for the replacement of old and damaged cells. In some unicellular eukaryotic organisms, e.g. yeast cells, mitosis is also responsible for reproduction.

▸ Two important processes must occur in order for new cells to be produced. The first is the duplication of the genetic material (DNA). The second is the division (splitting) of the parent cell into two identical daughter cells. The two daughter cells have the same genetic material as the parent cell.

▸ In multicellular animals, mitosis only occurs in body cells (somatic cells). Sperm and egg cells (gametes) are produced by a different type of cell division called meiosis.

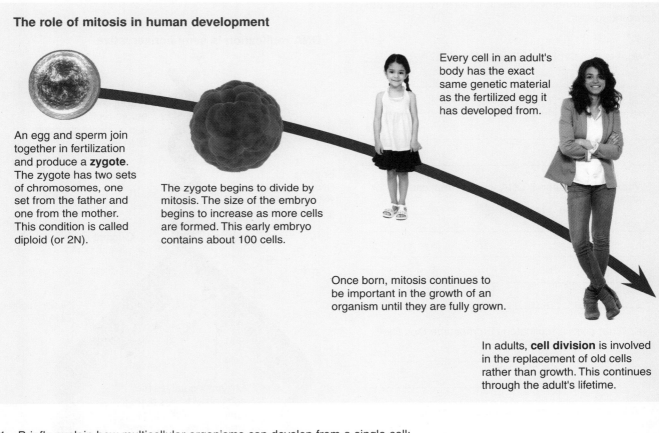

The role of mitosis in human development

An egg and sperm join together in fertilization and produce a **zygote**. The zygote has two sets of chromosomes, one set from the father and one from the mother. This condition is called diploid (or 2N).

The zygote begins to divide by mitosis. The size of the embryo begins to increase as more cells are formed. This early embryo contains about 100 cells.

Every cell in an adult's body has the exact same genetic material as the fertilized egg it has developed from.

Once born, mitosis continues to be important in the growth of an organism until they are fully grown.

In adults, **cell division** is involved in the replacement of old cells rather than growth. This continues through the adult's lifetime.

1. Briefly explain how multicellular organisms can develop from a single cell: _____

_____

_____

2. What two things must occur for a new cell to be produced? _____

_____

_____

3. Explain the role of mitosis in:

(a) A developing embryo: _____

(b) An adult: _____

©2022 **BIOZONE** International
ISBN: 978-1-98-856692-4
Photocopying Prohibited

LS1.B

86 DNA Replication

- Before a cell can divide, its DNA must be copied (replicated). **DNA replication** ensures that the two daughter cells receive identical genetic information.

- In eukaryotes, DNA exists in the cell nucleus and is organized into structures called chromosomes.

- DNA replication takes place in the time between **cell divisions**. After the DNA has replicated, each chromosome is made up of two chromatids which are joined at the centromere.

- The process of DNA replication is known as semi-conservative, meaning that each chromatid contains half original (parent) DNA and half new (daughter) DNA. The two chromatids will become separated during cell division to form two separate chromosomes.

1. What is the purpose of DNA replication? _____

_____

_____

_____

_____

2. What would happen if DNA was not replicated prior to cell division?

_____

_____

_____

_____

3. (a) What does a replicated chromosome look like?

_____

_____

   (b) What is the purpose of the centromere?

_____

_____

_____

4. Explain what semi-conservative replication means:

_____

_____

_____

_____

_____

DNA replication duplicates chromosomes

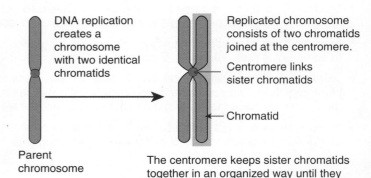

DNA replication creates a chromosome with two identical chromatids

Parent chromosome

Replicated chromosome consists of two chromatids joined at the centromere.

Centromere links sister chromatids

Chromatid

The centromere keeps sister chromatids together in an organized way until they are separated before nuclear division.

DNA replication is semi-conservative

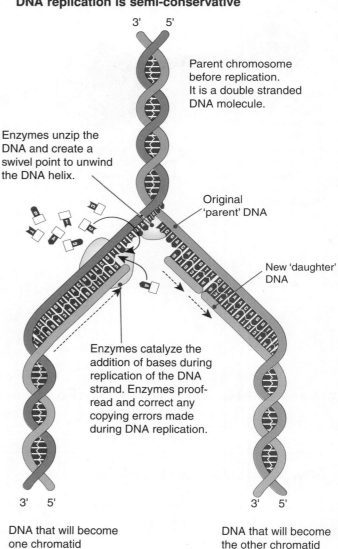

Parent chromosome before replication. It is a double stranded DNA molecule.

Enzymes unzip the DNA and create a swivel point to unwind the DNA helix.

Original 'parent' DNA

New 'daughter' DNA

Enzymes catalyze the addition of bases during replication of the DNA strand. Enzymes proof-read and correct any copying errors made during DNA replication.

DNA that will become one chromatid

DNA that will become the other chromatid

DNA replication is called semi-conservative. This is because each resulting DNA molecule is made up of one parent strand and one daughter strand of DNA.

©2022 **BIOZONE** International
ISBN: 978-1-98-856692-4
Photocopying Prohibited

   LS1.B SSM

87 Details of DNA Replication

Key Question: How does DNA unwind for replication happen and what enzymes are involved?

▸ The individual units that make up the DNA molecule are called **nucleotides**. During **DNA replication**, new nucleotides are added at a region called the replication fork. This replication fork moves along the chromosome as replication progresses.

▸ Nucleotides are added in by complementary base-pairing. The base pairing rule ensures that nucleotide A is always paired with nucleotide T and nucleotide C is always paired with nucleotide G.

▸ The DNA strands can only be replicated in one direction, so one strand has to be copied in short segments which are joined together later.

▸ This whole process occurs simultaneously for each chromosome of a cell and the entire process is tightly controlled by enzymes.

1. How are the new strands of DNA lengthened?

_____

_____

_____

_____

2. What rule ensures that the two new DNA strands are identical to the original strand?

_____

_____

3. Why does one strand of DNA need to be copied in segments?

_____

_____

_____

_____

_____

4. Describe three activities carried out by enzymes during DNA replication:

(a) _____

_____

(b) _____

_____

(c) _____

_____

Stages in DNA replication

Parent DNA is made up of two **anti-parallel** strands coiled into a double helix.

The two strands are joined by base pairing

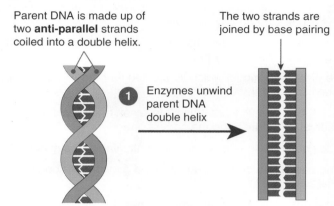

1 Enzymes unwind parent DNA double helix

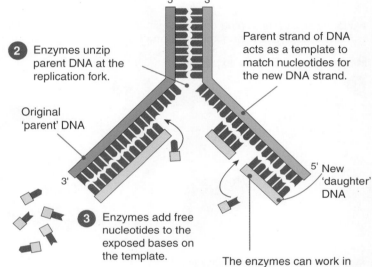

5'   3'

2 Enzymes unzip parent DNA at the replication fork.

Parent strand of DNA acts as a template to match nucleotides for the new DNA strand.

Original 'parent' DNA

3'

3 Enzymes add free nucleotides to the exposed bases on the template.

5' New 'daughter' DNA

The enzymes can work in only one direction and the strands are anti-parallel, so one strand is made in fragments that are later joined by other enzymes.

Nucleotide symbols

G   C   A   T

DNA base pairing rule

G pairs with C
A pairs with T

4 Two new double-stranded DNA molecules

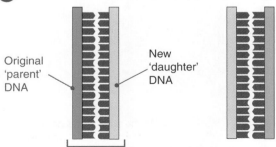

Original 'parent' DNA

New 'daughter' DNA

Enzymes are involved at every step of DNA replication. They unzip the parent DNA, add the free nucleotides to the 3' end of each single strand, join DNA fragments, and check and correct the new DNA strands.

©2022 **BIOZONE** International
ISBN: 978-1-98-856692-4
Photocopying Prohibited

 SSM   LS1.B

88 Modeling DNA Replication

Key Question: How do we know that DNA replication is semi-conservative?

Initially, three models were proposed to explain how DNA replicated:

1 – a semi-conservative model in which each DNA strand served as a template, forming a new DNA molecule that was half old and half new DNA;

2 – a conservative model which suggested that the original DNA served as a complete template so that the resulting DNA was completely new; and

3 – a dispersive model which suggested that the two new DNA molecules had part new and part old DNA interspersed throughout them.

In 1958, two scientists, Matthew Meselson and Franklin Stahl, carried out an experiment that proved the semi-conservative model to be the correct one.

▶ They grew bacteria in a solution containing a heavy nitrogen isotope (^{15}N) until their DNA contained only ^{15}N.

▶ The bacteria were then placed into a growth solution containing the nitrogen isotope (^{14}N), which is lighter than ^{15}N.

▶ After a set number of generation times, the DNA was extracted and centrifuged in a solution that provides a density gradient.

▶ Heavy DNA (containing only ^{15}N) sinks to the bottom, light DNA (containing only ^{14}N) rises to the top, and intermediate DNA (one light and one heavy strand) settles in the middle, as shown below:

Meselson and Stahl's experiment

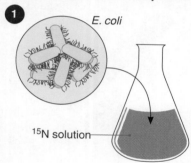

E. coli were grown in a nutrient solution containing ^{15}N. After 14 generations all the bacterial DNA contained ^{15}N. A sample was removed. This was generation 0.

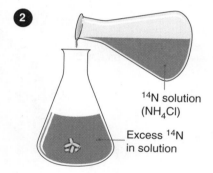

Generation 0 was added to a solution with excess 14N (as NH_4Cl). During replication, new DNA would incorporate 14N and be 'lighter' than the original DNA (which had only ^{15}N).

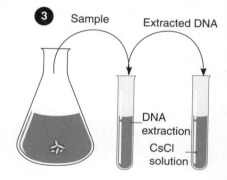

Each generation (~ 20 minutes), a sample was taken and treated to release the DNA. The DNA was placed in a CsCl solution which provided a density gradient for separating the DNA.

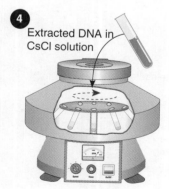

Samples were spun in a high speed ultracentrifuge at 140,000 *g* for 20 hours. Heavier ^{15}N DNA moved closer to the bottom of the test tube than light 14N DNA or 14N/ ^{15}N intermediate DNA.

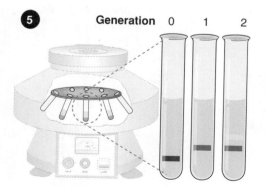

All the DNA in the generation 0 sample moved to the bottom of the test tube. All the DNA in the generation 1 sample moved to an intermediate position. At generation 2, half the DNA was at the intermediate position and half was near the top of the test tube. In subsequent generations, more DNA was near the top and less was in the intermediate position.

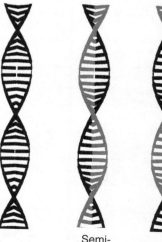

Models for DNA replication

Conservative | Semi-conservative | Dispersive

©2022 **BIOZONE** International
ISBN: 978-1-98-856692-4

1. In this activity, you will model the **semi-conservative** model of DNA replication. 1(a) and (d (i)) are done for you:

(a) The DNA model below represents the DNA of the bacteria after growing in the solution containing the heavy nitrogen (Generation 0). The relative mass of the DNA can be modelled by adding together the nitrogen masses.

(b) In the space below (center) (Generation 1), split the DNA from Generation 0 along its centre, then write in the complementary base pairs to form two DNA strands (i and ii). This represents the DNA after it has been grown in the solution containing ^{14}N for one generation.

(c) In the space below (right) (Generation 2), split the two DNA chains along their centres, then write the complementary base pairs to form four new DNA strands (i, ii, iii, and iv). This represents the DNA after it has been growing in the solution containing the ^{14}N for two generations.

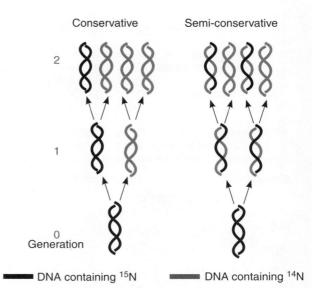

Conservative   Semi-conservative

DNA containing ^{15}N DNA containing ^{14}N

Generation 0   **Generation 1**   **Generation 2**

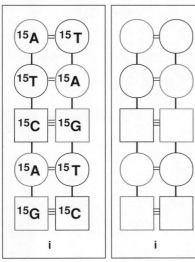 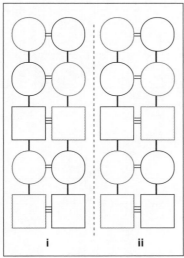   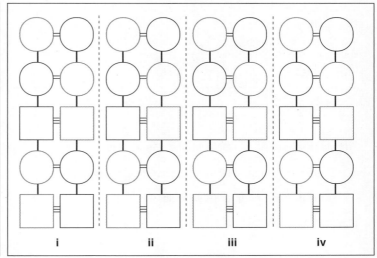

(d) Mass of nitrogen in DNA strand, Generation 0:   i: _150_____

Mass of nitrogen in DNA strands, Generation 1:   i: _____ ii: _____

Mass of nitrogen in DNA strands, Generation 2:   i: _____ ii: _____

iii: _____ iv: _____

(e) On the test tubes (right), mark a bar representing the mass of each of the DNA strands from (a), (b), and (c).

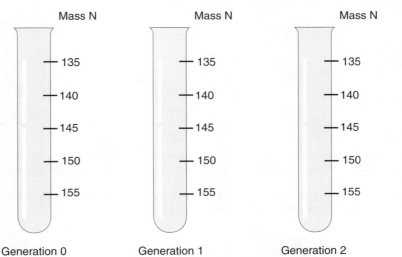

Mass N   Mass N   Mass N

Generation 0   Generation 1   Generation 2

2. In this part of the activity, you will model the conservative model of **DNA replication**:

   (a) The DNA model below represents the DNA of the bacteria after it has grown in the solution containing the heavy nitrogen (Generation 0). The relative mass of the DNA can be modelled by adding together the nitrogen masses.

   (b) In the box below (center), recreate the original DNA strand. Beside it, create a matching strand with bases using the light nitrogen. This represents the DNA after it has been replicated in the solution with the ^{14}N for one generation.

   (c) In the space below (right), recreate the original DNA strand and the first generation strand. Beside these, create matching strands with bases using the ^{14}N. This represents the DNA after it has been growing in the solution containing the ^{14}N for two generations.

Generation 0   **Generation 1**   **Generation 2**

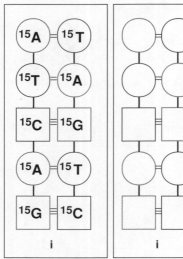
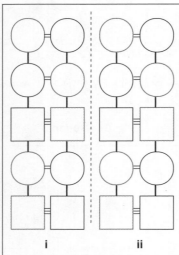
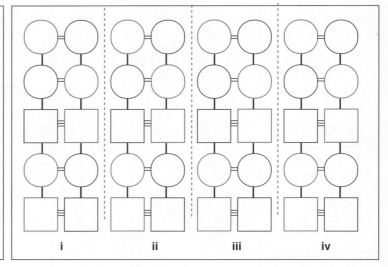

   (d) Mass of nitrogen in DNA strand, Generation 0:   i: _____

   Mass of nitrogen in DNA strands, Generation 1:   i: _____   ii: _____

   Mass of nitrogen in DNA strands, Generation 2:   i: _____   ii: _____

   iii: _____   iv: _____

   (e) On the appropriate test tubes (right), mark a bar representing the mass of each of the DNA strands from Generation 0, Generation 1, and Generation 2

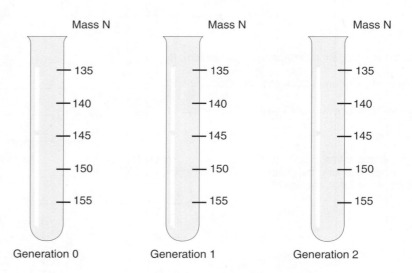

3. In their experiment, Meselson and Stahl obtained the following results: Generation 0 = 100% 'heavy DNA'; **Generation 1**= 100% 'intermediate DNA'; Generation 2 = 50% 'intermediate DNA', 50% 'light DNA'.

   From the results of your two modelling exercises, decide which matches the result of Meselson and Stahl. How does DNA replicate (conservatively or semi-conservatively)?

   _____

   _____

© 2022 **BIOZONE** International
ISBN: 978-1-98-856692-4
Photocopying Prohibited

89 The Functions of Mitosis

Key Question: What are the three primary functions of mitosis?

Mitotic **cell division** has three purposes

▶ Growth: Multicellular organisms grow from a single fertilized cell into a mature organism. Depending on the organism, the mature form may consist of several thousand to several trillion cells. These cells that form the building blocks of the body are called somatic cells.

▶ Repair: Damaged and old cells are replaced with new cells.

▶ Asexual reproduction: Some unicellular eukaryotes such as yeasts and some multicellular organisms, e.g. *Hydra,* reproduce asexually by mitotic division.

Fertilized egg cell   *Embryo*   *Adult*

Matthias Zepper

Asexual reproduction
Some simple eukaryotic organisms reproduce asexually by **mitosis**. Yeasts, such as baker's yeast, can reproduce by budding. The parent cell buds to form a daughter cell (right). The daughter cell continues to grow, and eventually separates from the parent cell.

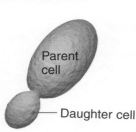

Parent cell

Daughter cell

Growth
Multicellular organisms develop from a single fertilized egg cell and grow by increasing in cell numbers. Cells complete a **cell cycle** in which the cell copies its DNA and then divides to produce two identical cells. During the period of growth, the production of new cells is faster than the death of old ones. Organisms, such as the 12 day old mouse embryo (above, middle), grow by increasing their total cell number and the cells become specialized as part of development. Cell growth is highly regulated and once the mouse reaches its adult size (above, right), physical growth stops and the number of cell deaths equals the number of new cells produced.

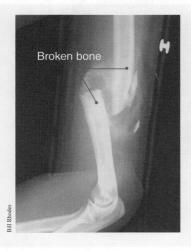

Broken bone

Bill Rhodes

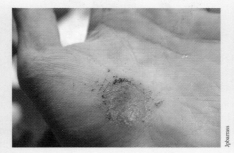

Jpbarrass

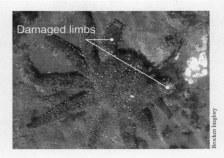

Damaged limbs

Brocken Inaglory

Repair
Mitosis is vital in the repair and replacement of damaged cells. When you break a bone or graze your skin, new cells are generated to repair the damage. Some organisms, like this sea star (above right), are able to generate new limbs if they are broken off.

1. Use examples to explain the role of **mitosis** in:

(a) Growth of an organism: _____

_____

(b) Replacement of damaged cells: _____

_____

(c) Asexual reproduction: _____

_____

LS1.B

90 The Eukaryotic Cell Cycle

Key Question: What are the phases of the eukaryotic cell cycle, and what specific cellular events occur in each phase?

The life cycle of a eukaryotic cell is called the **cell cycle**. The cell cycle can be divided into two broad phases; **interphase** and **M phase**. Specific activities occur in each phase.

Interphase

Cells spend most of their time in interphase. Interphase is divided into three stages:

▸ The first gap phase (G1).

▸ The S-phase (S).

▸ The second gap phase (G2).

During interphase the cell increases in size, carries out its normal activities, and replicates its DNA in preparation for **cell division**. Interphase is not a stage in **mitosis**.

Mitosis and cytokinesis (M-phase)

Mitosis and cytokinesis occur during M-phase. During mitosis, the cell nucleus containing the replicated DNA divides in two equal parts. Cytokinesis occurs at the end of M-phase. During cytokinesis the cell cytoplasm divides, and two new daughter cells are produced.

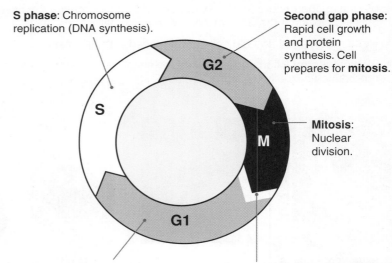

S phase: Chromosome replication (DNA synthesis).

Second gap phase: Rapid cell growth and protein synthesis. Cell prepares for **mitosis**.

Mitosis: Nuclear division.

First gap phase: Cell increases in size and makes the mRNA and proteins needed for DNA synthesis.

Cytokinesis: The cytoplasm divides and the two cells separate. Cytokinesis is part of M phase but distinct from nuclear division.

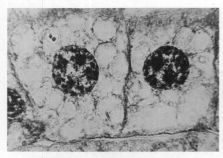

During interphase, the cell grows and acquires the materials needed to undergo mitosis. It also prepares the nuclear material for separation by replicating it.

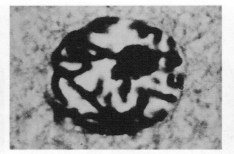

During interphase, the nuclear material is unwound. As mitosis approaches, the nuclear material begins to reorganize in readiness for nuclear division.

During mitosis, the chromosomes are separated. Mitosis is a highly organized process and the cell must pass "checkpoints" before it proceeds to the next phase.

1. Briefly outline what occurs during the following phases of the cell cycle:

(a) Interphase: _____

_____

_____

(b) Mitosis: _____

_____

_____

_____

(c) Cytokinesis: _____

_____

_____

  LS1.B SSM

©2022 **BIOZONE** International
ISBN: 978-1-98-856692-4

91 Mitosis

Key Question: Mitosis is an important part of the eukaryotic cell cycle in which the replicated chromosomes are separated and the cell divides, producing two new cells.

Mitosis is a stage in the cell cycle

▶ M-phase (**mitosis** and cytokinesis) is the part of the **cell cycle** in which the parent cell divides in two to produce two genetically identical daughter cells, shown in the diagram (right).

▶ Mitosis results in the separation of the nuclear material and division of the cell. It does not result in a change of chromosome number.

▶ Mitosis is one of the shortest stages of the cell cycle. When a cell is not undergoing mitosis, it is said to be in **interphase**.

▶ In animals, mitosis takes place in the somatic (body) cells. Somatic cells are any cell of the body except sperm and egg cells.

▶ In plants, mitosis takes place in the meristems. Meristems are specific regions of growth where new cells are produced, such as the tips of roots and shoots.

Mitosis produces identical daughter cells

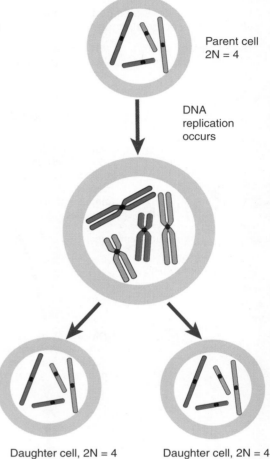

Parent cell
2N = 4

DNA replication occurs

Daughter cell, 2N = 4   Daughter cell, 2N = 4

The cell divides forming two identical daughter cells. The chromosome number remains the same as the parent cell.

Onion cells

At any one time, only a small proportion of the cells in an organism will be undergoing mitosis. The majority of the cells will be in interphase.

The meristematic **tissue** (M) at the growing tip is the site of mitosis in this plant root. The root cap below the meristem protects the dividing cells.

1. Briefly outline the events in mitosis: _____

_____

_____

2. Where does mitosis take place in:

   (a) Animals: _____

   (b) Plants: _____

3. A cell with 10 chromosomes undergoes mitosis.

   (a) How many daughter cells are created: _____

   (b) How many chromosomes does each daughter cell have? _____

   (c) The genetic material of the daughter cells is the same as / different to the parent cell (delete one).

92 Mitosis and Cytokinesis

Key Question: What happens in the different stages of mitosis leading up to the formation of two daughter cells, and is it different for plant and animal cells?

The cell cycle and stages of mitosis

▸ **Mitosis** is continuous, but is divided into stages for easier reference (1-6 below). Enzymes are critical at key stages. The example below illustrates the **cell cycle** in an animal cell.

▸ In animal cells, centrioles (located in the centrosome) form the spindle. During cytokinesis (division of the cytoplasm) a constriction forms dividing the cell in two. Cytokinesis is part of M-phase, but it is distinct from mitosis.

▸ Plant cells lack centrioles, and the spindle is organized by structures associated with the plasma membrane. In plant cells, cytokinesis involves formation of a cell plate in the middle of the cell. This will form a new cell wall.

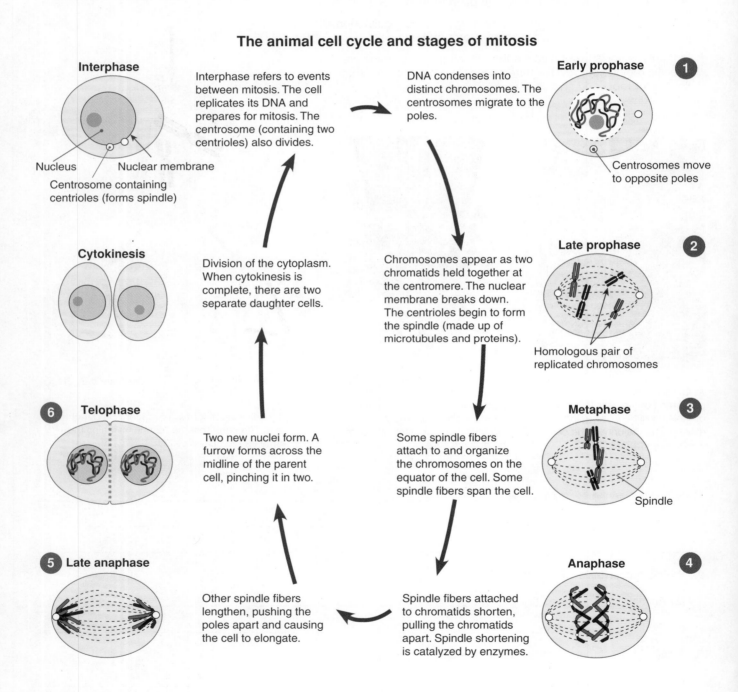

The animal cell cycle and stages of mitosis

Interphase

Nucleus — Nuclear membrane
Centrosome containing centrioles (forms spindle)

Interphase refers to events between mitosis. The cell replicates its DNA and prepares for mitosis. The centrosome (containing two centrioles) also divides.

DNA condenses into distinct chromosomes. The centrosomes migrate to the poles.

Early prophase 1

Centrosomes move to opposite poles

Late prophase 2

Chromosomes appear as two chromatids held together at the centromere. The nuclear membrane breaks down. The centrioles begin to form the spindle (made up of microtubules and proteins).

Homologous pair of replicated chromosomes

Cytokinesis

Division of the cytoplasm. When cytokinesis is complete, there are two separate daughter cells.

6 **Telophase**

Two new nuclei form. A furrow forms across the midline of the parent cell, pinching it in two.

Metaphase 3

Some spindle fibers attach to and organize the chromosomes on the equator of the cell. Some spindle fibers span the cell.

Spindle

5 **Late anaphase**

Other spindle fibers lengthen, pushing the poles apart and causing the cell to elongate.

Spindle fibers attached to chromatids shorten, pulling the chromatids apart. Spindle shortening is catalyzed by enzymes.

Anaphase 4

1. What must occur before mitosis takes place? _____

_____

  LS1.B SSM

©2022 **BIOZONE** International
ISBN: 978-1-98-856692-4
Photocopying Prohibited

Cytokinesis

In plant cells (below right), cytokinesis (division of the cytoplasm) involves construction of a cell plate (a precursor of the new cell wall) in the middle of the cell. The cell wall materials are delivered by vesicles derived from the Golgi. The vesicles join together to become the plasma membranes of the new cell surfaces. Animal cell cytokinesis (below left) begins shortly after the sister chromatids have separated in anaphase of mitosis. A ring of microtubules assembles in the middle of the cell, next to the plasma membrane, constricting it to form a cleavage furrow. In an energy-using process, the cleavage furrow moves inwards, forming a region of separation where the two cells will separate.

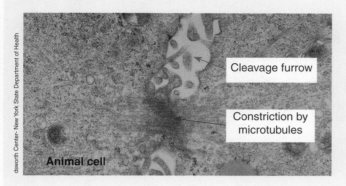

Cleavage furrow

Constriction by microtubules

Animal cell

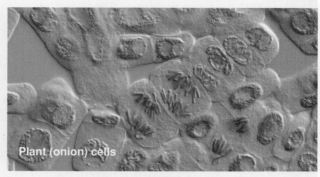

Plant (onion) cells

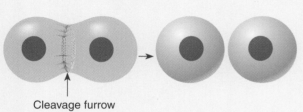

Cleavage furrow

Cytokinesis in an animal cell

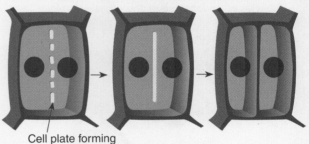

Cell plate forming

Cytokinesis in a plant cell

2. Summarize what happens in each of the following phases:

(a) Prophase: _____

_____

(b) Metaphase: _____

_____

(c) Anaphase: _____

_____

(d) Telophase: _____

_____

3. (a) What is the purpose of cytokinesis? _____

_____

_____

(b) Describe the differences between cytokinesis in an animal cell and a plant cell: _____

_____

_____

_____

_____

_____

© 2022 **BIOZONE** International
ISBN: 978-1-98-856692-4
Photocopying Prohibited

93 Modeling Mitosis

Key Question: How can I model the stages of mitosis to help to visualize and understand the process?

Investigation 4.1   Modeling mitosis

See appendix for equipment list.

1. You can work in pairs for this activity if you wish.

2. Use the information on the previous pages to model **mitosis** in an animal cell using pipe cleaners and string. Work in pairs and use four chromosomes for simplicity (2N = 4). Photograph or film each stage.

3. Photo 1 (below) can be used as a starting point for your model. It represents a cell in interphase before mitosis begins. The circular structures are the replicated centrosomes.

4. Before you start, identify the structures A–C in photo

    A: _____   B: _____   C: _____

5. Remember to label your photos as you place them on the page.

Photo 1

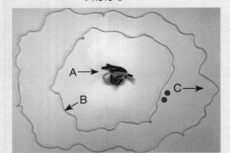

©2022 **BIOZONE** International
ISBN: 978-1-98-856692-4
Photocopying Prohibited

94 Differentiation of Cells

Key Question: How do many different cell types arise during development of the embryo?

▶ When a cell divides by **mitosis**, it produces genetically identical cells. However, a multicellular organism is made up of many different types of cells, each specialized to carry out a particular role. How can it be that all of an organism's cells have the same genetic material but the cells have a wide variety of shapes and functions? The answer is through **cellular differentiation** (transformation) of unspecialized cells called **stem cells**.

▶ Although each cell has the same genetic material (genes), different genes are turned on (activated) or off in different patterns during development in particular types of cells. The differences in gene activation controls what type of cell forms (below). Once the developmental pathway of a cell is determined, it cannot alter its path and change into another cell type.

How stem cells give rise to different cell types

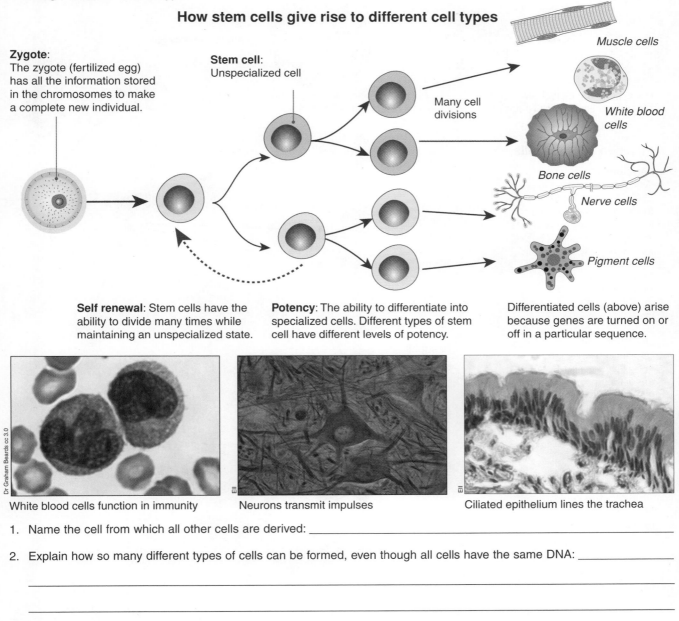

Zygote:
The zygote (fertilized egg) has all the information stored in the chromosomes to make a complete new individual.

Stem cell:
Unspecialized cell

Many cell divisions

Muscle cells

White blood cells

Bone cells

Nerve cells

Pigment cells

Self renewal: Stem cells have the ability to divide many times while maintaining an unspecialized state.

Potency: The ability to differentiate into specialized cells. Different types of stem cell have different levels of potency.

Differentiated cells (above) arise because genes are turned on or off in a particular sequence.

White blood cells function in immunity

Neurons transmit impulses

Ciliated epithelium lines the trachea

Dr Graham Beards cc 3.0

EII

EII

1. Name the cell from which all other cells are derived: _____

2. Explain how so many different types of cells can be formed, even though all cells have the same DNA: _____
   _____
   _____

3. (a) What are stem cells? _____
   _____

   (b) What are the two defining properties of stem cells? _____

   i _____

   ii _____

SSM | LS1.B | |

95 Stem Cells Give Rise to Other Cells

Key Question: How do stem cells, which are undifferentiated, develop into many different cell types, and how do related cell types come together to form tissues such as blood?

Totipotent stem cells
These **stem cells** can differentiate into all the cells in an organism. Example: in humans, the **zygote** and its first few divisions. The **tissue** at the root and shoot tips of plants is also totipotent.

Pluripotent stem cells
These stem cells can give rise to any cells of the body, except extra-embryonic cells, e.g. placenta and chorion. Example: embryonic stem cells.

Multipotent stem cells
These adult stem cells can give rise to a limited number of cell types, related to their tissue of origin. Example: bone marrow stem cells (below), skin stem cells, bone stem cells, umbilical cord blood.

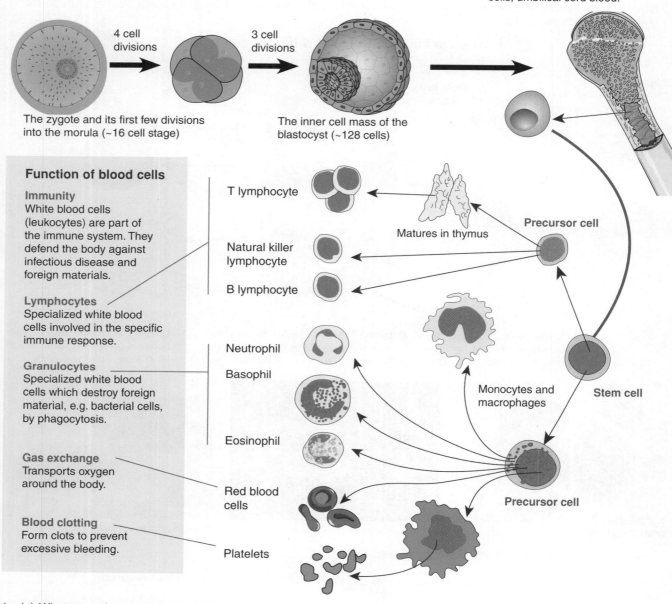

4 cell divisions

3 cell divisions

The zygote and its first few divisions into the morula (~16 cell stage)

The inner cell mass of the blastocyst (~128 cells)

Function of blood cells

Immunity
White blood cells (leukocytes) are part of the immune system. They defend the body against infectious disease and foreign materials.

Lymphocytes
Specialized white blood cells involved in the specific immune response.

Granulocytes
Specialized white blood cells which destroy foreign material, e.g. bacterial cells, by phagocytosis.

Gas exchange
Transports oxygen around the body.

Blood clotting
Form clots to prevent excessive bleeding.

T lymphocyte

Natural killer lymphocyte

B lymphocyte

Neutrophil

Basophil

Eosinophil

Red blood cells

Platelets

Matures in thymus

Precursor cell

Monocytes and macrophages

Stem cell

Precursor cell

1. (a) What type of stems cells are blood cells produced from? _____

   (b) What are the features of the stem cells you described in (a): _____
   _____
   _____

2. Describe the functional roles of blood cells: _____
   _____
   _____
   _____
   _____

  LS1.B SF

©2022 **BIOZONE** International
ISBN: 978-1-98-856692-4
Photocopying Prohibited

96 Tissues Work Together

Key Question: How do different tissue types work together to meet the body's needs efficiently?

A **tissue** is a collection of related cell types that work together to carry out a specific function. Different tissues come together to form organs. The cells, tissues, and organs of the body interact to meet the needs of the entire organism. This activity explains the role of the four tissue types (below) in humans.

Muscle tissue	Epithelial tissue	Nervous tissue	Connective tissue
▶ Contractile tissue.	▶ Lining tissue.	▶ Receives and responds to stimuli.	▶ Supports, protects, and binds other tissues.
▶ Produces movement of the body or its parts.	▶ Covers the body and lines internal surfaces.	▶ Makes up the structures of the nervous system.	▶ Contains cells in an extracellular matrix.
▶ Includes smooth, skeletal, and cardiac muscle.	▶ Can be modified to perform specific roles.	▶ Regulates function of other tissues.	▶ Can be hard or fluid.

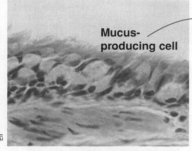

Mucus-producing cell

The upper respiratory tract is lined with ciliated epithelium to move irritants before they reach the lungs. The lungs and cardiovascular system work together to respond to changes in oxygen demand.

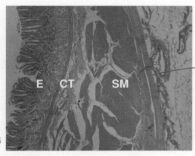

E CT SM

The digestive tract is lined with epithelial tissue (E) and held in place by connective tissue (CT). It is moved by smooth muscle (SM) in response to messages from neurons.

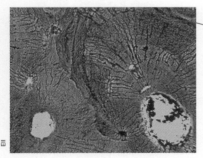

Bone is a type of connective tissue. It provides shape to the body and works with muscle to produce movement. Ligaments are also connective tissue structures. They hold bones together.

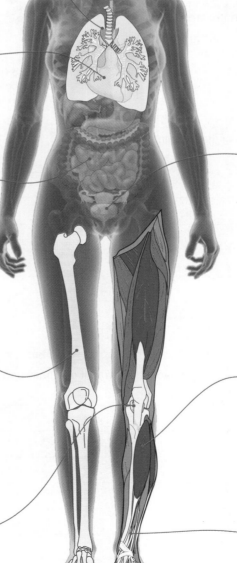

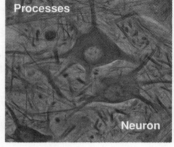

Processes

Neuron

Nervous tissue is made up of nerve cells (neurons) and supporting cells. The long processes of neurons control the activity of muscles and glands.

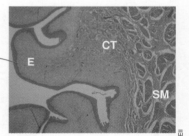

CT

E

SM

Epithelial tissue (E) lines organs such as the bladder. Connective tissue (CT) supports the organ. This epithelium is layered so that it can stretch. The bladder's activity is controlled by smooth muscle (SM) and neurons.

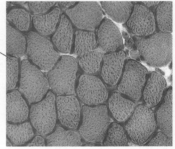

Skeletal muscle tissue contracts to pull on the rigid bones of the skeleton to bring about movement of the body. Tendons are connective tissue structures that attach muscles to bones.

SF   LS1.B

Tissues work together and make up organs, which perform specific functions

The body's **tissues** work together in order for the body to function. Tissues also group together to form organs. For example, epithelial tissues are found associated with other tissues in every organ of the body. Other examples include:

Nerves, muscles, and movement

▶ Nerves stimulate muscles to move.

▶ Connective tissue binds other tissues together and holds them in place (e.g. skeletal muscle tissue is held together by connective tissue sheaths to form discrete muscle, neurons are bundled together by connective tissue to form nerves).

▶ Bones are held together by connective tissue ligaments at joints, allowing the skeleton to move. Skeletal muscles are attached to bone by connective tissue tendons. Muscle contraction causes the tendon to pull on the bone, moving it.

Heart, lungs, blood vessels, and blood

▶ Cardiac muscle pumps blood (a specialized connective tissue) around the body within blood vessels.

▶ In the lungs, blood vessels surround the epithelium of the tiny air sacs to enable the exchange of gases between the blood and the air in the lungs.

▶ Neurons regulate the activity of heart and lungs to respond to changes in oxygen demand, as when a person is exercising.

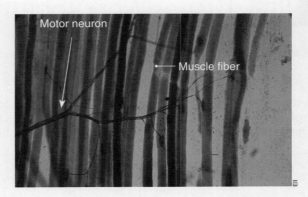

This image shows a neuron branching to supply muscle fibers. Impulses from the neuron will cause the muscle to contract. Neurons are bundled together by connective tissue wrappings to form nerves.

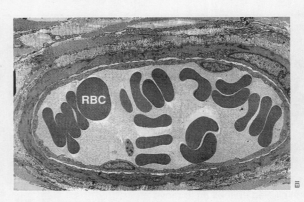

This image shows red blood cells within a vein. Veins return blood to the heart. When oxygen demand increases, heart and breathing rates increase and blood is delivered more quickly to working tissues.

1. Describe the main function of each of the following types of tissues:

   (a) Epithelial tissue: _____

   _____

   (b) Connective tissue: _____

   _____

   (c) Nervous tissue: _____

   _____

   (d) Muscle tissue: _____

   _____

2. Describe how different tissues interact to bring about movement of a body part: _____

   _____

   _____

   _____

   _____

   _____

   _____

97 Review Your Understanding

Key Question: How can the axolotl have such superpowers of regeneration, with the ability to regrow amputated limbs, damaged spinal cords, and even parts of its brain?

▶ At the beginning of this chapter you were asked to consider how a skin wound healed and think about the ability of the axolotl to heal from injuries, compared to how humans do this.

▶ Now that you have learned more about cell division and cell differentiation you should have a better understanding of how the axolotl is able to regrow complex tissues.

1. Briefly describe the stages that skin cells go through, in order to divide and replicate correctly, to heal after an injury?

_____

_____

_____

_____

_____

2. How do cells surrounding an axolotl's injured limb "know" to develop into skin, bone, brain tissue, blood etc?

_____

_____

_____

3. When an axolotl loses its limb, initial healing begins with blood clotting and then cells from nearby tissues move to the wound site. These cells undergo a change so that they have the ability to become any other tissue. What do we call unspecialized cells that have the ability to differentiate into any cell type

_____

4. Use the information that you have learned to describe the stage of mitosis shown in each of the photographs below. Underneath your identification, briefly state the reason for your choice.

(a) _____

_____

_____

_____

(b) _____

_____

_____

_____

(c) _____

_____

_____

_____

(d) _____

_____

_____

_____

5. Where are the different locations of mitosis in animal cells and plant cells?

_____

_____

_____

_____

98 Summing Up

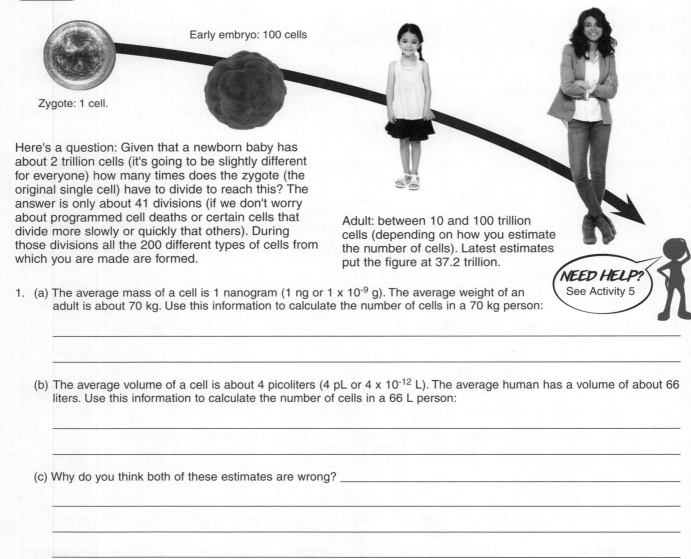

Early embryo: 100 cells

Zygote: 1 cell.

Here's a question: Given that a newborn baby has about 2 trillion cells (it's going to be slightly different for everyone) how many times does the zygote (the original single cell) have to divide to reach this? The answer is only about 41 divisions (if we don't worry about programmed cell deaths or certain cells that divide more slowly or quickly that others). During those divisions all the 200 different types of cells from which you are made are formed.

Adult: between 10 and 100 trillion cells (depending on how you estimate the number of cells). Latest estimates put the figure at 37.2 trillion.

NEED HELP?
See Activity 5

1. (a) The average mass of a cell is 1 nanogram (1 ng or 1×10^{-9} g). The average weight of an adult is about 70 kg. Use this information to calculate the number of cells in a 70 kg person:

_____

_____

(b) The average volume of a cell is about 4 picoliters (4 pL or 4×10^{-12} L). The average human has a volume of about 66 liters. Use this information to calculate the number of cells in a 66 L person:

_____

_____

(c) Why do you think both of these estimates are wrong? _____

_____

_____

_____

Differentiating cells

Every cell in your body has the same genetic information. How is it that there are so many different types of cells in your body? The answer is that during different cell divisions only certain parts of the genetic information is used (much like reading only some books in the library), thus producing the many different cell types.

2. A hypothetical cell has 11 genes in its DNA. Each gene initiates certain processes in the cell as shown in the table below. Use these "genes" to fill in the boxes opposite showing the cell as it progresses through two divisions. Note if a gene is switched on, the cell follows the instruction for that gene.

Gene number	Instructions
1	Grow spikes (overrides gene 2)
2	Lose spikes
3	Grow by 20% in volume
4	Grow by 30% in volume
5	Keep single nucleus (overrides gene 6)
6	Divide nucleus in two to produce a cell with more than one nucleus
7	Grow 50% longer along y axis
8	Grow 50% longer along x axis
9	Remain as a singular cell (overrides genes 10 and 11)
10	Produce connections to join to parent cell
11	Grow microvilli along unattached border (only if gene 10 is on)

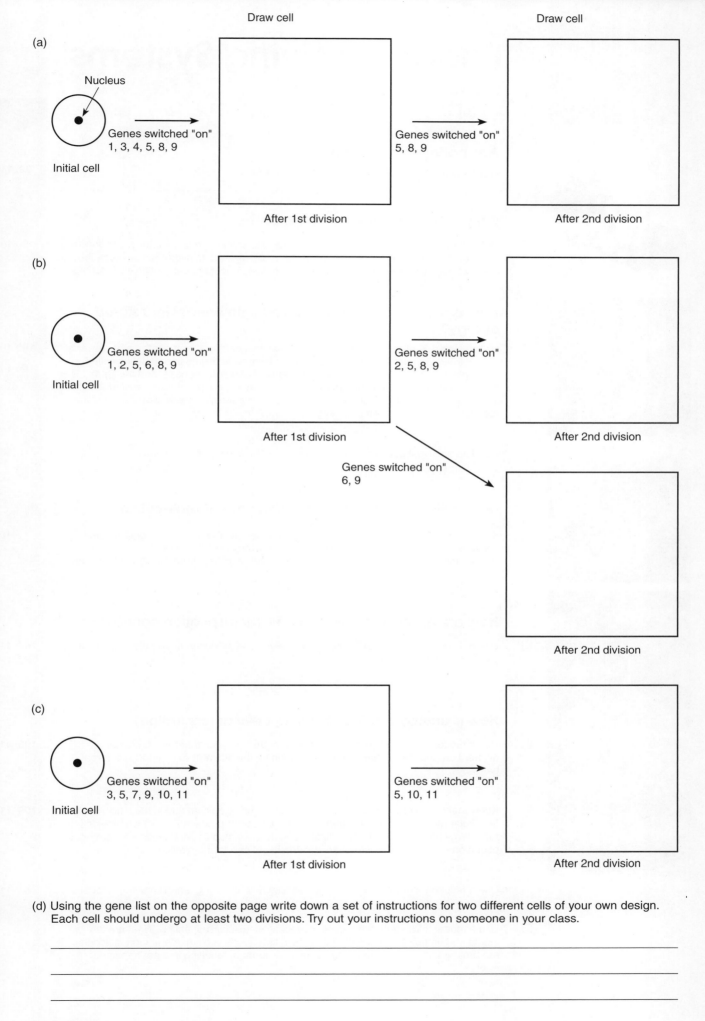

(a)

Nucleus

Initial cell

Draw cell

Genes switched "on"
1, 3, 4, 5, 8, 9

After 1st division

Draw cell

Genes switched "on"
5, 8, 9

After 2nd division

(b)

Initial cell

Genes switched "on"
1, 2, 5, 6, 8, 9

After 1st division

Genes switched "on"
2, 5, 8, 9

After 2nd division

Genes switched "on"
6, 9

After 2nd division

(c)

Initial cell

Genes switched "on"
3, 5, 7, 9, 10, 11

After 1st division

Genes switched "on"
5, 10, 11

After 2nd division

(d) Using the gene list on the opposite page write down a set of instructions for two different cells of your own design. Each cell should undergo at least two divisions. Try out your instructions on someone in your class.

_____

_____

_____

CHAPTER 5

Energy in Living Systems

Anchoring Phenomenon

Mouse Trap: Can an animal survive in a sealed system? 99 111

___

What is ATP and why is it important?

☐ 1 Understand that living organisms use cellular respiration to release energy and form molecules of ATP. ATP is an energy transfer molecule and its hydrolysis provides the energy to do cellular work. Show how ATP provides energy to carry out life's functions. 100 101

How does photosynthesis convert light energy into stored energy?

☐ 2 Understand that photosynthesis is the process that captures light energy and converts it into stored chemical energy. In plants, photosynthesis occurs in organelles called chloroplasts, and consists of two stages; the light dependent phase, and the light independent phase. Construct an explanation for how photosynthesis transforms light energy into stored chemical energy. During photosynthesis, carbon dioxide and water are converted into glucose and oxygen. 102 103 104 112

☐ 3 Conduct an investigation to demonstrate that light drives photosynthesis. 105

How is glucose used to make other macromolecules?

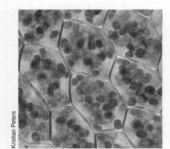

☐ 4 Know that glucose consists of carbon, oxygen, and hydrogen atoms. Construct and revise an explanation based on evidence for how organisms build a wide range of organic molecules, e.g. amino acids, nucleic acids, and fats, from the atoms in glucose. 106

How are photosynthesis and cellular respiration connected?

☐ 5 Use a diagrammatic model to show how the glucose produced during photosynthesis is used in cellular respiration. 107 112

How is energy produced during cellular respiration?

☐ 6 Use a model to compare ATP production from aerobic and anaerobic pathways in different organisms. Construct an explanation for the different ATP yields in different pathways. 108 112

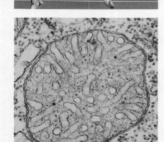

☐ 7 Know that cellular respiration takes place in the cell cytoplasm and in the mitochondrion. In cellular respiration, glucose and oxygen are used to produce ATP, which provides the energy needed to perform cellular work, such as muscle contraction. Use a model based on evidence to construct an explanation of cellular respiration. 109 112

☐ 8 Use a respirometer to investigate cellular respiration in living organisms, e.g. seeds. 110

☐ 9 Use a model to illustrate the chemical process of cellular respiration. What are the inputs and outputs? Use the model to show that bonds in oxygen and food molecules are broken and bonds in new compounds are formed, resulting in a net transfer of energy.

Kristian Peters

99 Mouse Trap

Key Question: Under what conditions can an animal survive in a sealed system?

Mouse in a jar

▸ Around 1772, Joseph Priestley carried out a series of interesting experiments. He wanted to see if there was a relationship between the survival of plants and animals in a closed system. One of his experiments is shown below.

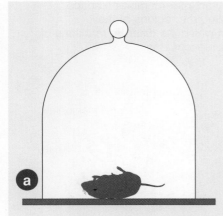

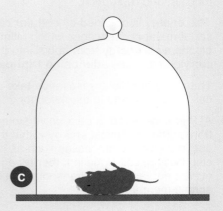

Priestley placed a mouse in a sealed, empty bell jar. The mouse quickly collapsed and died.

Priestley placed a mint plant in a sealed bell jar and left it for several days before adding a mouse. After several minutes the mouse was still alive.

Priestley removed the mouse from jar (b) and placed it into a sealed, empty bell jar. The mouse died very quickly.

1. (a) Can you explain why the mouse died in jars (a) and (c), but not in jar (b)? _____

   _____

   _____

   _____

   _____

   (b) What metabolic or chemical processes might explain the results Joseph Priestley obtained? _____

   _____

   _____

   _____

   _____

2. Draw a very simple model to show what is happening in jar (b):

3. In another experiment, Joseph Priestley left a plant covered with a bell jar for many days. He then placed a candle with a glowing wick into the jar. The wick ignited and began to burn. What was present to allow the wick to ignite?

   _____

100 Energy in Cells

Key Question: How does the ATP produced from cellular respiration provide the energy needed to perform essential life functions?

Energy for metabolism

▶ All organisms require energy to be able to perform the metabolic processes required for them to function and reproduce.

▶ This energy is obtained by **cellular respiration**, a set of metabolic reactions which ultimately convert biochemical energy from "food" into the energy-carrying molecule **adenosine triphosphate** (**ATP**).

▶ The steps of cellular respiration take place in the cell cytoplasm and in the mitochondria.

▶ ATP is considered to be a universal energy carrier, transporting chemical energy within the cell for use in metabolic processes such as biosynthesis, cell division, cell signaling, thermoregulation, cell movement, and active transport of substances across membranes.

The mitochondrion

A mitochondrion is bounded by a double membrane. The inner and outer membranes are separated by an inter-membrane space, compartmentalizing the regions of the mitochondrion in which the different reactions of cellular respiration occur.

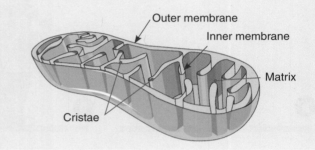

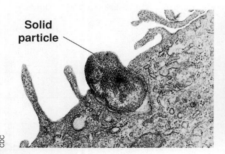

Energy is needed to actively transport molecules across the cellular membrane. The above image shows phagosytosis a large particle being engulfed).

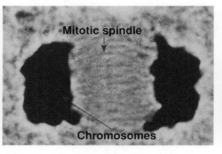

Cell division (mitosis) (above), requires energy to proceed. ATP provides energy for the mitotic spindle formation and chromosome separation.

The maintenance of body temperature requires energy. Both heating and cooling the body require energy as it either shivers or produces sweat.

1. What process produces usable energy in cells? _____

2. How is energy carried around the cell? _____

_____

3. (a) Describe the general role of mitochondria in the cell: _____

_____

_____

_____

(b) What is the purpose of the folded inner membrane in mitochondria? _____

_____

_____

_____

4. (a) What energy-using process helps warm the body? _____

(b) What energy-using process helps cool the body? _____

LS1.C EM SF

©2022 **BIOZONE** International
ISBN: 978-1-98-856692-4
Photocopying Prohibited

101 ATP

Adenosine triphosphate (ATP)

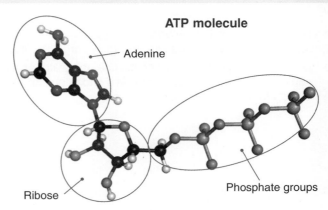

ATP molecule

▸ The **ATP molecule** (right) is a nucleotide derivative. It has three components;

- A purine base (adenine)
- A pentose sugar (ribose)
- Three phosphate groups.

▸ ATP acts as a store of energy within the cell. The bonds between the phosphate groups contain electrons in a high energy state, which store a large amount of energy that is released during a chemical reaction. The removal of one phosphate group from ATP results in the formation of **adenosine diphosphate (ADP)**.

Note: Adenine + ribose = adenosine

How does ATP provide energy?

▸ The bonds between the phosphate groups of ATP are unstable and very little energy is needed to break them. The energy in the ATP molecule is transferred to a target molecule, e.g. a protein, by a hydrolysis reaction. Water is split during the reaction and added to the terminal phosphate on ATP, forming ADP and an inorganic phosphate molecule (Pi).

▸ When the Pi molecule combines with a target molecule, energy is released. Most of the energy (about 60%) is lost as heat (this helps keep you warm). The rest of the energy is transferred to the target molecule, allowing it to do work, e.g. joining with another molecule (right).

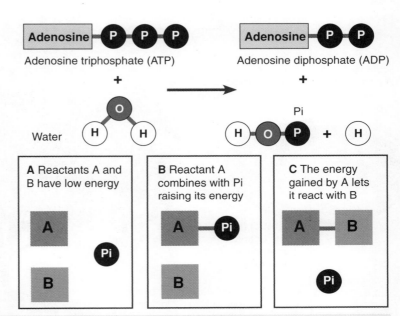

Note! The phosphate bonds in ATP are often referred to as being high energy bonds. This can be misleading. The bonds contain *electrons* in a high energy state (making the bonds themselves relatively weak). A small amount of energy is required to break the bonds, but when the intermediaries recombine and form new chemical bonds a large amount of energy is released. The final product is less reactive than the original reactants.

1. What are the three components of ATP? _____

2. (a) What is the biological role of ATP? _____

 _____

 (b) Where is the energy stored in ATP? _____

 _____

 (c) What products are formed during hydrolysis of ATP? _____

 _____

3. Why does the conversion of ATP to ADP help keep us warm? _____

 _____

©2022 **BIOZONE** International
ISBN: 978-1-98-856692-4
Photocopying Prohibited

102 Introduction to Photosynthesis

Key Question: How does photosynthesis convert sunlight, carbon dioxide, and water into glucose and oxygen?

▸ Plants, algae, and some bacteria are photoautotrophs. They use pigments called chlorophylls to absorb light of specific wavelengths and capture light energy. The light energy is used in a process called photosynthesis.

▸ During photosynthesis, carbon dioxide and water are converted into glucose and oxygen. The reaction requires sunlight energy which is transformed into chemical energy within the bonds of the glucose molecule. This chemical energy fuels life's essential processes.

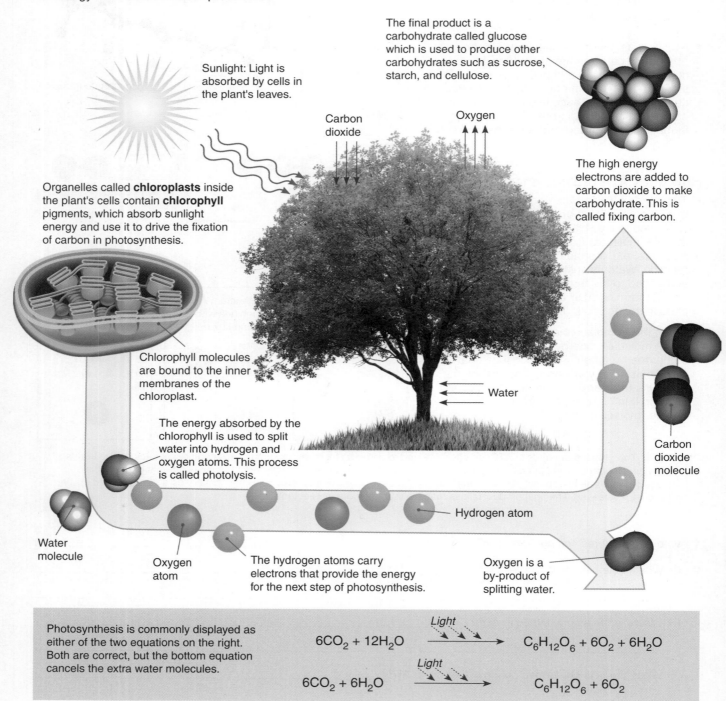

Sunlight: Light is absorbed by cells in the plant's leaves.

The final product is a carbohydrate called glucose which is used to produce other carbohydrates such as sucrose, starch, and cellulose.

Carbon dioxide

Oxygen

The high energy electrons are added to carbon dioxide to make carbohydrate. This is called fixing carbon.

Organelles called **chloroplasts** inside the plant's cells contain **chlorophyll** pigments, which absorb sunlight energy and use it to drive the fixation of carbon in photosynthesis.

Chlorophyll molecules are bound to the inner membranes of the chloroplast.

Water

Carbon dioxide molecule

The energy absorbed by the chlorophyll is used to split water into hydrogen and oxygen atoms. This process is called photolysis.

Hydrogen atom

Water molecule

Oxygen atom

The hydrogen atoms carry electrons that provide the energy for the next step of photosynthesis.

Oxygen is a by-product of splitting water.

Photosynthesis is commonly displayed as either of the two equations on the right. Both are correct, but the bottom equation cancels the extra water molecules.

$$6CO_2 + 12H_2O \xrightarrow{\text{Light}} C_6H_{12}O_6 + 6O_2 + 6H_2O$$

$$6CO_2 + 6H_2O \xrightarrow{\text{Light}} C_6H_{12}O_6 + 6O_2$$

1. Write the word equation for photosynthesis: _____

   _____

2. Where does the oxygen released during photosynthesis come from? _____

LS1.C   EM

©2022 **BIOZONE** International
ISBN: 978-1-98-856692-4

Requirements for photosynthesis

Plants need only a few raw materials to make their own food:

▸ *Light energy* from the Sun.

▸ *Chlorophyll* absorbs light energy.

▸ CO_2 *gas* is reduced to carbohydrate.

▸ *Water* is split to provide the electrons for the fixation of carbon as carbohydrate.

Photosynthesis is not a single process but two complex processes (the light dependent and light independent reactions) each with multiple steps.

Production of carbohydrate occurs in the fluid stroma of the **chloroplast**. Commonly called carbon "fixation", it does not require sunlight (light independent phase).

Energy capture occurs in the inner membranes (thylakoids) of the chloroplast and requires sunlight (light dependent phase).

The photosynthesis of marine algae, such as these diatoms, supplies a substantial portion of the world's oxygen. The oceans also act as sinks for absorbing large amounts of CO_2.

Macroalgae, like this giant kelp, are important marine producers. Algae living near the ocean surface get access to light for photosynthesis.

On land, vascular plants, such as trees with transport vessels, are the main producers of food. Plants at different levels in a forest receive different intensities and quality of light.

3. (a) What form of energy is used to drive photosynthesis? _____

   (b) What is the name of the molecule that captures this energy? _____

   (c) Where in the plant cell does photosynthesis take place? _____

   (d) What form of energy is your answer to (a) converted into? _____

   (e) What happens in each of the two phases of photosynthesis? _____

   _____

4. (a) Primary production is the production of carbon compounds from carbon dioxide, generally by photosynthesis. Study the graph (right) showing primary production in the oceans. Describe what the graph is showing:

   _____

   _____

   _____

   (b) Explain the shape of the curves described in (a): _____

   _____

   _____

   _____

   (c) About 90% of all marine life lives in the photic zone (the depth to which light penetrates). Suggest why this is so:

   _____

   _____

   _____

   _____

Ocean primary production

Primary production
(mgC/m³/day)

Depth (m)

2011 ——
2012 ·······
2013 ——

©2022 **BIOZONE** International
ISBN: 978-1-98-856692-4
Photocopying Prohibited

103 Chloroplasts

Key Question: How does the structure of a chloroplast relate to its photosynthetic function?

Photosynthesis takes place in disk-shaped organelles called **chloroplasts** (4-6 µm in diameter). The inner structure of chloroplasts is characterized by a system of membrane-bound compartments called thylakoids arranged into stacks called grana, linked together by stroma lamellae. The light dependent reactions of photosynthesis occur in the thylakoids.

Pigments on these membranes, called chlorophylls, capture light energy by absorbing light of specific wavelengths. Chlorophylls reflect green light, giving leaves their green color.

Chloroplasts are usually aligned with their broad surface parallel to the cell wall to maximize the surface area for light absorption.

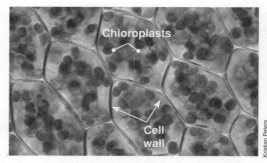
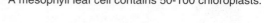

A mesophyll leaf cell contains 50-100 chloroplasts.

Chloroplast structure

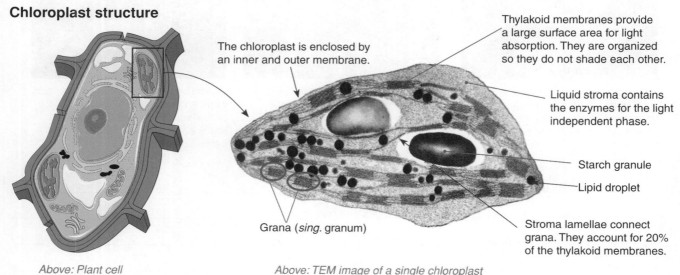

The chloroplast is enclosed by an inner and outer membrane.

Thylakoid membranes provide a large surface area for light absorption. They are organized so they do not shade each other.

Liquid stroma contains the enzymes for the light independent phase.

Starch granule

Lipid droplet

Stroma lamellae connect grana. They account for 20% of the thylakoid membranes.

Grana (*sing.* granum)

Above: Plant cell

Above: TEM image of a single chloroplast

1. The photo below is a transmission electron micrograph (TEM) of a chloroplast. Use the information above to label the parts:

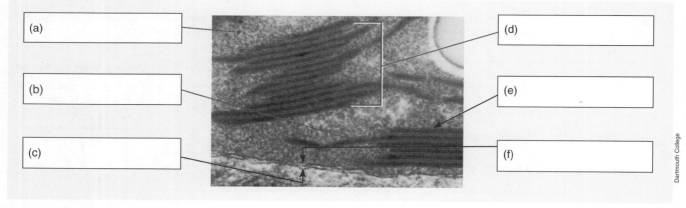

(a)

(b)

(c)

(d)

(e)

(f)

2. What does chlorophyll do? _____

_____

3. What features of chloroplasts help maximize the amount of light that can be absorbed? _____

_____

_____

_____

LS1.C   EM   SF

©2022 **BIOZONE** International
ISBN: 978-1-98-856692-4
Photocopying Prohibited

104 Stages in Photosynthesis

Key Question: What are the two main reactions in photosynthesis?

▸ **Photosynthesis** has two phases, the light dependent phase and the light independent phase.

▸ In the reactions of the light dependent phase, light energy is converted to chemical energy in the form of ATP and NADPH. This phase occurs in the thylakoid membranes of the chloroplasts.

▸ In the reactions of the light independent phase, the chemical energy is used to make carbohydrate. This phase occurs in the stroma of chloroplasts.

▸ An overview of the two stages of photosynthesis is shown in the diagram below.

Light dependent phase (LDP):

Location: Thylakoid membranes of the grana.
Process: In the first phase of photosynthesis, chlorophyll captures light energy, which is used to split water, producing O_2 gas (expelled as a waste product), electrons and H^+ ions, which are transferred to the molecule NADPH. ATP is also produced.

Light independent phase (LIP):

Location: Stroma.
Process: The second phase of photosynthesis uses the NADPH and ATP produced in the LDP to drive a series of enzyme-controlled reactions (the Calvin cycle) that fix carbon dioxide to produce triose phosphate. This phase does not need light to proceed.

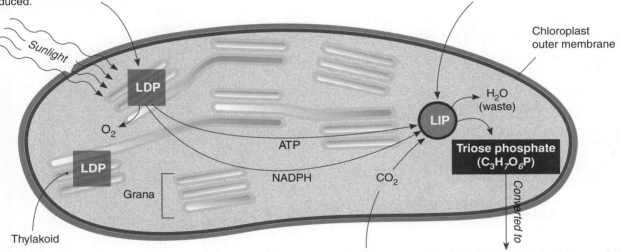

Above: Drawing of a chloroplast

CO$_2$ from the air provides raw materials for glucose production.

Monosaccharides, e.g. glucose, and other carbohydrates, lipids, and amino acids.

1. (a) Where does the light dependent phase of photosynthesis occur? _____

   (b) Where does the light independent phase of photosynthesis occur? _____

2. How are the light dependent and light independent phases linked? _____

   _____

   _____

3. In two experiments, radioactively-labeled oxygen (shown in blue in the equations below) was used to follow oxygen through the photosynthetic process. The results of the experiment are shown below:

   Experiment A: $6CO_2 + 12H_2O + $ sunlight energy $\rightarrow C_6H_{12}O_6 + 6O_2 + 6H_2O$

   Experiment B: $6CO_2 + 12H_2O + $ sunlight energy $\rightarrow C_6H_{12}O_6 + 6O_2 + 6H_2O$

   From these results, what would you conclude about the source of the oxygen in:

   (a) The carbohydrate produced? _____

   _____

   (b) The oxygen released? _____

105 Investigating Photosynthetic Rate

Key Question: How does light intensity affect photosynthesis rate?

Investigation 5.1 Measuring bubble production in *Cabomba*

See appendix for equipment list.

1. Fill a boiling tube 2/3 full with a 20°C solution of 1% sodium hydrogen carbonate ($NaHCO_3$).

2. Cut ~ 7 cm long piece of *Cabomba* stem (cut underwater). Place the *Cabomba* into the boiling tube (cut end up). Carefully push the *Cabomba* down.

3. Place the boiling tube in a rack and position a lamp so that it will shine on the tube when switched on.

4. To test the set-up, switch on the lamp for one minute to check that bubbles emerge freely from the stem. If they don't, you may have to recut the stem to open it.

5. When you have checked your set-up, switch off the lamp and, **after 5 minutes**, use a stopwatch to record the number of bubbles emerging from the stem in one minute. Repeat.

6. Use a ruler to mark out distances 0, 5, 10, 15, 20, and 25 cm from the boiling tube.

7. Starting at 25 cm, move the lamp to each of the distances in turn and use a stopwatch to record the number of bubbles emerging from the stem in one minute. Run two tests at each distance and allow 5 minutes after moving to a new distance before recording (this allows for acclimation).

8. Record your results in the table (right). Calculate the mean rate of gas production for each distance (and lamp OFF).

9. After you have finished recording, remove the stopper from the tube and test the gas with a glowing splint. What happens?

_____

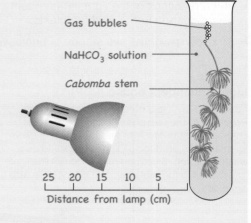

Gas bubbles

$NaHCO_3$ solution

Cabomba stem

25 20 15 10 5
Distance from lamp (cm)

NEED HELP?
See Activity 23

Distance (cm)	Bubbles per minute		
	Test 1	Test 2	Mean
OFF			
25			
20			
15			
10			
5			
0			

1. Use your calculated means to draw a graph gas production vs light intensity (distance).

2. What did your splint test tell you about the gas produced by the *Cabomba* plant?

NEED HELP?
See Activities 17 & 18

_____

3. From this experiment what can you say about photosynthesis, light, and the gas produced?

_____

_____

_____

4. How could you improve the design of this investigation?

_____

_____

_____

_____

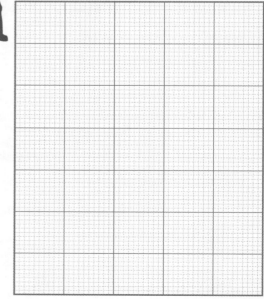

    LS1.C EM

©2022 **BIOZONE** International
ISBN: 978-1-98-856692-4
Photocopying Prohibited

106 The Fate of Glucose

Key Question: How do living organisms use glucose to produce a wide range of other molecules?

Glucose is a multipurpose molecule

▸ **Glucose** is a versatile biological molecule. It contains the elements carbon, oxygen, and hydrogen, which are used to build many other molecules required by plants, animals, and other living organisms.

▸ Plants make their glucose directly through the process of **photosynthesis** and use it to build all the molecules they require. Animals obtain their glucose (as carbohydrates) by consuming plants or other animals. Other molecules, e.g. amino acids and fatty acids, are also obtained by animals this way.

▸ Glucose has three main fates: immediate use to produce ATP molecules (available energy for work), storage for later ATP production, or for use in building other molecules. The elements present in each type of molecule are identified below, in blue.

The fate of glucose

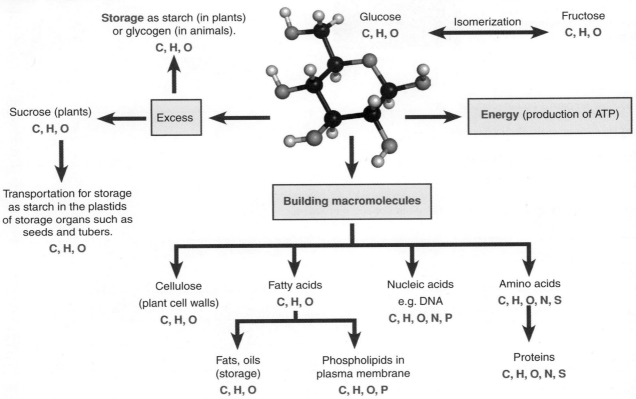

Storage as starch (in plants) or glycogen (in animals).
C, H, O

Glucose **C, H, O**

Isomerization

Fructose **C, H, O**

Sucrose (plants) **C, H, O**

Excess

Energy (production of ATP)

Transportation for storage as starch in the plastids of storage organs such as seeds and tubers. **C, H, O**

Building macromolecules

Cellulose (plant cell walls) **C, H, O**

Fatty acids **C, H, O**

Nucleic acids e.g. DNA **C, H, O, N, P**

Amino acids **C, H, O, N, S**

Fats, oils (storage) **C, H, O**

Phospholipids in plasma membrane **C, H, O, P**

Proteins **C, H, O, N, S**

How do we know how glucose is used?

▸ Labeling the carbon atoms in a glucose molecule with isotopes shows how glucose is incorporated into other molecules.

▸ An isotope is an element, e.g. carbon, whose atoms have a particular number of neutrons in their nucleus. The different number of neutrons allows the isotopes to be identified by their density, e.g. a carbon atom with 13 neutrons is denser than a carbon atom with 12 neutrons.

▸ Some isotopes are radioactive. These radioactive isotopes can be traced using X-ray film or devices that detect the disintegration of the isotopes, such as Geiger counters.

The carbon atom

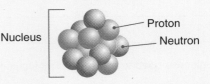

Nucleus

Proton

Neutron

The nucleus of an atom is made up of neutrons and protons. For any element, the number of protons remains the same, but the number of neutrons can vary. Electrons (not shown) are found outside the nucleus.

Naturally occurring C isotopes

^{12}C
6 protons
6 neutrons
Stable. 99.9% of all C isotopes.

^{13}C
6 protons
7 neutrons
Stable

^{14}C
6 protons
8 neutrons
Radioactive

©2022 **BIOZONE** International
ISBN: 978-1-98-856692-4
Photocopying Prohibited

Isotope experiments with animals

Experiments using ^{13}C isotopes to identify the fate of glucose in guinea pigs showed that 25% of the glucose intake was used as fuel for cellular respiration. The rest of the glucose was incorporated into proteins, fats, and glycogen.

Corals are small sea anemone-like organisms that live in a symbiotic relationship with algae. The algae transfer sugars to the coral in return for the safe environment provided by the coral. Experiments with ^{13}C showed that the major molecule being transferred to the coral was glucose.

Isotope experiments with plants

^{13}C isotopes were used to trace the movement of glucose in plant leaves. It was found that some glucose is converted to fructose (a sugar molecule similar to glucose). Fructose molecules can be joined together to be stored as fructan in plant vacuoles. Fructose is also added to glucose to produce sucrose. Sucrose is transported out of the leaf.

Four molecule model of glucose use in a plant

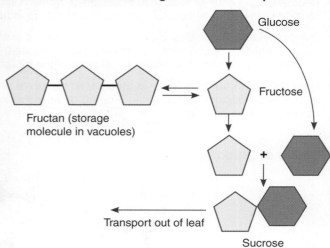

Fructan (storage molecule in vacuoles)

Glucose

Fructose

Transport out of leaf

Sucrose

1. (a) How do plants obtain glucose? _____

   (b) How do animals obtain glucose? _____

2. What are the three main fates of glucose? _____
   _____

3. Identify a use for glucose in a plant that does not occur in animals: _____
   _____
   _____
   _____

4. (a) How can isotopes of carbon be separated? _____
   _____
   _____

   (b) How can this help trace how glucose is used in an organism? _____
   _____
   _____
   _____

5. How is glucose used to make other molecules needed by an organism? _____
   _____
   _____
   _____

6. Describe the fate of glucose in the glucose, fructose, sucrose system shown above: _____
   _____
   _____
   _____

©2022 **BIOZONE** International
ISBN: 978-1-98-856692-4
Photocopying Prohibited

107 Energy Transfer Between Systems

Key Question: How is the stored energy in glucose used to power the chemical reactions which occur in living organisms?

▶ During **photosynthesis,** light energy is converted into chemical energy in the form of glucose. **Glucose** is used by plants and animals to provide the energy for **cellular respiration**.

▶ During cellular respiration, ATP is formed through a series of chemical reactions. The ATP provides the energy to drive life's essential processes.

▶ Heterotrophs (organisms that cannot make their own food) obtain their glucose by eating plants or other organisms.

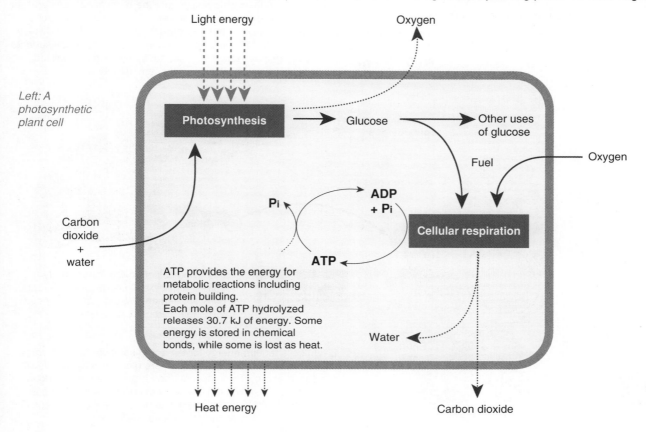

Left: A photosynthetic plant cell

1. Use the diagram above to explain where heterotrophs join the energy transfer system: _____

_____

_____

2. Complete the schematic diagram (below) of the transfer of energy and production of macromolecules using the following word list: *water, ADP, protein, carbon dioxide, amino acid, glucose, ATP*.

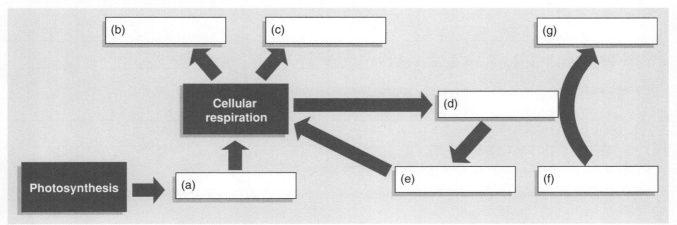

EM   LS1.C

108 Energy From Glucose

Key Question: How is energy released from glucose during the process of cellular respiration?

▶ Energy is released in cells by the breakdown of sugars and other substances in **cellular respiration**. During aerobic respiration, oxygen is consumed and carbon dioxide is released. These gases need to be exchanged with the environment by diffusion. Diffusion gradients are maintained by transport of gases away from the gas exchange surface, e.g. the cell surface. In **anaerobic** pathways, ATP is generated but oxygen is not used.

The overall equation for cellular respiration is:

Glucose + Oxygen ⟶ Carbon dioxide + Water + Energy

$$C_6H_{12}O_6 + 6O_2 \longrightarrow 6CO_2 + 6H_2O + Energy$$

To carry out **aerobic** respiration, the body needs oxygen and must remove carbon dioxide. Gas exchange surfaces provide a way for these respiratory gases to enter and leave the body by diffusion. Some organisms use the body surface as the gas exchange surface, but many have specialized gas exchange structures, e.g. lungs, gills, or stomata.

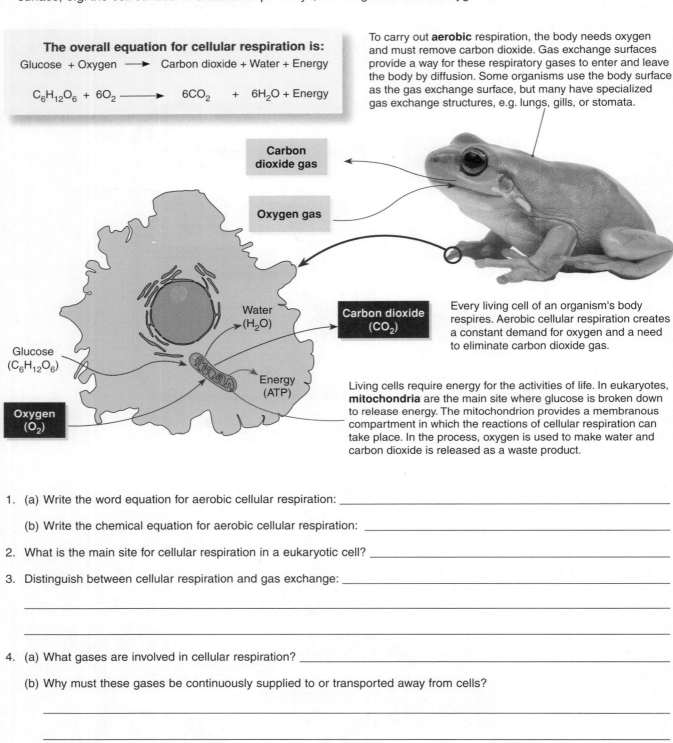

Carbon dioxide gas

Oxygen gas

Water (H_2O)

Carbon dioxide (CO_2)

Glucose ($C_6H_{12}O_6$)

Energy (ATP)

Oxygen (O_2)

Every living cell of an organism's body respires. Aerobic cellular respiration creates a constant demand for oxygen and a need to eliminate carbon dioxide gas.

Living cells require energy for the activities of life. In eukaryotes, **mitochondria** are the main site where glucose is broken down to release energy. The mitochondrion provides a membranous compartment in which the reactions of cellular respiration can take place. In the process, oxygen is used to make water and carbon dioxide is released as a waste product.

1. (a) Write the word equation for aerobic cellular respiration: _____

    (b) Write the chemical equation for aerobic cellular respiration: _____

2. What is the main site for cellular respiration in a eukaryotic cell? _____

3. Distinguish between cellular respiration and gas exchange: _____

    _____

    _____

4. (a) What gases are involved in cellular respiration? _____

    (b) Why must these gases be continuously supplied to or transported away from cells?

    _____

    _____

    (c) By which transport process do these gases move through the gas exchange surface? _____

5. What is the main difference between aerobic and anaerobic pathways for ATP generation? _____

    _____

   LS1.C EM

©2022 **BIOZONE** International
ISBN: 978-1-98-856692-4
Photocopying Prohibited

Aerobic and anaerobic pathways for ATP production

A Aerobic respiration

B Lactic acid fermentation

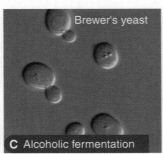

Brewer's yeast

C Alcoholic fermentation

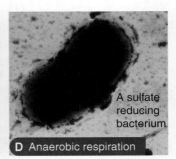

A sulfate reducing bacterium

D Anaerobic respiration

Aerobic respiration produces the energy (as ATP) needed for metabolism. The rate of aerobic respiration is limited by the amount of oxygen available. In animals and plants, most of the time the oxygen supply is sufficient to maintain aerobic metabolism. Aerobic respiration produces a high yield of ATP per molecule of glucose (path A below).

During maximum physical activity, when oxygen is limited, **anaerobic** metabolism provides ATP for working muscle. In mammalian muscle, metabolism of a respiratory intermediate produces lactate, which provides fuel for working muscle and produces a low yield of ATP. This process is called lactic acid fermentation (path B below).

The process of brewing utilizes the anaerobic metabolism of yeasts. Brewers' yeasts preferentially use anaerobic metabolism in the presence of excess sugars. This process, called alcoholic fermentation, produces ethanol and CO_2 from the respiratory intermediate pyruvate. It is carried out in vats that prevent entry of O_2 (path C below).

Many bacteria and archaea are anaerobic, using molecules other than oxygen, e.g. nitrate or sulfate, as a terminal electron acceptor of their electron transport chain. These electron acceptors are not as efficient as oxygen (less energy is released per oxidized molecule) so the energy (ATP) yield from anaerobic respiration is generally quite low (path D below).

In most energy-yielding pathways the initial source of chemical energy is glucose. The first step, glycolysis, is an almost universal pathway. The paths differ in what happens after **glucose** has been converted to pyruvate.

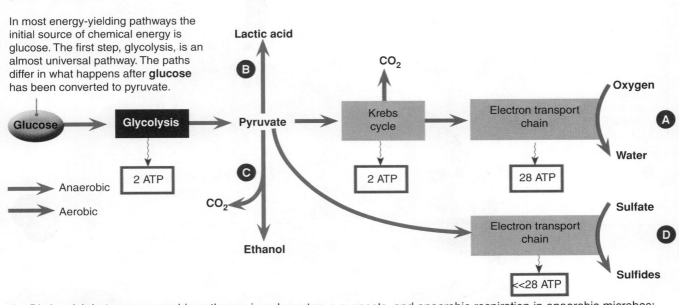

6. Distinguish between anaerobic pathways in eukaryotes, e.g. yeasts, and anaerobic respiration in anaerobic microbes:

_____

_____

7. When brewing alcohol, why is it important to prevent entry of oxygen to the fermentation vats? _____

_____

_____

8. Explain why aerobic respiration is energetically more efficient than fermentation and anaerobic respiration: _____

_____

_____

9. (a) How many ATP molecules are produced from one glucose molecule during aerobic respiration? _____

   (b) How many ATP molecules are produced during lactic acid or alcoholic fermentation? _____

   (c) Calculate the efficiency of fermentation compared to aerobic respiration: _____

NEED HELP?
See Activity 6

   _____

109 Aerobic Cellular Respiration

Key Question: How does aerobic cellular respiration convert the chemical energy in glucose into usable energy (ATP), carbon dioxide, and water?

▶ **Cellular respiration** is the process of extracting the energy stored in the chemical bonds in **glucose** and storing it in ATP molecules. The process includes many chemical reactions, some of which produce ATP molecules and some that prepare molecules for further chemical reactions.

▶ Cellular respiration can be divided into four major steps, each with its own set of chemical reactions. The four steps are: glycolysis, the link reaction, the Krebs cycle, and the electron transport chain (ETC). Every step, except the link reaction, produces ATP.

▶ Glycolysis occurs in the cytoplasm, the other steps take place within the **mitochondrion**.

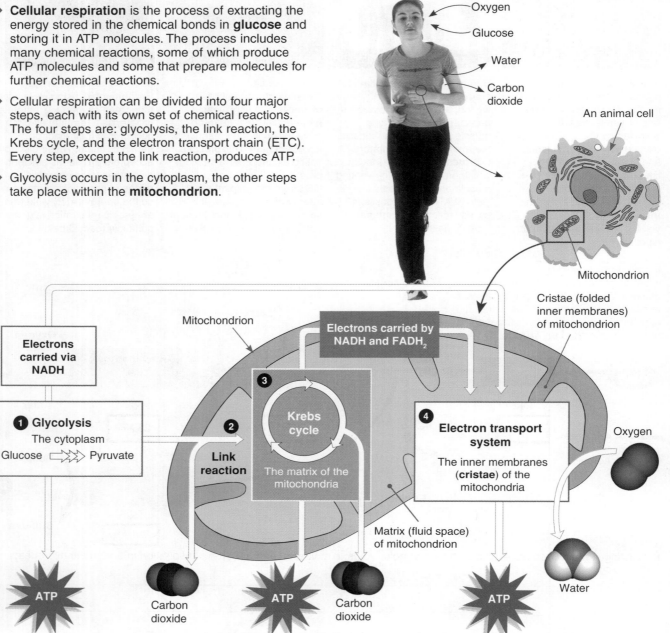

1. Which of the four steps in cellular respiration yield ATP? _____

_____

2. (a) What are the main reactants (inputs) for cellular respiration? _____

_____

(b) What are the main products (outputs) for cellular respiration? _____

_____

3. Which of the four steps of cellular respiration occur in the mitochondria? _____

_____

   LS1.C EM SF

©2022 **BIOZONE** International
ISBN: 978-1-98-856692-4
Photocopying Prohibited

How does cellular respiration provide energy?

▸ A molecule's energy is contained in the electrons within its chemical bonds. During a chemical reaction, energy, e.g. heat, can break the bonds of the reactants.

▸ When the reactants form products, the new bonds within the product will contain electrons with less energy, making the bonds more stable. The difference in energy is usually lost as heat. However, some of the energy can be captured to do work.

▸ Glucose contains 16 kJ of energy per gram (2870 kJ/mol). The step-wise breakdown of glucose through a series of chemical reactions yields ATP. In total, 32 ATP molecules can be produced from 1 glucose molecule.

A model for ATP production and energy transfer from glucose

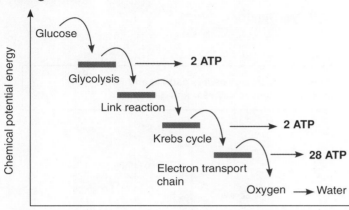

A model for ATP use in the muscles

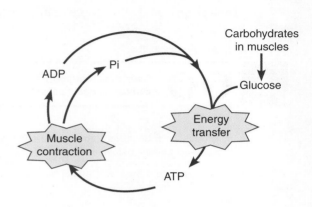

4. Explain how the energy in glucose is converted to useful energy in the body. Use the example of muscle contraction to help illustrate your ideas:

_____

_____

_____

_____

_____

5. (a) One mole of glucose contains 2870 kJ of energy. The hydrolysis of one mole of ATP releases 30.7 kJ of energy. Calculate the percentage of energy that is transformed to useful energy in the body. Show your working.

_____

_____

_____

_____

_____

(b) Use your calculations above to explain why shivering keeps you warm and extreme muscular exertion causes you to get hot:

_____

_____

_____

_____

_____

©2022 **BIOZONE** International
ISBN: 978-1-98-856692-4
Photocopying Prohibited

110 Measuring Respiration

Key Question: How can a respirometer be used to measure the rate of cellular respiration in germinating seeds?

▸ A respirometer can be used to measure the amount of oxygen consumed by an organism during cellular respiration and so can be used to measure respiration rate.

▸ A simple respirometer is shown in the diagram below. The carbon dioxide produced during respiration is absorbed by the potassium hydroxide. As the oxygen is used up, the colored bubble in the glass tube moves. Measuring the movement of the bubble, e.g. with a ruler or taped graph paper, allows us to estimate the change in volume of gas and therefore the rate of cellular respiration.

Investigation 5.2   Measuring respiration in germinating seeds

See appendix for equipment list.

   Caution is required when handling potassium hydroxide as it is caustic and can cause chemical burns. You should wear protective eyewear and gloves.

1. Work in groups of four to set up three respirometers using the set-up shown below as a guide.

2. Collect three boiling tubes and place two cotton balls in the bottom of each.
   Label the tubes A, B, and C.

3. Use a dropper to add 15% potassium hydroxide (KOH) solution on the cotton balls until they are saturated (there should be no liquid in the boiling tube). Add the same amount of KOH to the cotton balls in each boiling tube.

4. Place gauze on top of the cotton balls in each tube. This prevents the KOH coming into contact with the seeds and killing them.

5. Quarter fill tube A with germinated bean seeds. These seeds will be wet because they have been germinated under wet paper towels for four days.

6. Quarter fill tube B with ungerminated (dry) seeds.

7. Quarter fill tube C with glass beads.

8. Place a two-hole stopper firmly in each boiling tube. In one hole insert a bent glass tube or bent pipette. In the second hole insert a tube that can be clamped shut using a screw clip.

9. Use a dropper or fine pipette to place a drop of colored liquid into the bent tube/pipette of each set up. Attach a syringe to the clamped tube. Open the screw clip and use the syringe to draw the colored bubble into the middle of the bent tube/pipette.

10. Place all three tubes in a water bath at 25°C. Secure them with a clamp stand or in racks.

11. Leave the apparatus to acclimatize for 10 minutes.

12. At the end of the acclimation period, close the screw clip on the boiling tubes. Mark the position of the bubble with a marker pen. This is your time zero position. Start the timer.

13. Use a ruler or the pipette's scale (if there is one) to measure the distance the colored bubble moved at 5, 10, 15, 20, and 25 minutes.

14. Record your results on the table at the top of the next page.

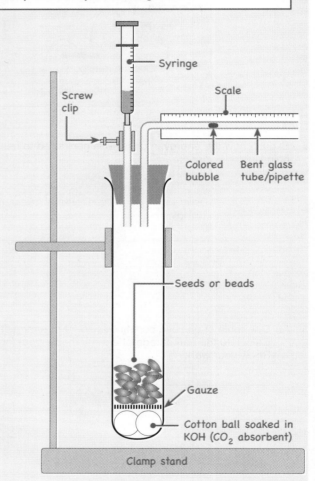

Respirometers of this sort are very sensitive to poor procedure because the volumes involved are so small.

Be very careful with your set-up and when taking readings. Have one person responsible taking the measurements of the bubble movement.

Time (minutes)	Distance bubble moved (mm)		
	Germinated seeds	Ungerminated seeds	Glass beads
0			
5			
10			
15			
20			
25			

5. What is the purpose of the test tube with the beads? _____

_____

_____

6. (a) Calculate the corrected distance the bubble moved in tubes A and B by subtracting the distance moved in tube C from each value. Record these values in the table below.

(b) Use the corrected distance the bubble moved to calculate the rate of respiration. Record this in the table below:

NEED HELP? See Activity 6

Time (minutes)	Corrected distance bubble moved (mm)		Rate (mm/min)	
	Germinated seeds	Ungerminated seeds	Germinated seeds	Ungerminated seeds
0				
5				
10				
15				
20				
25				

NEED HELP? See Activities 17 & 18

(c) Plot the rate of respiration on the grid (right). Include appropriate titles and axis labels:

(d) What does your plot show? _____

_____

_____

_____

_____

_____

_____

©2022 **BIOZONE** International
ISBN: 978-1-98-856692-4
Photocopying Prohibited

7. Why does the bubble in the capillary tube move? _____

_____

_____

_____

8. What conclusion can you make about cellular respiration in germinated and ungerminated seeds?

_____

_____

_____

_____

_____

9. How would you have to modify the experiment if you were measuring respiration in a plant instead of seeds?

_____

_____

_____

_____

_____

10. Explain the purpose of:

(a) KOH: _____

_____

(b) The acclimation period: _____

_____

_____

(c) The ungerminated seeds: _____

_____

_____

11. A student decided to repeat the respirometer experiment but used maggots instead of seeds. Their results are shown on the table (right).

(a) Calculate the rates and record them in the table:

NEED HELP? See Activities 6 & 17

(b) Graph the rates:

(c) Describe the results: _____

_____

_____

_____

_____

_____

_____

Time (minutes)	Distance bubble moved (mm)	Rate (mm/min)
0	0	
5	25	
10	65	
15	95	
20	130	
25	160	

©2022 **BIOZONE** International
ISBN: 978-1-98-856692-4
Photocopying Prohibited

111 Review Your Understanding

Key Question: Under what conditions can an animal survive in a sealed system?

Man in a box

▶ In 2012, researchers carried out a larger version of Joseph Priestley's famous mouse in a jar experiment.

▶ 274 plants were placed in a sealed container with oxygen-depleted air (12.4% oxygen). A healthy 47 year man entered the container for 48 hours. Gas levels were monitored throughout the experiment. The container was kept in constant light.

▶ The experiment was run to completion with no harm to the person within the box.

▶ Ethics approval was obtained beforehand and medical staff were on hand during the experiment.

Martin D, Thompson A, Stewart I, et al. A paradigm of fragile Earth in Priestley's bell jar. Extrem Physiol Med. 2012;1(1):4. Published 2012 Sep 4. doi:10.1186/2046-7648-1-4

1. Use your understanding from the information in this chapter to identify:

   (a) The two gases primarily being monitored in the experiment: _____

   _____

   (b) The two metabolic processes involved in this experiment: _____

   _____

2. (a) The graph on the right shows the change in oxygen concentration over the course of the experiment. Describe the trend in oxygen levels over time:

   _____

   _____

   _____

   _____

   (b) Explain why this change occurred (your answer should make reference to the gases and metabolic pathways involved):

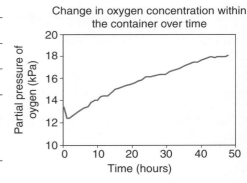

Change in oxygen concentration within the container over time

   _____

   _____

   _____

   _____

   _____

3. Revisit the model you produced in activity 99. Refine it and add more detail to explain the relationship between cellular respiration and photosynthesis:

112 Summing up

Investigation 5.3   Modeling photosynthesis and cellular respiration

See appendix for equipment list.

1.   Work by yourself for this task. If you have beads or molecular models you could use these instead of the shapes on the next page.

2.   Cut out the atoms and shapes on the following page. They are color coded as follows:

Carbon   Hydrogen   Oxygen

3.   Use the cutouts to model photosynthesis and cellular respiration by following the steps below.

1.   Write the equation for **photosynthesis** here: _____

   (a) State the starting reactants in photosynthesis: _____

   (b) State the total number of atoms of each type needed to make the starting reactants:

      Carbon: _____   Hydrogen: _____   Oxygen: _____

   (c) Use the atoms you have cut out to make the starting reactants in photosynthesis.

   (d) State the end products of photosynthesis: _____

   (e) State the total number atoms of each type needed to make the end products of photosynthesis:

      Carbon: _____   Hydrogen: _____   Oxygen: _____

   (f)   Use the atoms you have cut out to make the end products of photosynthesis.

   (g) What do you notice about the number of C, H, and O atoms on each side of the photosynthesis equation? _____

      _____

   (h) Name the energy source for this process and add it to the model you have made: _____

2.   Write the equation for **cellular respiration** here: _____

   (a) State the starting reactants in cellular respiration: _____

   (b) State the total number of atoms of each type needed to make the starting reactants:

      Carbon: _____   Hydrogen: _____   Oxygen: _____

   (c) Use the atoms you have cut out to make the starting reactants in cellular respiration.

   (d) State the end products of cellular respiration: _____

   (e) State the total number of atoms of each type needed to make the end products of cellular respiration:

      Carbon: _____   Hydrogen: _____   Oxygen: _____

   (f)   Use the atoms you have cut out to make the end products of cellular respiration.

   (g) Name the end products of cellular respiration that are utilized in photosynthesis: _____

LS1.C   EM

©2022 **BIOZONE** International
ISBN: 978-1-98-856692-4

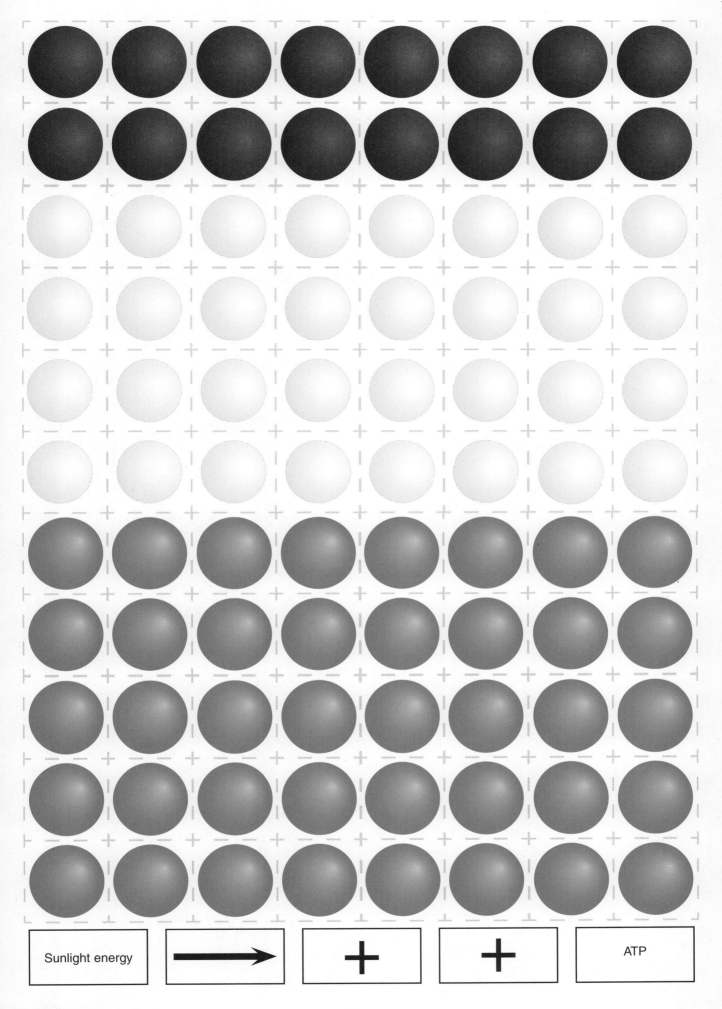

| Sunlight energy | ⟶ | + | + | ATP |

This page is left blank deliberately

©2022 **BIOZONE** International
ISBN: **978-1-98-856692-4**

3. **Photosynthesis** is the process in which carbon dioxide is fixed to form glucose. In the space below, draw a diagram to show the process of photosynthesis, including reactants and products and the location of the reactions involved:

4. Cellular respiration is a continuous, integrated process. A simple diagram of the process in a eukaryote is shown below.
   (a) In the diagram, fill in the rectangles with the process and the ovals with the substance used or produced.
       Use the following word list (some words can be used more than once): *pyruvate, glycolysis, glucose, oxygen (O₂), link reaction, electron transport chain (ETC), Krebs cycle, ATP, carbon dioxide (CO₂), water (H₂O)*
   (b) Add in a pathway to show fermentation. Write the two possible products of this pathway in eukaryotes.

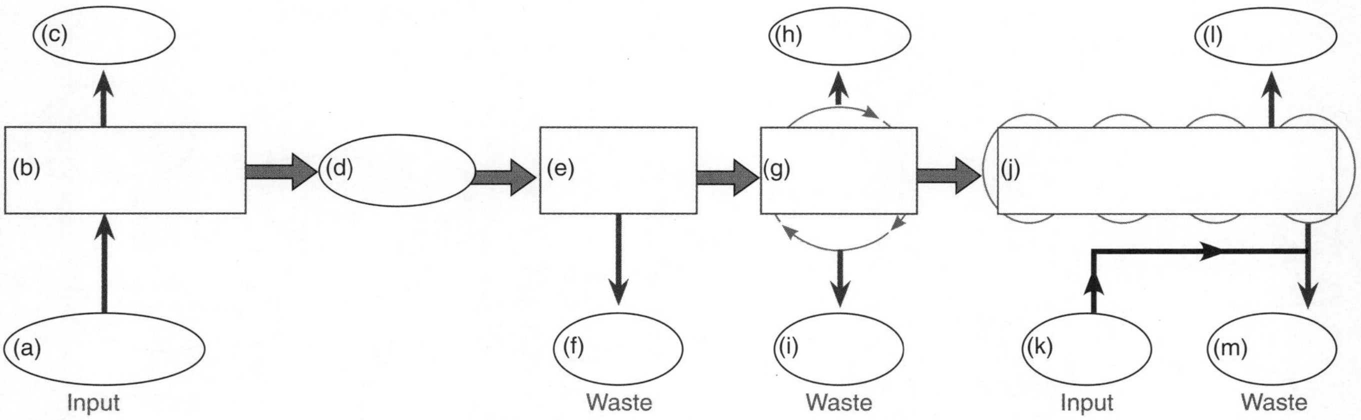

(c) Use the completed diagram to explain the difference in ATP yield between aerobic and anaerobic pathways:

_____

_____

_____

_____

_____

_____

©2022 **BIOZONE** International
ISBN: 978-1-98-856692-4
Photocopying Prohibited

Ecosystems: Interactions, Energy, and Dynamics

Concepts and connections
Use arrows to make connections between related concepts in this section of the book

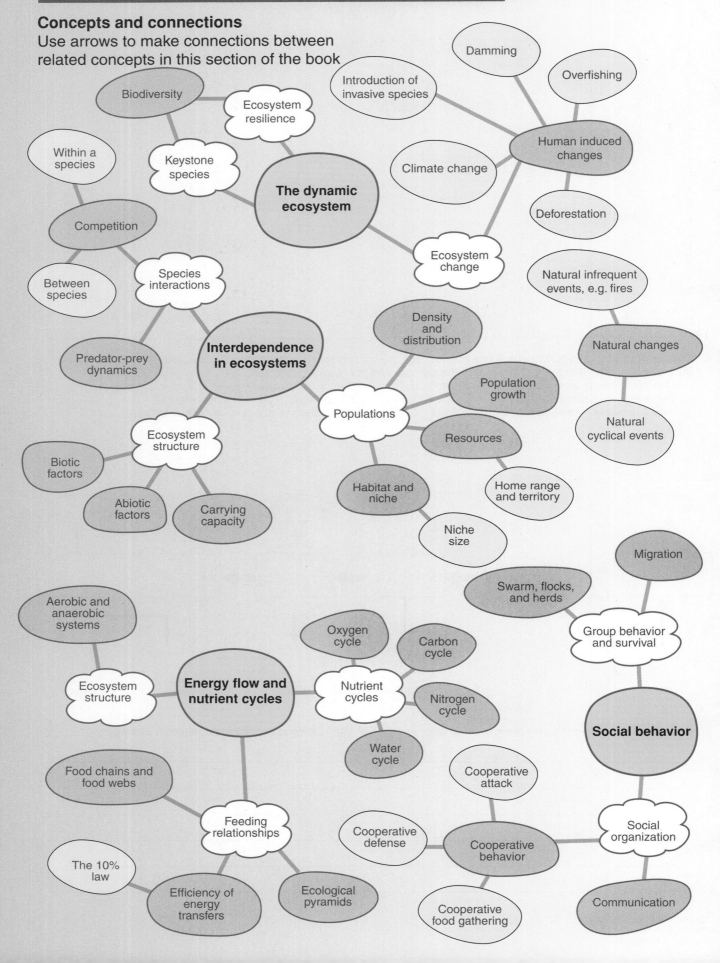

CHAPTER 6

Interdependence in Ecosystems

Anchoring Phenomenon

A plague of mice: What causes explosive population growth?   113 134

What species interactions occur in an ecosystem?

☐ 1 Describe an ecosystem as an area that can be can be large or small, as defined by   114
boundaries, e.g. a forest ecosystem or the ecosystem in a puddle of water. Know that it
includes all the living organisms (biotic) and physical factors (abiotic) in an area.

☐ 2 Understand that each organism in an ecosystem has a niche. Explain that an organism's   115 -117
ecological niche describes its functional position in the ecosystem, including its   128
habitat and its relationships with other species. Explain the difference between an
organism's fundamental niche and its realized niche, and that competition plays a role in
determining the realized niche.

☐ 3 Describe the different distribution of populations in an ecosystem. Examine models   118
of population distribution and why these distributions occur. Understand that population
distribution and density depend on factors such as competition and the availability
of resources.

☐ 4 Species interact in ways that may be beneficial to both species, e.g. mutualism, or   119
harmful to at least one species, e.g. competition, predation, and parasitism. Explore
the different types of species interaction. Identify interactions in specific cases, and the
species that benefit and that are harmed in each type.

☐ 5 Competition occurs when species exploit the same limited resources. Competition may   120-124
be between individuals, or between large groups. Explore different aspects of intraspecific
competition. Explain how competition can limit a species population size in a specific area.
Describe the differences between inter- and intraspecific competition. Examine evidence
and explain how different species reduce resource competition.

How does the environment affect population growth?

☐ 6 The number of organisms and populations an ecosystem can support is its carrying   125 -128
capacity. Abiotic and biotic factors determine the carrying capacity. Use case studies   127
to explain the effect of ecosystem size and diversity on carrying capacity. Carry out a
simulation of population growth to investigate how resource limitations affect carrying
capacity and population growth.

☐ 7 The population growth of a species is limited by the resources of its environment.   129 -131
Investigate exponential and logistic growth rate, and describe the shape of a graph   129
of exponential growth and a graph of logistic growth. Explain why exponential growth
cannot continue indefinitely. Identify factors that lead to each type of growth. Use a
spreadsheet to write formulas to model logistic growth. Use data to investigate bacterial
population growth and determine if it is exponential or logistical growth.

☐ 8 Use simulation software (Populus) to simulate population growth. Explore the effect   131 132
of changing certain starting values, e.g. population number, intrinsic growth rate, on
population growth rate. Explain the relationship between prey and predator populations.

Michael L. Baird CC 2.0

113 A Plague of Mice

Key Question: What causes explosive population growth?

▶ In 1924, the Bureau of Biological Survey (now the US Fish and Wildlife Service) launched a project to eliminate coyotes, hawks, and various other predators from Kern County, California.

▶ Two years later, in 1926, an autumn rainstorm triggered an invasion by 50 million mice from the surrounding (flooding) fields of the Kern County city of Taft. This particular mouse plague has been described as the largest in North American history.

▶ But this mouse plague pales in comparison to the mouse plagues experienced by rural areas of Australia. The most recent occurred in 2021, in New South Wales. In this particular plague, it is estimated at least US$700 million of crops and equipment was ruined by hundreds of millions of mice.

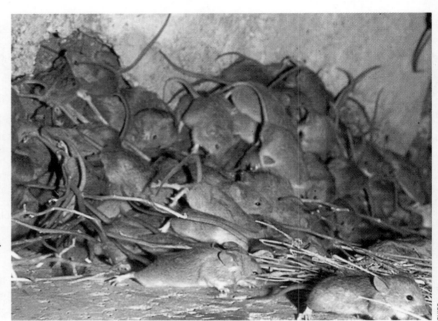

▶ New South Wales is a particularly good environment for growing wheat and other grains, with mild springs and long, hot summers. Mice were introduced by accident in the 1700s and are preyed upon by various native predators. Mice breed extremely rapidly in good conditions, often experienced in New South Wales. During mouse plagues, they can reach **population** densities of 3000 mice per hectare.

1. What do you think caused the mouse plague in Taft, 1924?_____

_____

_____

2. The graph below shows the success of mouse traps in Southern Australia. The greater the trap success, the greater the mouse population:

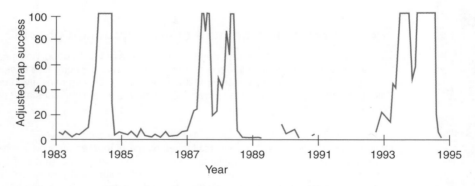

Rate of increase as a function of rainfall for house mouse *Mus domesticus* populations in a cereal-growing region in southern Australia
Peter R. Brown, Journal of Applied Ecology, 2001

Suggest why the mouse population, as related to the trap success, rises and falls so steeply:

_____

_____

_____

3. Other kinds of animals also produce plagues. Can you think of another animal that can reach plague-like populations? Briefly describe the reasons why they do:

_____

_____

LS2.A   CE

©2022 **BIOZONE** International
ISBN: 978-1-98-856692-4

114 What is an Ecosystem?

Key Question: What are the components of an ecosystem? How are these components linked through nutrient cycles and energy flow?

Ecosystems

▸ **Ecosystems** are natural units made up of all the living organisms (**biotic factors**) and the physical (**abiotic factors**) in an area.

▸ Abiotic factors (non-living physical factors) include the soil, water, atmosphere, temperature, and sunlight (SWATS). Biotic factors are all the living organisms, e.g. plants, animals, fungi, protists, and microorganisms.

▸ The interactions of living organisms with each other and with the physical environment help determine the features of an ecosystem. The components of an ecosystem are linked to each other, and to other ecosystems, through nutrient cycles and energy flows.

BIOTIC FACTORS	ABIOTIC FACTORS		

| The living organisms in the environment, including their interactions, e.g. as competitors, predators, or symbionts.
 • Plants
 • Animals
 • Microorganisms
 • Fungi
 • Protists, e.g. algae, protozoa | **Atmosphere (air)**
 • Wind speed
 • Wind direction
 • Humidity
 • Light intensity/quality
 • Precipitation
 • Temperature | **Hydrosphere (water)**
 • Dissolved nutrients
 • pH
 • Salinity
 • Dissolved oxygen
 • Precipitation
 • Temperature | **Geosphere (rock/soil)**
 • Nutrient availability
 • Soil moisture
 • pH
 • Composition
 • Temperature
 • Depth |

1. (a) What are biotic factors? _____

_____

(b) Give an example of a biotic factor: _____

2. (a) What is an abiotic factor? _____

_____

(b) Give an example of an abiotic factor: _____

3. How are the components of an ecosystem linked? _____

_____

4. How do the particular characteristics of an ecosystem arise?_____

_____

LS2.A

115 Habitat and Tolerance Range

Key Question: How do we define habitat? How does the tolerance range of an organism determine the optimum position of organisms in their habitat?

The habitat is where an organism lives

▶ The natural environment in which an organism lives is its **habitat**. It includes all the physical and **biotic** factors in that occupied area. Habitats vary widely in scale, as do the physical factors that influence them.

A habitat may be vast and relatively homogeneous, as is the open ocean. Predatory barracuda (above) occur around reefs and in the open ocean.

For sessile organisms like this fungus, a suitable habitat may be defined by the environment in a small area, such as on a decaying log.

For microbial organisms, such as those in the ruminant gut, the habitat is defined by the chemical environment within the rumen (R) of the host, in this case, a cow.

Tolerance range determines distribution in the habitat

▶ Each species has a **tolerance range** for factors in its environment (below). However, the members of a population are individually different and vary in their tolerance. Organisms are usually most abundant where the conditions for their survival and reproduction are optimal. Outside this optimal ecological space, the environment is less favorable for survival and there are fewer individuals. For any species, there will also be environments that are unavailable to them.

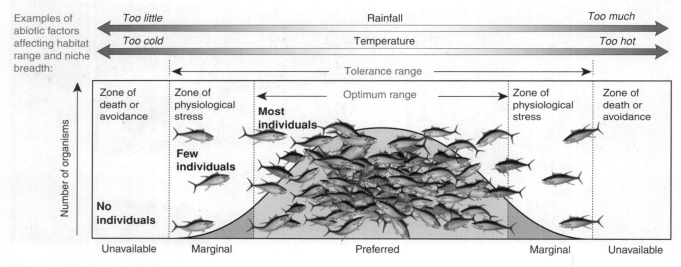

Examples of abiotic factors affecting habitat range and niche breadth:

Too little — Rainfall — Too much

Too cold — Temperature — Too hot

Tolerance range

| Zone of death or avoidance | Zone of physiological stress | Optimum range | Zone of physiological stress | Zone of death or avoidance |

Most individuals

Few individuals

No individuals

Number of organisms

Unavailable — Marginal — Preferred — Marginal — Unavailable

1. In which part of an organism's range will competition for resources be most intense, and why? _____

_____

Species tolerant of large environmental variations tend to be more widespread than organisms with a narrow tolerance range. The Atlantic blue crab (left) is widespread along the Atlantic coast from Nova Scotia to Argentina. Adults tolerate a wide range of water salinity, ranging from almost fresh to highly saline. This species is an omnivore and eats anything, from shellfish to carrion and animal waste.

2. Suggest an advantage to being able to tolerate variations in a wide range of environmental factors?

_____

_____

_____

Wendy Kaveney

  LS2.A SPQ

©2022 **BIOZONE** International
ISBN: 978-1-98-856692-4
Photocopying Prohibited

116 The Ecological Niche

Key Question: What is an organism's niche? How is it influenced by interactions with other species?

The niche is the functional role of an organism

▸ The **ecological niche** (or niche) of an organism describes its functional position in its environment. The full range of environmental conditions under which an organism can exist describes its fundamental niche.

▸ Interactions with other species, e.g. **competition**, usually force organisms to occupy a space that is narrower than this. This is called the realized niche.

▸ Central to the niche concept is the idea that two species with exactly the same niche cannot coexist, because they would compete for the same resources and one would exclude the other. More often, species compete for only some of the same resources. These competitive interactions limit population sizes and influence **distributions**.

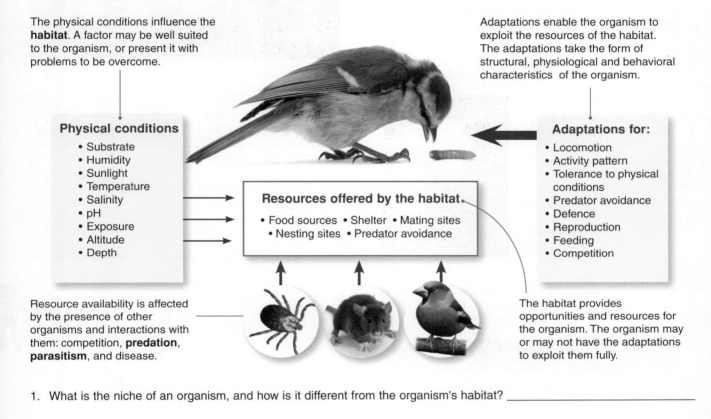

The physical conditions influence the **habitat**. A factor may be well suited to the organism, or present it with problems to be overcome.

Physical conditions
- Substrate
- Humidity
- Sunlight
- Temperature
- Salinity
- pH
- Exposure
- Altitude
- Depth

Resources offered by the habitat.
- Food sources • Shelter • Mating sites
- Nesting sites • Predator avoidance

Adaptations enable the organism to exploit the resources of the habitat. The adaptations take the form of structural, physiological and behavioral characteristics of the organism.

Adaptations for:
- Locomotion
- Activity pattern
- Tolerance to physical conditions
- Predator avoidance
- Defence
- Reproduction
- Feeding
- Competition

Resource availability is affected by the presence of other organisms and interactions with them: competition, **predation**, **parasitism**, and disease.

The habitat provides opportunities and resources for the organism. The organism may or may not have the adaptations to exploit them fully.

1. What is the niche of an organism, and how is it different from the organism's habitat? _____

_____

_____

2. (a) In what way is the size of the realized niche flexible? _____

_____

_____

   (b) How does competition with another species affect the size of an organism's niche? Explain: _____

_____

_____

3. The diagram (right) shows the resource-use curves of two bird species. They overlap in the size of the food items they exploit. On the diagram, use A, B, and C to mark:

   (a) The food item sizes most exploited by species A.
   (b) The food item sizes most exploited by species B.
   (c) The regions where competition is most intense.

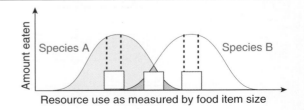

 LS2.A

117 Dingo Habitats

Key Question: How does the amount of resources in a habitat influence an organism's population density?

Habitats provide resources

▸ As we have seen, species may tolerate wide variations in a range of physical and **biotic** factors. As a result of this tolerance range, the **habitat** that is occupied by members of a species may be quite variable.

▸ It is important to know that the habitat provides resources for the organisms that live there. These resources include water, food, shelter, and places to raise offspring.

▸ Some habitats can be richer in resources than others and are usually described with reference to their main features. For example, riverine habitats (rivers and creeks containing water and thick vegetated cover) provide water, food, and cover (right).

Dingo habitats

Dingoes (right) are wild dogs found throughout Australia. The table on the far right gives information about five dingo packs at one location, including how much of their territory is made up of riverine areas. Kangaroos are the main prey for these dingoes.

Dingo pack name	Territory area (km^2)	Pack size	Dingo density per 100 km^2	% of territory made up of riverine areas
Pack A	113	12	10.6	10
Pack B	94	12		14
Pack C	86	3		2
Pack D	63	6		12
Pack E	45	10		14

1. Calculate the density of each of the dingo packs per 100 km^2 using the equation below and record it in the table above. The first one has been done for you.

NEED HELP?
See Activity 19

> Density = pack size ÷ territory area x 100

2. (a) Plot a scatter graph of dingo density versus how much of their territory is made up of riverine areas for each pack.

   (b) Describe the relationship between dingo density and amount of riverine area:

   _____

   _____

   _____

   (c) Can you explain why this relationship might occur?

   _____

   _____

   _____

   _____

   _____

   LS2.A

©2022 **BIOZONE** International
ISBN: 978-1-98-856692-4
Photocopying Prohibited

118 Population Density and Distribution

Key Question: What are population density and population distribution? Why do these vary between different species of organisms?

Population density

▸ A **population** refers to all the organisms of the same species in a particular area. The **density** of a population is the number of individuals of that species per unit area (for land organisms) or volume (for aquatic organisms).

▸ Populations can exist naturally at different densities. Social insects, such as termites, exist naturally at high densities, whereas the population density of solitary or territorial species is naturally lower (below). Some species may occur at high densities at certain times of the year, e.g. during breeding.

▸ The density of populations is also affected by the availability of resources. Population density is higher where resources are plentiful and lower where they are scarce or highly variable.

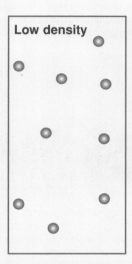
Low density

In low density populations, individuals are spaced well apart. There are only a few individuals per unit area. Highly territorial or solitary animal species, such as tigers, leopards, and bears, occur at low densities.

Plant population density is strongly correlated with the availability of water and nutrients. The density of plant populations, especially of larger species, is therefore lower in deserts, where water is limited.

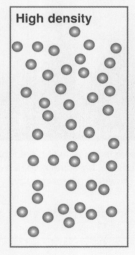
High density

In high density populations, there are many individuals per unit area. This can be a natural feature of highly social species, e.g. ants and termites, or species that reproduce asexually to form large colonies, e.g. corals.

Human populations reach their highest densities in large cities, which are often centers of commerce. Cities were originally established in areas where resources, such as water or fuel, were plentiful.

1. (a) How would you express the population density of a terrestrial species? _____

   (b) How would you express the population density of an aquatic species? _____

2. Explain how the distribution and availability of resources might influence population density? _____

   _____

   _____

   _____

©2022 **BIOZONE** International
ISBN: 978-1-98-856692-4
Photocopying Prohibited

SPQ LS2.A

Population distribution

▶ Population **distribution** describes how organisms are distributed in the environment, relative to each other.

▶ Three distribution patterns are usually recognized: random, clumped (or aggregated), and uniform, described below. In the examples, the circles represent individuals of the same species.

Random distribution

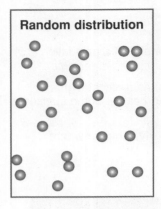

Dune grass

Oyster bed

In random distributions, the spacing between individuals is unpredictable, i.e. the position of one individual is independent of the other individuals. Random distribution is uncommon but can occur in homogeneous environments where unpredictable factors determine distribution, e.g. dandelion seeds germinating after being blown by the wind or oyster larvae settling after being carried by ocean currents.

Clumped distribution

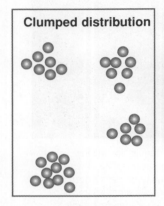

Musk oxen herd

Elephant herd

In clumped (or aggregated) distributions, individuals are grouped in patches (often around a resource). Clumped distributions are the most common type of distribution pattern in nature and are typical of herding and other highly social species, and in environments where resources are patchy.

Uniform distribution

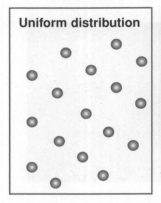

Gannet colony

Allocasuarina in Australia

In uniform (regular) distributions, individuals are evenly spaced and the distance between neighboring individuals is maximized. Uniform distributions occur in territorial species, e.g. breeding colonies of seabirds, but also occur in plants that produce chemicals to inhibit the growth of others nearby.

3. Explain how the behavior of a species might influence the population density: _____

_____

4. What factors might influence the distribution of individuals in their environment? _____

_____

_____

5. What type of distribution pattern would you expect to see when:

(a) Resources are not evenly spread out: _____ (c) Animals are social: _____

(b) Resources are evenly spread out: _____ (d) Animals are territorial: _____

6. Why do you think random distributions are uncommon in nature? _____

_____

©2022 **BIOZONE** International
ISBN: 978-1-98-856692-4
Photocopying Prohibited

119 Species Interactions

Key Question: How do interactions such as predation, competition, and parasitism between species influence the size and distribution of their populations?

Species interact in ways that limit the size of populations

▶ Within **ecosystem**s, each species interacts with others in their community. In many of these interactions, at least one of the parties in the relationship is disadvantaged. Predators eat prey, parasites and pathogens exploit their hosts, and species compete for limited resources. These interactions contribute to the **abiotic** factors that limit the number of organisms in a population and prevent any one population from becoming too large.

▶ Not all relationships involve exploitation. Some species form relationships that are mutually beneficial, e.g. some flowering plants have mutualistic relationships with their pollinating insects. Mutualistic relationships can enable two species to exist in greater numbers than either would alone.

	Type of interaction between species				
Mutualism	**Exploitation**				**Competition**
	Predation	**Herbivory**	**Parasitism**		
A ⇌ B Benefits Benefits	A → B Benefits Harmed	A → B Benefits Harmed	A → B Benefits Harmed		A ⇌ B Harmed Harmed
Both species benefit. **Example**: Flowering plants and their insect pollinators. The flowers are pollinated and the insect gains food. **Population effects**: Flower population spreads by producing seeds. Bees use pollen to make honey and feed larvae, ensuring the hive's survival.	Predator kills the prey outright and eats it. **Examples**: Praying mantis consuming insect prey. Canada lynx consuming snowshoe hare. **Population effects**: Invertebrate predators may control the population numbers of their prey. Numbers of vertebrate predators are often limited by prey availability.	Herbivore eats parts of a plant. Plants often have defences to limit damage. **Example**: Giraffes browsing acacia trees. Browsing stimulates the acacia to produce toxic alkaloids, which cause the giraffe to move to another plant. **Population effects**: Browser damage is self limiting, so the plant is able to recover.	Parasite lives in or on the host, taking all its nutrition from it. The host is harmed but usually not killed. **Examples**: Tapeworm in a pig's gut. **Population effects**: Parasite numbers generally stay at a level that is tolerated by the host. High parasite loads may weaken the host and reduce survival.		Species, or individuals, compete for the same resources. Both parties suffer, especially when resources are limited. **Examples**: Plants growing close to each other compete for light and soil nutrients. **Population effects**: Competition reduces the maximum number of any one species in an area as resources are limited.

Honeybee and flower   Mantid eats cricket   Giraffe browses acacia   Pork tapeworm   Forest plants

1. Plants are not defenceless and they have evolved physical and chemical defences to deter herbivores. In some cases, as in grasses, grazing stimulates growth in the plant.

(a) What is the acacia's response to giraffe browsing? _____

(b) How might this response prevent over-browsing?_____

(c) How might the acacia's adaptations contribute to the sustainability of its browser population? _____

LS2.A

2. Although hyenas attack and kill large animals, such as wildebeest, they will also scavenge carrion or drive other animals off their kills.

   (a) Identify the interaction pictured here: _____

   (b) How many species are involved in the interaction? _____

   (c) Describe how each species is affected by the presence of the other (benefits/harmed/no effect):

   _____

   _____

3. Ticks are obligate blood feeders and must obtain blood to pass from one life stage to the next. Ticks attach to the outside of hosts, in this case a cat, where they suck blood and fluids and cause irritation.

   (a) Identify this type of interaction: _____

   (b) Describe how each species is affected (benefits/harmed/no effect):

   _____

   (c) How would the tick population be affected if the host became rare?

   _____

   _____

NJR ZA CC 3.0

4. It is tempting to assume that large mammalian predators, such as big cats and wolves, control the numbers of their prey species. However, population studies have shown that predator-prey interactions are very complex and the outcomes depend on factors such as the food available to the prey and the availability of alternative prey for the predator. Sometimes, predators can control the numbers of prey, but sometimes the opposite is true.

   How could availability of food for the prey affect the number of predators?

   _____

   _____

5. Many insect predators, e.g. ladybugs, are very effective at limiting the numbers of their invertebrate prey, e.g. aphids. As the prey become more abundant, the predators both take more prey and become more numerous. This is the basis of biological control programs in which an invertebrate predator is introduced to control the numbers of an insect pest. Once the numbers of the pest are reduced, both populations stabilize at a low level.

   Why do you think a successful biological control program relies on the predator having just one prey source (the pest)?

   _____

   _____

   _____

6. Many butterfly species breed within a relatively short time period, laying their eggs on suitable plants where the larvae will hatch and feed. Competition between the larvae will be intense as they all compete for the food available.

   (a) What is the interaction occurring here? _____

   _____

   (b) What is the likely outcome for eggs laid too late in the season?

   _____

   _____

   _____

©2022 **BIOZONE** International
ISBN: 978-1-98-856692-4
Photocopying Prohibited

120 Competition for Resources

Key Question: Why does competition within and between species occur?

▶ No organism exists in isolation. Each organism interacts with other organisms, and with the physical (**abiotic**) components of the environment.

▶ **Competition** occurs when two or more organisms are competing for the same limited resource, e.g. food or space.

▶ The resources available for growth, reproduction, and survival for each competitor are reduced relative to a situation of no competition. Competition, therefore, has a negative effect on both competitors and limits population numbers.

▶ Competition can occur between members of the same species (**intraspecific competition**), or between members of different species (**interspecific competition**).

A complex system of interactions occurs between the different species living on this coral reef in Hawaii. Population numbers will be limited by competition for limited resources, such as food and space on the reef.

Examples of limited resources

Space can be a limited resource
These sea anemones are competing for space in a tidal pool. Some species defend areas, called territories, which have resources they need.

Suitable mates can be hard to find
Within a species, individuals may compete for a mate. These male red deer are fighting to determine which one will mate with the females.

Food is usually a limited resource
In most natural systems, there is competition for food between individuals of the same species, and between different species with similar diets.

1. (a) What is competition? _____

_____

_____

(b) Why does competition occur? _____

_____

_____

(c) Why does competition have a negative effect on both competitors? _____

_____

_____

_____

LS2.A

121 Intraspecific Competition

Key Question: Why does intraspecific competition occur? How does intraspecific competition regulate population size?

- **Intraspecific competition** occurs when individuals of the same species compete for the same limited resources. In addition to food, space, nutrients, and light, intraspecific competition also includes competition for mating partners and breeding sites.

- In most cases, intraspecific competition is more intense that competition between different species because individuals are all competing for the same resources, e.g. same food and mates. It is an important factor in limiting the **population** size of many species.

- Strategies such as territoriality and social hierarchies can reduce the conflict associated with resource competition and can be important in determining which individuals in the population will breed.

How does intraspecific competition limit population size?

- Most resources are limited and this is a major factor in determining how large a population can grow.

- As population numbers increase, the demands on the resources are higher. The resources are used up more quickly and some individuals receive fewer resources than others. High population densities may also increase the influence of population limiting factors such as disease. Populations respond by decreasing their numbers. This occurs by:

- Reduced survival (more individuals die).

- Reduced birth rates (fewer individuals are born).

- If resources increase, e.g. food increases, population numbers can increase. The relationship between resources and population numbers is shown on the right.

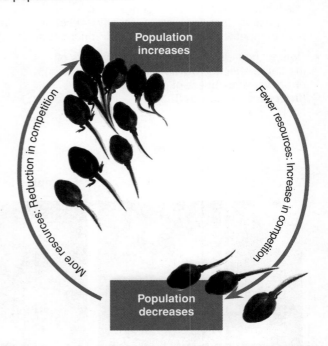

Population increases

Fewer resources: Increase in competition

More resources: Reduction in competition

Population decreases

Scramble competition
Direct competition between members of the same species for a finite resource is called scramble competition. These silkworm caterpillars are all competing for the same food. When it is insufficient, none of the individuals may survive.

Mike Baird

Contest competition
In contest competition, there is a winner and a loser and resources are obtained completely or not at all. For example, male elephant seals fight for territory and mates. Unsuccessful males may not mate at all.

Competition in social species
In some animals, strict social orders ensure that dominant individuals will have priority access to resources. Lower ranked individuals must contest what remains. If food is very limited, only dominant individuals may receive enough to survive.

1. (a) What is intraspecific competition? _____

_____

(b) How does scramble competition differ from contest competition? _____

_____

_____

LS2.A   CE   SPQ

©2022 **BIOZONE** International
ISBN: 978-1-98-856692-4
Photocopying Prohibited

Territories and limitations of population size

▶ Territoriality in birds and other animals is usually a result of intraspecific competition. A territory is a defended area containing the resources required by an individual or breeding pair to survive and reproduce. Territories space organisms out in the **habitat** according to the availability of resources. Those without territories usually do not breed.

▶ In the South American rufous-collared sparrow, males and females occupy small territories (below). These birds make up 50% of the population. The remaining 50% or the population, called floaters, occupy **home ranges** (which are undefended areas) within the territory boundaries. They are tolerated by the territory owners, but these floaters do not breed.

▶ By using tagging studies and removal of birds, researchers found that when a territory owner (male or female) dies or disappears, it is replaced by a floater of the appropriate sex. This is shown for the females in the left diagram as a darker region.

▶ Territoriality can limit population size in some circumstances. If there is no lower limit to territory size and all individuals or pairs gain a territory, then the population becomes spaced out but not limited. However, if territories have a lower size limit, then only a limited number of individuals or pairs can claim a territory. Those that fail to do so must leave and this limits population numbers.

Females

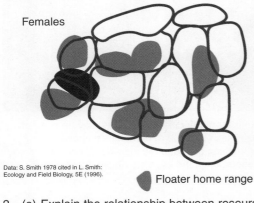

Males

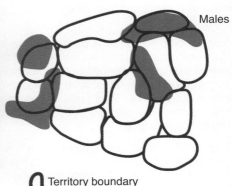

Data: S. Smith 1978 cited in L. Smith:
Ecology and Field Biology, 5E (1996).

⬤ Floater home range

◗ Territory boundary

2. (a) Explain the relationship between resource availability and intraspecific competition: _____

_____

_____

(b) What happens to population numbers as intraspecific competition increases? _____

_____

_____

(c) Intraspecific competition is considered to be a density-dependent process. What do you think this means?

_____

_____

3. Territoriality is a way to ensure that at least some individuals have the resources to survive and reproduce. The territories are established and maintained by direct conflict and by calls and displays.

(a) Identify the benefits of possessing a territory: _____

_____

_____

(b) Can you think of a cost of having a territory? _____

_____

_____

(c) What evidence is there from the rufous-collared sparrow study to show that territoriality can limit population size?

_____

_____

_____

_____

_____

122 Interspecific Competition

Key Question: Why does interspecific competition occur, and how does it affect the species involved?

Interspecific competition involves individuals of different species competing for the same limited resources. They may do this by:

Interfering directly with the ability of others to gain access to the resource.

Exploiting the resource before other individuals can get access to it (exploitative competition).

▸ Interspecific competition is usually less intense than competition between members of the same species because competing species have different requirements for at least some resources, e.g. different **habitat** or food preferences. In other words, their **niches** are different even if they exploit some of the same resources.

▸ Interspecific competition can have a role in limiting a population's size and determining the species present or their **distribution**. However, in naturally occurring populations, it is generally less effective at limiting population size than intraspecific competition, especially in animals. This is because each species usually has alternative resources it can exploit to avoid competition.

▸ Interspecific competition in natural plant communities is very dependent on nutrient availability and will be greater when soil nutrients are low. Fast growing plants with large, dense root systems can absorb large amounts of nitrogen, depleting soil nitrogen so that other plants cannot grow close to them. Similarly, fast growing plants may quickly grow tall enough to intercept the available light and prevent the germination of plants nearby. Pest plants often have this strategy and become very difficult to control.

▸ Sometimes, humans may introduce a species with the same resource requirements as a native species. The resulting competition can lead to the decline of the native species.

In some communities, many different species may be competing for the same resource. This type of competition is called interference competition because the individuals interact directly over a scarce resource. In the example above, three species compete for what remains of a carcass.

The plant species in the forest community above compete for light, space, water, and nutrients. A tree that can grow taller than those around it will be able to absorb more sunlight, and grow more rapidly to a larger size, than the plants in the shade below.

1. (a) What is interspecific competition? Describe an example: _____

_____

_____

(b) Why is interspecific competition usually less intense than intraspecific competition? _____

_____

_____

_____

(c) Why is interspecific competition generally less effective at limiting population size than intraspecific competition?

_____

_____

_____

©2022 **BIOZONE** International
ISBN: 978-1-98-856692-4

Change in distributions of red and gray squirrels in the UK

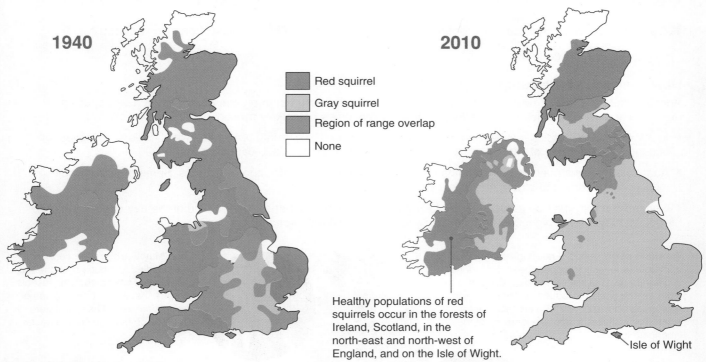

1940

2010

	Red squirrel
	Gray squirrel
	Region of range overlap
	None

Healthy populations of red squirrels occur in the forests of Ireland, Scotland, in the north-east and north-west of England, and on the Isle of Wight.

Isle of Wight

Paul Whippey cc 3.0

Red squirrel

The European red squirrel was the only squirrel species in Britain until the introduction of the American gray squirrel in 1876. Regular distribution surveys (above) have recorded the reducing range of the reds, with the larger, more aggressive gray squirrel displacing populations of reds over much of England. Gray squirrels can exploit tannin-rich foods, which are unpalatable to reds. In mixed woodland and in competition with grays, reds may not gain enough food to survive the winter and breed. Reds are also very susceptible to several viral diseases, including squirrelpox, which is transmitted by grays.

Whereas red squirrels once occupied a range of forest types, they are now almost solely restricted to coniferous forest. The data suggest that the gray squirrel is probably responsible for the red squirrel decline, but other factors, such as **habitat** loss, are also likely to be important.

BirdPhotos.com cc 3.0

Gray squirrel

2. (a) What evidence is there that competition with gray squirrels is responsible for the decline in red squirrels in the UK?

_____

_____

_____

(b) Is the evidence conclusive? If not why not?_____

_____

_____

_____

_____

3. The ability of red and gray squirrels to coexist appears to depend on the diversity of habitat type and availability of food sources (reds appear to be more successful in regions of coniferous forest). Suggest why careful habitat management is thought to offer the best hope for the long term survival of red squirrel populations in Britain:

_____

_____

_____

123 Reducing Competition Between Species

Key Question: How can competition between species with similar resources be reduced?

How species reduce competition

Species exploiting similar resources have adaptations (evolved features) to reduce **competition**. Each species exploits a smaller proportion of the entire spectrum of resources potentially available to it. For example:

▸ In a forest, each species may feed on a different part of a tree, e.g. trunk, branches, twigs, flowers, or leaves, or occupy different areas of vertical air-space, e.g. ground, understorey, sub-canopy, or canopy.

▸ Aquatic organisms may also inhabit different zones to reduce competition for resources. Some organisms will inhabit the bottom, and others may occupy surface waters.

▸ Competition may also be reduced by exploiting the same resources at a different time of the day or year, e.g. one species may feed at night and another may exploit the same resource in the morning.

Warblers of the genus *Setophaga* spend a lot of their time feeding in conifer trees. By feeding in different parts of the tree and on different food sources, the warblers reduce competition. The blue areas shown on the tree indicate areas where the warblers spent 50% or more of their time feeding.

Blackburnian warblers forage in new needles and buds in upper branches of trees. They tend to move horizontally through the tree, feeding on insects or spiders.

Bay-breasted warblers are found on older needles and lichen near middle branches. They feed on insects, particularly the spruce budworm. These birds will also feed on berries and nectar.

Cape May warblers feed around new needles and buds near the top of the tree. They catch flying insects and pick insects up from the tips of conifer branches. These warblers also feed on berry juice and nectar.

Black-throated warblers forage in new needles and buds, as well as older needles and branches in the upper to middle parts of the tree. They feed mainly on insects, sometimes hovering-(gleaning), or catching insects in flight-(hawking). Berries will occasionally be consumed.

Myrtle warblers forage on lower trunks and middle branches. These birds are insectivorous, but will readily take wax-myrtle berries in winter, a habit which gives the species its name. They make short flights in search of insects.

All bird photos Cephas CC 3.0

1. (a) How do the species of warbler avoid competition? _____

_____

(b) What evidence is there that their adaptations for doing this are largely behavioral? _____

_____

2. Cape May warblers and Blackburnian warblers appear to occupy the same habitat in the tree. Explain how they are able to avoid competition and coexist:

_____

_____

_____

   LS2.A CE

©2022 **BIOZONE** International
ISBN: 978-1-98-856692-4
Photocopying Prohibited

Adaptations reduce competition in foraging bumblebees

▶ Studies on bumblebee foraging have shown that when bumblebees forage in the presence of other bumblebee species, they tend to spend the majority of their time on particular flower types. In many cases, the length of the corolla (the length of the flower petals) of flowers visited correlates with the length of the bumblebee's proboscis (mouthparts).

▶ Bumblebee species in the mountains of Colorado (graph right) compete for nectar from flowers. Species with a long proboscis take nectar from flowers with long petals. Species with a short proboscis take nectar from flowers with short petals. This reduces competition for food between the bumblebee species.

Bumblebee species

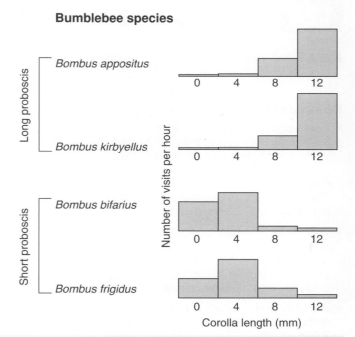

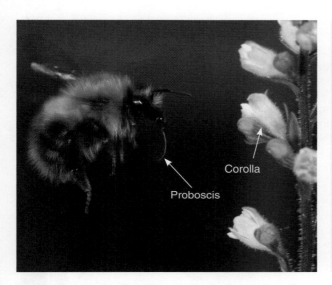

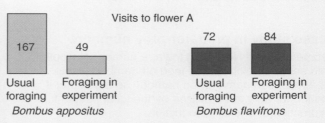

Visits to flower A

Bombus appositus
Usual foraging: 167
Foraging in experiment: 49

Bombus flavifrons
Usual foraging: 72
Foraging in experiment: 84

The bumblebees *Bombus appositus* and *Bombus flavifrons* normally show a preference for particular flower species (call these A and F respectively for reference). However, in the absence of competition, they will forage on either flower species. This was shown in an experiment in which visits of *Bombus appositus* to its usual forage flower A were restricted. *Bombus flavifrons*, which usually forages on flower F, responded by increasing its visits to flower A and decreasing its visits to flower F.

3. (a) How do the *Bombus* species in Colorado reduce competition for flower resources? _____

_____

_____

_____

(b) Are the differences between the Colorado species mainly structural, physiological, or behavioral? Explain:

_____

_____

_____

_____

4. What evidence is there that competition restricts bumblebee species to certain flower types? _____

_____

_____

_____

_____

_____

©2022 **BIOZONE** International
ISBN: 978-1-98-856692-4
Photocopying Prohibited

124 Predator-Prey Relationships

Key Question: Are the populations of predators and prey related and how do they change over time?

Do predators limit prey numbers?

▸ It was once thought that predators always limited the numbers of their prey populations. While this is often true for invertebrate predator-prey systems, prey species are very often regulated more by factors, such as climate and the availability of food, than by **predation**.

▸ In contrast, predator populations can be strongly affected by the availability of prey, especially when there is little opportunity for prey switching, i.e hunting another prey if the preferred one becomes scarce.

▸ Predator and prey populations may settle into a stable oscillations, where the predator numbers follow those of the prey, with a time lag (right).

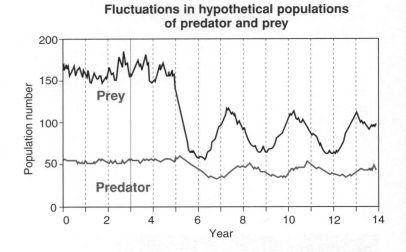

Fluctuations in hypothetical populations of predator and prey

A case study in predator-prey numbers

In some areas of Northeast India, a number of woolly aphid species colonize and feed off bamboo plants. The aphids can damage the bamboo so much that it is no longer able to be utilized by the local people for construction and textile production.

Giant ladybug beetles (*Anisolemnia dilatata*) feed exclusively off the woolly aphids of bamboo plants. There is some interest in using them as biological control agents to reduce woolly aphid numbers, and limit the damage woolly aphids do to bamboo plants.

The graph below shows the relationship between the giant ladybug beetle and the woolly aphid, when grown in controlled laboratory conditions.

Bamboo plants are home to many insect species, including ladybugs and aphids.

Aphids feed off the bamboo sap, and the ladybugs are predators of the aphids (below).

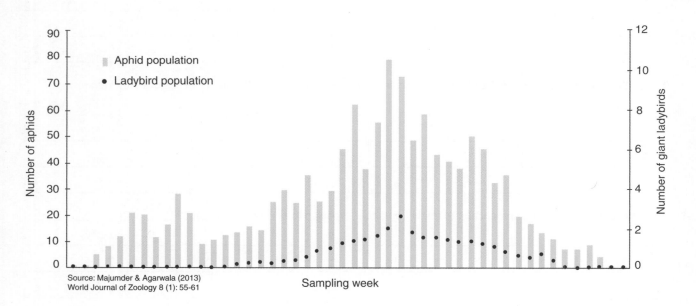

Source: Majumder & Agarwala (2013)
World Journal of Zoology 8 (1): 55-61

Sampling week

1. (a) On the graph above, mark the two points (using different colored pens) where the peak numbers of woolly aphids and giant ladybugs occur.

    LS2.A CE

(b) Do the peak numbers for both species occur at the same time? _____

(c) Why do you think this is? _____

_____

_____

2. (a) What is the response of the ladybug population when their prey decline? _____

_____

(b) Although this was a laboratory situation, what features of the ladybug predator suggest it would be a good choice to control woolly aphids?:

_____

_____

_____

3. A census of a deer population on an island forest reserve indicated a population of 2000 animals in 1960. In 1961, ten wolves (natural predators of deer) were brought to the island in an attempt to control deer numbers. The numbers of deer and wolves were monitored over the next nine years. The results of these population surveys are presented (right).

(a) Plot a line graph for the results. Use one scale (on the left) for numbers of deer and another scale (on the right) for the number of wolves. Use different symbols or colors to distinguish the lines and include a key.

(b) What does the plot show?_____

_____

_____

_____

Island population surveys (1961-1969)		
Year	Wolf numbers	Deer numbers
1961	10	2000
1962	12	2300
1963	16	2500
1964	22	2360
1965	28	2244
1966	24	2094
1967	21	1968
1968	18	1916
1969	19	1952

NEED HELP?
See Activities
17 & 18

(c) Suggest a possible explanation for the pattern in the data: _____

_____

_____

_____

125 The Carrying Capacity of an Ecosystem

Key Question: What does the carrying capacity of an environment mean, and what environmental factors affect an environment's carrying capacity?

▶ The **carrying capacity** is the maximum number of organisms of a given species a particular environment can support indefinitely. An **ecosystem's** carrying capacity, and therefore the maximum population size it can sustain, is limited by its resources and is affected by both **biotic** factors, e.g. food and **abiotic** factors, e.g. water, temperature. It is determined by the most limiting factor and can change over time, e.g. as a result of a change in food availability or a climate shift.

The graph (right) shows how the carrying capacity of a forest environment varies, based on changes to the limiting factors:

1 A population moves into the forest and rapidly increases in numbers due to abundant resources.

2 The population overshoots the carrying capacity.

3 Large numbers damage the environment and food becomes more limited, lowering the original carrying capacity.

4 The population becomes stable at the new carrying capacity.

5 The forest experiences a drought and the carrying capacity is reduced as a result.

6 The drought breaks, and the carrying capacity rises but is less than before because of **habitat** damage during the drought.

Environmental resistance

Limiting factors
Water, space, food

Overshoot

Carrying capacity

Population

Population size

Time

Factors affecting population size

Density dependent factors
The effect of these on population size is influenced by population density. They include:

▶ Competition

▶ Predation

▶ Disease

Density dependent factors tend to be biotic and are less important when population density is low.

They regulate population size by decreasing birth rates and increasing death rates.

Density independent factors
The effect of these on population size does not depend on population density. They include catastrophic events such as:

▶ Volcanic eruption, fire

▶ Drought, flood, tsunami

▶ Earthquake

Density independent factors tend to be abiotic.

They regulate population size by increasing death rates.

1. What is carrying capacity? _____

_____

2. How does carrying capacity limit population numbers? _____

_____

_____

3. What limiting factors have changed at points 3, 5, and 6 in the graph above, and how have they changed?

(a) 3: _____

(b) 5: _____

(c) 6: _____

LS2.A   SPQ

©2022 **BIOZONE** International
ISBN: 978-1-98-856692-4

126 A Case Study in Carrying Capacity

Key Question: How does the environment influence predator-prey interactions?

What happened when wolves were introduced to Coronation Island?

▶ Coronation Island is a small island (116 km²) off the Alaskan coast. A resident black tailed deer **population** had overgrazed the island and, as a result, very little forest understorey remained and many common plant species were absent. The forest was quite open and park-like and not dense like a typical South Alaskan forest. Researchers noted that the deer on the island were smaller than those in other populations and that several died each year from malnutrition.

▶ In 1960, the Alaska Department of Fish and Game released two breeding pairs of timber wolves onto the island. Their aim was to control the black-tailed deer (top right), which had been overgrazing the land.

▶ Initially, wolves fed off the deer (lower right), bred successfully, and deer numbers fell. The island's vegetation began to return and by 1964 the vegetation was quite abundant. However, within a few years the deer population crashed. The wolves ran out of food (deer) and began eating each other, causing a drop in wolf numbers. Within 8 years of the wolves being introduced, only one wolf remained on the island.

▶ By 1983, wolves were absent and the deer were once again abundant.

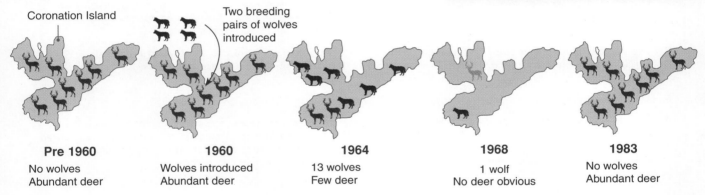

Pre 1960	1960	1964	1968	1983
No wolves Abundant deer	Wolves introduced Abundant deer	13 wolves Few deer	1 wolf No deer obvious	No wolves Abundant deer

Coronation Island

Two breeding pairs of wolves introduced

1. Work in pairs to evaluate the introduction of wolves to Coronation Island. What did it show? Was it a success?

_____

_____

_____

_____

2. If the experiment was carried out at a larger scale on a bigger island with more resources, do you think the outcome would have been different? What is your reasoning?

_____

_____

_____

_____

_____

_____

©2022 **BIOZONE** International
ISBN: 978-1-98-856692-4
Photocopying Prohibited

SPQ LS2.A

127 Carrying Capacity Simulation

Key Question: How does competition for resources limit population growth?

▸ Any environment has limited resources, such as water, food, or nesting sites. Some of these will be continuously renewed, but at a limited rate. This affects how quickly a **population** can grow.

▸ If a new species moves into an **ecosystem** with a small population, at first it may find excess resources and be able to expand quickly, but as resources are used up **population growth** will eventually slow down.

Investigation 6.1   Investigating carrying capacity

See appendix for equipment list.

1. This activity is best done outside for ease of movement.

2. Each student needs a spoon and a cup. For the class, divide 100 dried beans (or peas, beads, or small marbles, etc.) into two sets of 50.

3. Place each set of 50 dried beans onto a separate tray with an upturned edge (so the beans don't roll off), wide enough so that the beans are single layered and have a small amount of space between them.

4. The teacher chooses a student (or has a volunteer) to be the data recorder. They will need a pen or pencil and a copy of Table 1 below.

5. Students arrange themselves in a circle about 10 m in diameter. Place the trays in the center of the circle, but about a meter apart.

6. The teacher will choose one student to start the first round of "feeding". The student places their cup on the ground to mark their position in the circle. The teacher starts the timer and tells the student. This student has 15 seconds to run to either tray and using only their spoon, pick up as many beans as they can. They then run back to their place and tip the beans into their cup (if beans fall off the spoons, they are not picked up until the end of this step). They keep doing this until the 15 seconds is up.

7. The number of beans in the cup are then counted. For each 7 beans in the cup, another student will participate in the next round. The teacher will choose this student. The data recorder notes down the generation (1) and the population of feeding students.

8. All the beans are returned to the trays so that there are again 50 beans on each tray.

9. The second generation of students is then given 15 seconds to collect beans as before. Again, for each 7 beans in a cup another student to added to the next generation (it is not the total number of beans that is counted but the number in each cup e.g. if there are 2 cups and one has 11 beans and the other has 17 beans then 3 extra students are added, not 4). Again, the generation and student number is recorded.

10. This continues for 10 generations (teacher may decide longer if time and beans allow).

11. Repeat the experiment, but this time the teacher will decided to raise or lower the number of beans needed to spawn another feeding student, e.g. 5 or 9 beans.

Table 1

Generation	Number of students
0	1
1	
2	
3	
4	
5	
6	
7	
8	
9	
10	

Table 2

Generation	Number of students
0	1
1	
2	
3	
4	
5	
6	
7	
8	
9	
10	

©2022 **BIOZONE** International
ISBN: 978-1-98-856692-4
Photocopying Prohibited

1. Plot a line graph of the results for both table 1 and 2 on the grid below. Include a key:

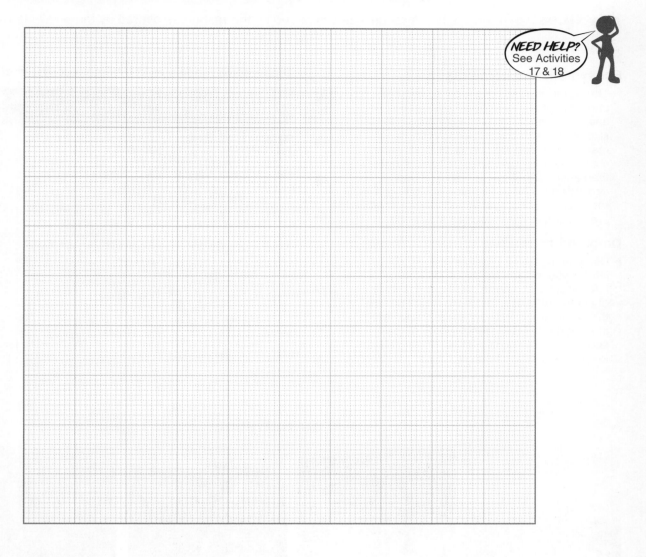

NEED HELP?
See Activities
17 & 18

2. Describe the shape of the graphs: _____

_____

3. Which population grows the fastest? _____

4. What factor is limiting the rate of growth of the populations? _____

5. The population growth rate can be calculated by: **Growth rate = change in population number ÷ change in time** (in this case, generations).

   (a) For table 1, what is the growth rate of the population from generation 0 to 3? _____

   (b) For table 1, what is the growth rate of the population from generation 4 to 7? _____

   (c) For table 1, what is the growth rate of the population from generation 8 to 10? _____

6. (a) What is the approximate carrying capacity for the first experiment? _____

   (b) What is the approximate carrying capacity of the second experiment? _____

7. Explain why the population growth rates for the graphs change the way they do: _____

_____

_____

128 Home Range Size in Dingoes

Key Question: How is home range size influenced by the resources offered by the ecosystem?

Ecosystem and home range

▶ The **home range** is the area in which an animal normally lives and moves about. An animal's home range can vary greatly in size. Animals that live in **ecosystems** rich in resources, e.g. good supply of food, water, shelter, tend to have smaller home ranges than animals that live in resource-poor ecosystems. This is because animals in a resource-poor ecosystem must cover a wider area to obtain the resources they need.

Dingo home ranges

▶ Dingoes are found throughout Australia, in ecosystems as diverse as the tropical rainforests of the north, to the arid deserts of central Australia. The table (right) shows the home range sizes for dingo packs living in a variety of ecosystems. Some of the ecosystems in Australia in which dingoes are found are described below.

Dingo home range size in different ecosystems

	Location (study site)	Ecosystem	Range (km²)
1	Fortescue River, North-west Australia	Semi-arid, coastal plains and hills	77
2	Simpson Desert, Central Australia	Arid, stony and sandy desert	67
3	Kapalga, Kakadu N.P., North Australia	Tropical, coastal wetlands and forests	39
4	Harts Ranges, Central Australia	Semi-arid, river catchment and hills	25
5	Kosciusko N.P., South-east Australia	Moist, cool forested mountains	21
6	Georges Creek N.R., East Australia	Moist, cool forested tablelands (plateaux)	18
7	Nadgee N.R., South-east Australia	Moist, cool coastal forests	10

Australian ecosystems

Arid: Little or no rain, and very dry. Very little, or no, vegetation grows. Often desert regions.

Semi arid: Rainfall is low, but sufficient to support some scrubby vegetation and grasses.

Cool forests: Moderate temperatures. Adequate water and abundant vegetation.

Tropical forests: Warm regions with high rainfall. Abundant lush vegetation, including large trees.

1. Using the information on dingo home range size from the table above:

   (a) Name the two regions where home ranges were largest: _____

   (b) Name the two regions where home ranges were smallest: _____

   (c) Use the information provided on Australian ecosystems to explain how ecosystem type influences the home range of the dingo packs identified in (a) and (b):

   _____

   _____

   _____

   _____

   _____

   _____

©2022 **BIOZONE** International
ISBN: 978-1-98-856692-4
Photocopying Prohibited

129 Population Growth

Population growth

▶ Births, deaths, immigrations (movements into the population), and emigrations (movements out of the population) are events that determine the numbers of individuals in a population. **Population growth** depends on the number of individuals added to the population from births and immigration, minus the number lost through deaths and emigration.

▶ Scientists usually measure the rate of these events. These rates are influenced by environmental factors, such as the availability of resources, and by the characteristics of the organisms themselves.

▶ In population studies, the per capita rate of population increase (also called the **biotic** potential) is often used. Ignoring migration, B – D / N gives the per capita rate of increase (denoted by the italicised letter r).

| Population growth | = B – D + I – E |
| Per capita growth rate, r = B – D / N | |

Population growth curves

▶ The change in population numbers over time is often presented as a population growth curve. Two basic population growth curves exist. Exponential growth (left) is unconstrained by the environment. Logistic growth (following page) is limited by **carrying capacity**.

Exponential growth

Exponential growth occurs when resources are unlimited (this rarely occurs). Growth is extremely rapid, but not sustainable. It produces a J-shaped growth curve.

When speaking of populations, exponential growth is expressed mathematically as: **dN/dt = rN**, where dN is the change in population (N), dt the change in time (t), and r is the population intrinsic (per capita) growth rate (the number of births minus the number of deaths per generation time).

It is an elaboration and rearrangement of the exponential equation $N = e^{rt}$.

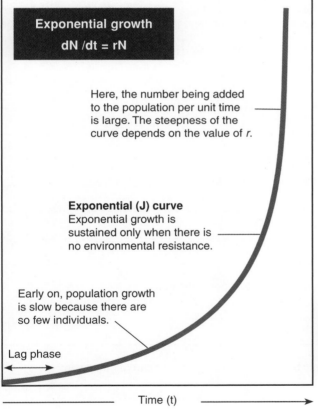

Exponential growth

$$dN/dt = rN$$

Here, the number being added to the population per unit time is large. The steepness of the curve depends on the value of r.

Exponential (J) curve
Exponential growth is sustained only when there is no environmental resistance.

Early on, population growth is slow because there are so few individuals.

Lag phase

Population numbers (N)

Time (t)

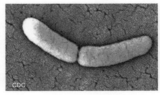

In bacterial populations, each cell divides in two, so the population doubles with every generation. Bacteria growing without environmental constraint show exponential growth.

1. Using the terms, B, D, I, and E (above), construct equations to express the following (the first is completed for you):

   (a) A population in equilibrium: _____B + I = D + E_____

   (b) A declining population: _____

   (c) An increasing population: _____

©2022 **BIOZONE** International
ISBN: 978-1-98-856692-4
Photocopying Prohibited

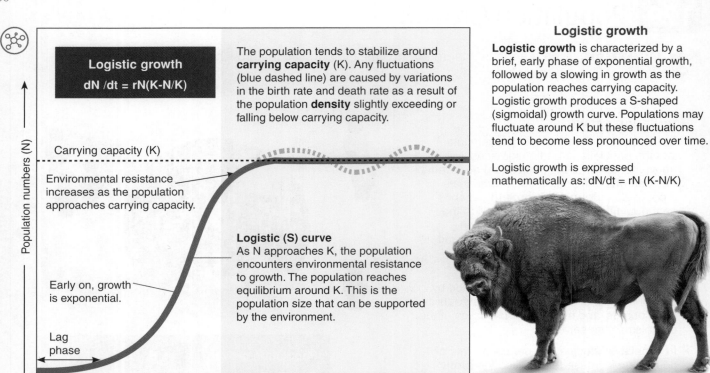

Logistic growth

$$dN/dt = rN(K-N/K)$$

The population tends to stabilize around **carrying capacity** (K). Any fluctuations (blue dashed line) are caused by variations in the birth rate and death rate as a result of the population **density** slightly exceeding or falling below carrying capacity.

Carrying capacity (K)

Environmental resistance increases as the population approaches carrying capacity.

Logistic (S) curve
As N approaches K, the population encounters environmental resistance to growth. The population reaches equilibrium around K. This is the population size that can be supported by the environment.

Early on, growth is exponential.

Lag phase

Population numbers (N)

Time (t)

Logistic growth

Logistic growth is characterized by a brief, early phase of exponential growth, followed by a slowing in growth as the population reaches carrying capacity. Logistic growth produces a S-shaped (sigmoidal) growth curve. Populations may fluctuate around K but these fluctuations tend to become less pronounced over time.

Logistic growth is expressed mathematically as: $dN/dt = rN(K-N/K)$

Populations of large mammals (above) show logistic growth and their populations exist at or near carrying capacity, which is usually determined by primary production (the amount of biomass produced by plants).

2. A population started with a total number of 100 individuals. Over the following year, population data were collected. Calculate birth rates, death rates, net migration rate, and rate of population change for the data below (as percentages):

(a) Births = 14: Birth rate = _____   (b) Net migration = +2: Net migration rate = _____

(c) Deaths = 20: Death rate = _____   (d) Rate of population change = _____

(e) State whether the population is increasing or declining: _____

3. (a) What are the features of exponential growth? _____

_____

(b) Why don't populations continue to increase exponentially in an environment? _____

_____

4. (a) Describe the features of logistic growth: _____

_____

(b) What is environmental resistance and what role does it have in limiting population growth? _____

_____

_____

(c) Explain why a population might overshoot carrying capacity before stabilizing around carrying capacity:

_____

_____

_____

5. What happens to population growth rate as K-N/K approaches 0? _____

_____

_____

©2022 **BIOZONE** International
ISBN: 978-1-98-856692-4
Photocopying Prohibited

A mathematical model of logistic growth

Plotting a logistic growth curve on a spreadsheet can help in understanding the effect of population size on the growth rate and how the logistic equation applies. In this investigation, you will create your own spreadsheet model of logistic growth for a hypothetical population of 2, where r is 0.15 and K 100. You can use Microsoft Excel or an equivalent spreadsheet program.

Investigation 6.2 Creating a model of logistic growth

See appendix for equipment list.

1. In cells **A1** to **F1**, add the headings **r, t(period), N, K, K–N/K,** and **dN/dt**, as shown in the image below.

	A	B	C	D	E	F
1	r_{max}	t (period)	N	K	K-N/K	dN/dt
2	0.15	0	2	100	=($D$2-C2)/$D$2	=$A$2*C2*E2
3		=B2+1	=C2+F2		=($D$2-C3)/$D$2	=$A$2*C3*E3
4		=B3+1	=C3+F3			=$A$2*C4*E4
5		=B4+1	=C4+F4			=$A$2*C5*E5
6		=B5+1	=C5+F5			=$A$2*C6*E6
7						
8						

Population at t_1 = population at t_0 + dN/dt (the amount of population change over 1 time period)

2. In cell **A2**, type **0.15**, the value for r.

3. In cell **B2**, type **0**.

4. In cell **C2**, type **2** (the initial population number).

5. In cell **D2**, type **100** (the carrying capacity).

6. In cell **E2**, type **=($D$2–C2)/$D$2**. This term, K–N/K, is the fraction of the carrying capacity that has not yet been "used up."

7. In cell **F2**, type **=$A$2*C2*E2**. This is the change in population number described by the logistic equation rN(K–N/K).

8. In cell **B3**, type **=B2+1**. In cell **C3**, type **=C2+F2**. Shift–select cells **B3** and **C3** and fill down about 65 cells.

9. Shift–select cells **E2** and **F2** and fill down to about 65 cells also. The first few cells are shown below.

	A	B	C	D	E	F	G	H
1	r_{max}	t (period)	N	K	K-N/K	dN/dt		
2	0.15	0	2.00	100	0.98	0.29		
3		1	2.29		0.98	0.34		
4		2	2.63		0.97	0.38		
5		3	3.01		0.97	0.44		
6		4	3.45		0.97	0.50		

10. When your time series is complete, select the data in columns B (time) and C (Numbers) and choose < **Insert** < **Chart** < **XY scatter** to create a plot of dN/dt.

11. Under **Chart Design** in the menu, you can choose **Add Chart Element** to add axis labels and a title.

6. (a) Describe the shape of the curve you have plotted: _____

   (b) Around which time period does the curve on the spreadsheet above begin to flatten out? _____

   (c) Use the logistic equation and mathematical reasoning to explain the changes in population growth rate (dN/dt):

   _____

   _____

   _____

   _____

   _____

   _____

   _____

130 Plotting Bacterial Growth

Key Question: How does a microbial population grow over time? Can growth be modeled or the population predicted?

Bacterial growth

▶ Bacteria normally reproduce by binary fission, a simple cell division that involves one cell dividing in two.

▶ When actively growing bacteria are inoculated into a liquid growth medium and the population is counted at intervals, a line can be plotted to show the growth of the cell population over time.

NEED HELP?
See Activities
8 & 17

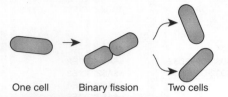

One cell   Binary fission   Two cells

Time (min)	Population size
0	1
20	2
40	4
60	8
80	
100	
120	
140	
160	
180	
200	
220	
240	
260	
280	
300	
320	
340	
360	

1. Complete the table above by doubling the number of bacteria for every 20 minute interval.

2. State how many bacteria were present after: 1 hour: _____ 3 hours: _____ 6 hours: _____

3. Graph the results on the grid above. Make sure that you choose suitable scales and labels for each axis.

4. (a) Predict the number of cells present after 380 minutes: _____

   (b) Plot this value on the graph above:

5. Why is a semi-log graph used to plot microbial growth? _____

_____

   LS2.A

©2022 **BIOZONE** International
ISBN: 978-1-98-856692-4
Photocopying Prohibited

131 Investigating Bacterial Growth

Key Question: How is a spectrophotometer used to measure the growth of microbial populations over time?

Background

The increase in cell numbers in a bacterial culture can be measured indirectly with a spectrophotometer as an increase in culture turbidity.

The aim

To investigate the growth rate of *E.coli* in two different liquid cultures: a minimal growth medium and a nutrient-enriched complex growth medium.

The method

Using aseptic technique, the students added 0.2 mL of a pre-prepared *E.coli* culture to two test tubes, one with 5.0 mL of a minimal growth medium and one with 5.0 mL of a complex medium. Both samples were immediately mixed, and 0.2 mL samples removed from each and added to a cuvette. The absorbance of the sample was measured using a spectrophotometer at 660 nm. This was the "time zero" reading. The test tubes were covered with parafilm, and placed in a 37°C water bath. Every 30 minutes, the test tubes were lightly shaken and 0.2 mL samples were taken from each and the absorbance measured. The results are shown on the right.

A spectrophotometer (left) is an instrument used to measure transmittance of a solution and can be used to quantify bacterial growth, where an increase in cell numbers results in an increase in turbidity.

In this experiment, students measured the absorbance of the solution. Absorbance measures the amount of light absorbed by the sample.

⚠ All bacteria should be treated as pathogenic and strict hygiene practices and aseptic techniques should be followed. This prevents infection or spread of bacteria into the environment.

Results

	Absorbance at 660 nm	
Incubation time (min)	Minimal medium	Complex medium
0	0.021	0.014
30	0.022	0.015
60	0.025	0.019
90	0.034	0.033
120	0.051	0.065
150	0.078	0.124
180	0.118	0.238
210	0.179	0.460
240	0.273	0.698
270	0.420	0.910
300	0.598	1.070

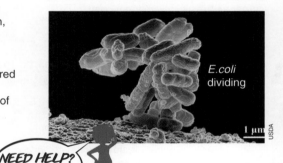

E.coli dividing

1 μm

NEED HELP?
See Activities 9 & 17

1. Why is it important to follow strict hygiene precautions when working with bacteria?

_____

_____

_____

2. (a) On the grid (right) plot the results for *E.coli* growth on the two media:

(b) What is the absorbance measuring? _____

_____

(c) Describe the effect of the complex medium on *E.coli* growth:

_____

_____

_____

(d) Can you see any differences in the shapes of the curves? Can you explain your observations? _____

_____

_____

LS2.A

132 Modeling Population Growth

Key Question: How does changing the starting position of populations affect the way a population grows?

Computer simulations can be used to model population growth

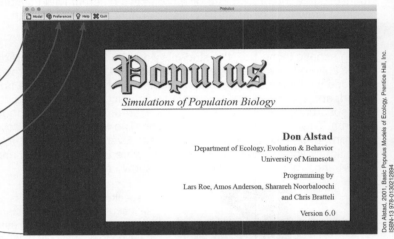

▶ **Population growth** can be simulated using spreadsheets or computer programs. In this investigation, you will use **Populus**, a Javascript program, which will run on Mac or Windows platforms. It models population growth, as well as the effects of **competition**.

▶ In this activity, you will model continuous density-independent (exponential) and density-dependent (logistic) growth.

Populus is shareware. Download it free:
https://cbs.umn.edu/populus/overview

(you can also download via **BIOZONE's Resource Hub**)

The opening screen looks like this.
▶ **Model** allows you to choose which type of simulation you want to run.
▶ **Preferences** lets you to load saved files and save new ones.
▶ **Help** loads a comprehensive PDF file covering all aspects of the program.

If it fills the entire screen, grab the lower corner and resize it with the mouse.

Investigation 6.3 Density independent growth

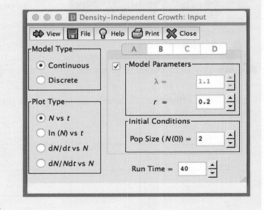

See appendix for equipment list.

1. Click on the Model in the menu bar.

2. Select "Single-Species-Dynamics" and then choose Density-Independent Growth.

3. Set the model type to continuous (as in continuous growth). This produces a single line in the output window.

4. Set plot type to N vs t. This models population vs time.

5. Up to four populations can be displayed on the one graph, using A, B, C, and D. Make sure the check box is ticked.

6. Set r to 0.2, population size N to 2, and run time to 40. Click View to see the graph. What shape is it?

   _____

7. Now increase r to 0.4. What happens to the shape of the graph?

   _____

8. Reset r to 0.2 and increase population size to 20. What happens to the shape of the graph now?

   _____

9. Leave the population size at 20 but decrease r to -0.2. What happens to the shape of the graph now?

   _____

10. Set the parameters back to N = 2 and r = 0.2. Set the plot type to dN/dt vs N and view the plot. Describe the shape of the graph and explain what it means.

   _____

11. Change the r value until the population doubles over one time period. What is the r value? _____

SAVE AND PRINT ALL YOUR SIMULATIONS AND ATTACH THEM TO THIS PAGE

     LS2.A CE SPQ SSM

Investigation 6.4 Density dependent growth (continuous)

See appendix for equipment list.

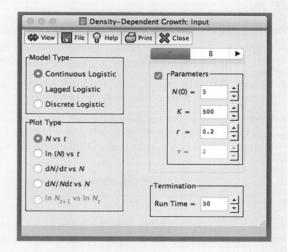

1. Click on the Model in the menu bar.

2. Select "Single-Species-Dynamics" and then choose Density-Dependent Growth.

3. As before, set the model type to continuous.

4. Set plot type to N vs t. This models population vs time.

5. Up to four populations can be displayed on one graph, using A, B, C, and D. Make sure the check box is ticked.

6. Set r to 0.2, population size N to 5, K to 500, and run time (t) to 50.

7. Click View to see the graph. What shape is it?

_____

8. Increase r to 0.4. What happens to the graph now?

_____

9. Reset r to 0.2 and increase population size to 50. What happens to the shape of the graph now?

_____

10. Reset the parameters (step 6). Set the plot type to dN/dt vs N and view the plot. Describe the shape of the graph and explain what it means.

_____

Investigation 6.5 Density dependent growth (continuous)

See appendix for equipment list.

The standard logistic growth curve assumes that population size immediately affects population growth rate. In real populations, there is a lag between change in population size and its effect on growth.

1. Set the graph type to **Lagged Logistic** to introduce a time lag (T). Set the parameters to N = 5, K = 500, r = 0.2, and run time (t) to 50. Set the time lag T to 4 and view the graph. What is the effect of the time lag on population growth?

_____

2. Now set r to 0.5 and t to 150. Describe the shape of the graph: _____

_____

3. What kinds of organisms would show this type of population growth? What features would they have?

_____

4. Keep T at 4 and set r to 0.2. View the graph. Describe the shape of the graph now:

_____

5. What kinds of organisms would show this type of population growth? What features would they have?

_____

6. Keeping r at 0.2, vary T between 1 and 10. What is the effect of increasing the lag?

_____

SAVE AND PRINT ALL YOUR SIMULATIONS AND ATTACH THEM TO THIS PAGE

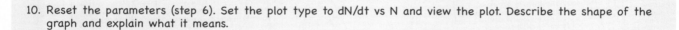

133 A Case Study in Population Growth

Key Question: How do populations fluctuate when a predator is dependent on a single prey species?

Population oscillations in a natural predator-prey system

▶ Snowshoe hares in the boreal forests of North America show cycles of population increase and decrease lasting 8-11 years. At the cycle's peak, they can reach densities that exceed the **carrying capacity** of the environment.

▶ Canada lynx are very dependent on the hares for food and they have little opportunity for prey switching. Consequently, their populations also rise and fall with fluctuations in populations of snowshoe hares.

▶ Regular trapping records of lynx and hare pelts over a 90 year period showed that their population numbers rise and fall with a similar pattern, with lynx numbers lagging behind those of the hare by 3-7 years. The lag is the response time of the lynx population to the change in hare numbers.

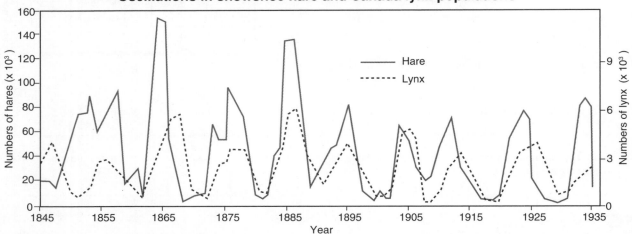

Oscillations in snowshoe hare and Canada lynx populations

Canada lynx are largely solitary and territorial. Territory size increases when food is scarce. They prey almost exclusively on snowshoe hares, which form 60-97% of their diet, and cannot maintain bodyweight successfully on alternative prey.

Snowshoe hares consume a variety of plant material depending on seasonal availability. Hare populations show regular cycles in numbers, with a 2-5 year period of abundance followed by a decline following over-exploitation of food sources.

1. Why is the lynx population so dependent on the fluctuations of the hare? _____

_____

2. (a) Explain how the availability of palatable food might regulate the numbers of hares: _____

_____

_____

(b) Explain how a decline in available palatable food might affect their ability to withstand predation pressure:

_____

_____

    LS2.A CE SPQ

©2022 **BIOZONE** International
ISBN: 978-1-98-856692-4
Photocopying Prohibited

134 Review Your Understanding

Key Question: What causes explosive population growth?

▸ At the beginning of this chapter, you were presented with episodes of mouse plagues in the USA and Australia. Plagues like this happen through the world and often seem to occur from nowhere. What causes them?

▸ You should now be able to explain why these explosions in mouse **populations** occur.

1. Originally you were asked what you thought caused the mouse plague in Taft city 1926? Does your original explanation match what you think now? Has it changed? Write down any changes in your explantation:

_____

_____

_____

2. The diagram below shows the density of various rodent populations throughout the world over time:

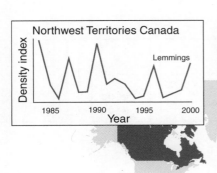

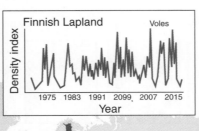

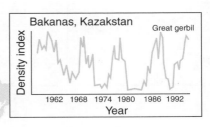

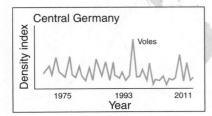

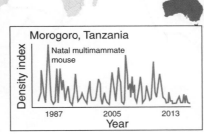

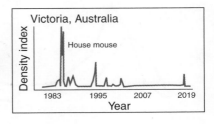

Andreassen, H.P., Sundell, J., Ecke, F. et al. Population cycles and outbreaks of small rodents: ten essential questions we still need to solve. Oecologia 195, 601–622 (2021). https://doi.org/10.1007/s00442-020-04810-w

The graphs all show a similar structure. What factors might be causing the populations to fluctuate as they do?

_____

_____

_____

_____

3. What might happen to the populations of predators that prey on the mice when populations explosions happen?

_____

_____

_____

135 Summing Up

Analyzing model predator-prey system

Mathematical models predict that predator and prey populations will form stable cycles of population increase and decrease. Early ecologists set out to verify these population oscillations in small model **ecosystems**. Two researchers, Gause and Huffaker, each worked on this question. Their results gave great insight into the nature of predatory prey interactions and the factors that control population size.

Gause's experiments

▸ Gause's experiments examined the interactions of two protists, *Paramecium* and its predator *Didinium* in simple test tube 'microcosms'. When *Didinium* was added to a culture of *Paramecium*, it quickly ate all the *Paramecium* and then died out. When sediment was placed in the microcosm, *Paramecium* could hide, *Didinium* died out and the *Paramecium* population recovered.

Huffaker's experiments

▸ Huffaker built on Gause's findings and attempted to design artificial systems that would better model a real world system. He worked on two mite species, the six spotted mite and its predator. Oranges provided both the habitat and the food for the prey.

▸ In a simple system, such as a small number of oranges grouped together, predators quickly ate all the prey and then died out.

▸ Huffaker then created a more complex system with arrays of 120 oranges (below). The amount of available food on each orange was controlled by sealing off parts of each orange with wax. Patchiness in the environment was created using balls (representing unsuitable habitat). Sticks aided dispersal of prey mites and vaseline was used to form barriers for predatory mite dispersal. In this system, the predator and prey coexisted for three full cycles (> year). In the diagram below, the arrays depict the distribution and **density** of the populations at the arrowed points. The circles represent oranges or balls and the dots the predatory mites.

○ Low density prey   ◔ Medium density prey   ● High density prey   ○ No prey

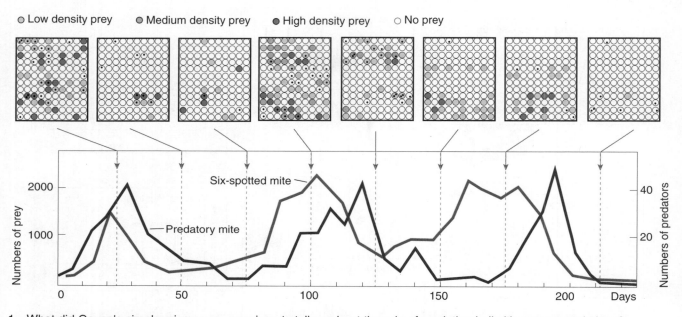

1. What did Gause's simple microcosm experiments tell us about the role of predation in limiting prey populations?

    _____

    _____

2. (a) Mark the three population cycles completed in Huffaker's experiment on the plot above.

    (b) In a different color, mark the lag in the predator population response to change in prey numbers.

    (c) What does the lag represent? _____

    _____

3. How well do you think Huffaker's model system approximated a real ecosystem? Use evidence from the arrays to discuss how variation in habitat makes it possible for populations to persist, despite periodic declines in their numbers.

    _____

    _____

    _____

    _____

©2022 **BIOZONE** International
ISBN: 978-1-98-856692-4
Photocopying Prohibited

136 Eat or be Eaten

▸ Over the time dinosaurs existed, from the Triassic period, 252 million years ago, to the end of the Cretaceous period, 65 million years ago, 66 species of carnivorous dinosaur and 185 species of herbivorous dinosaur were known to have existed in North America.

▸ *Tyrannosaurus rex* was an apex (top) predator of the late Cretaceous period, ending 65 million years ago. It was one of the largest land predators to have ever existed, measuring 12.3 meters long and weighing 8.4 tonnes.

▸ *T. rex* obtained its food by hunting herbivorous dinosaurs, and sometimes members of its own species. The herbivorous dinosaurs dominated the landscape and obtained food by eating a wide variety of plant-based materials such as ferns, horsetails, club-mosses, conifers, cycads, and ginkgos.

1. (a) How do you think we could represent the feeding relationships between the plants, herbivorous dinosaurs, and the carnivorous *T. rex* described above by a simple diagram?

    (b) All life on Earth needs energy to survive. If animals obtain energy from the food they eat, either from plants or by eating other animals, where do you think plants obtain their energy from?

    _____

2. What do you think would happen to the Cretaceous ecosystem described above if the number of plants fell significantly?

    _____

    _____

    _____

©2022 **BIOZONE** International
ISBN: 978-1-98-856692-4
Photocopying Prohibited

137 Energy in Ecosystems

Key Question: Where does the energy needed for essential life processes come from?

Photosynthesis and cellular respiration as the source of energy

▶ As matter and energy move through the biotic and abiotic environments, chemical elements are recombined in different ways. Each transformation results in storage and dissipation of energy into the environment as heat. Matter and energy are conserved at each transformation.

▶ The dissipation of energy as heat means that ecosystems must receive a constant input of new energy from an outside source to sustain themselves. Usually, the Sun is the ultimate source of energy in an ecosystem.

▶ **Photosynthesis** and **cellular respiration** provide most of the usable energy for life's essential processes, such as metabolism and growth. It is important to remember that all organisms, plants included, carry out cellular respiration.

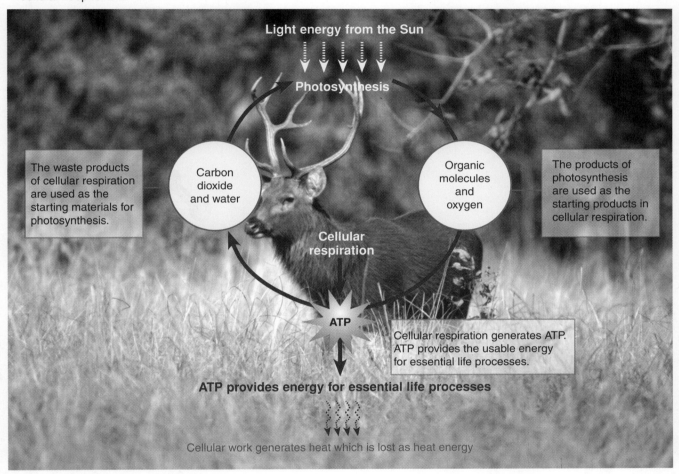

Light energy from the Sun

Photosynthesis

The waste products of cellular respiration are used as the starting materials for photosynthesis.

Carbon dioxide and water

Organic molecules and oxygen

The products of photosynthesis are used as the starting products in cellular respiration.

Cellular respiration

ATP

Cellular respiration generates ATP. ATP provides the usable energy for essential life processes.

ATP provides energy for essential life processes

Cellular work generates heat which is lost as heat energy

1. Why do ecosystems need a constant input of energy from an external source? _____

_____

2. How do photosynthesis and cellular respiration interact to cycle matter through an ecosystem? _____

_____

_____

_____

_____

3. What is the role of ATP in biological systems? _____

_____

_____

EM | LS2.B

138 Comparing Aerobic and Anaerobic Systems

Key Question: What are the differences in how matter cycles through aerobic systems and through anaerobic systems?

▶ Systems that operate **aerobically** (with oxygen) and **anaerobically** (without oxygen) both produce energy for use in living systems. Both systems are important in the cycling of matter through ecosystems.

▶ Most of the ATP made in respiration (aerobic or anaerobic) is generated through the transfer of electrons between electron carriers to a final electron acceptor. The energy released as a result of these transfers is captured in ATP and can be used to power essential chemical reactions in the cell.

An aerobic system: The breakdown of glucose

▶ Aerobic systems use oxygen as the final electron acceptor.

▶ Oxygen accepts electrons and joins with hydrogen to form water. Carbon dioxide is released earlier in the process.

▶ The example below shows **cellular respiration**.

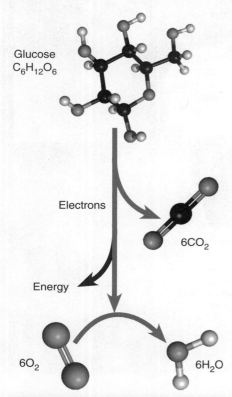

During cellular respiration, a series of chemical reactions breaks down glucose and uses the energy released to produce the energy carrier ATP, that is used to carry out the work of cells. Animals cannot make their own glucose and must obtain it by consuming other organisms.

An anaerobic system: Carbon dioxide to methane

▶ Anaerobic systems use a molecule other than oxygen as the terminal electron acceptor.

▶ There are many different anaerobic pathways that living organisms use to produce usable energy.

▶ Methanogenesis (below) is a form of anaerobic respiration found in methane-producing bacteria in the stomach of ruminants, e.g. sheep and cows.

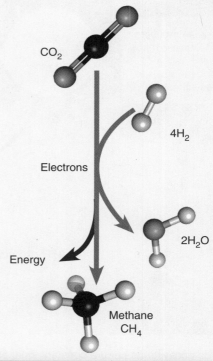

Methane is a greenhouse gas and contributes to global warming. Methane-producing bacteria in ruminants use the cellulose in vegetation as a source of energy and carbon. The methane generated is breathed out. The sheep (above) are fitted with devices to measure the exhaled methane.

    LS2.B EM

©2022 **BIOZONE** International
ISBN: 978-1-98-856692-4
Photocopying Prohibited

Matter cycles through the ecosystem

▶ The total amount of matter in a closed system is conserved. The Earth is, effectively, a closed system. This means that Earth's fixed supply of nutrients must be recycled to sustain life. Nutrients move between the various compartments on Earth: the atmosphere (air), hydrosphere (water), soils, and living organisms.

▶ Carbon, hydrogen, nitrogen, sulfur, phosphorus, and oxygen move through ecosystems in cycles, although the cycles may intersect and interact at various points. Energy drives the cycling of matter within and between systems.

▶ The rate of nutrient cycling can vary widely. Some nutrients, e.g. phosphorus, are cycled slowly. Others, such as nitrogen, are cycled more quickly. The type of environment and diversity of an ecosystem can also have a large effect on the rate at which nutrients are cycled.

Cycling nitrogen

Nitrogen must be in a form that can be utilized by plants. Some bacteria, called nitrogen fixers, can convert nitrogen gas to nitrate. Other bacteria convert nitrates back into nitrogen gas. Both processes are anaerobic.

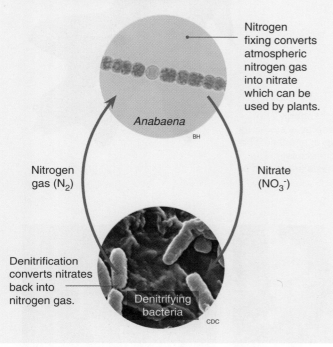

Nitrogen fixing converts atmospheric nitrogen gas into nitrate which can be used by plants.

Anabaena
BH

Nitrogen gas (N_2)

Nitrate (NO_3^-)

Denitrification converts nitrates back into nitrogen gas.

Denitrifying bacteria
CDC

Cycling carbon and oxygen

Photosynthesis and cellular respiration link the cycling of carbon and oxygen. Aerobic respiration, the conversion of glucose and O_2 into CO_2 and water, is part of the carbon and oxygen cycles. The CO_2 and water are converted back into glucose and O_2 by the anaerobic process of photosynthesis.

Cellular respiration (aerobic)

Oxygen and glucose

Carbon dioxide and water

Photosynthesis (anaerobic)

1. What is the main difference between an aerobic and anaerobic system? _____

_____

2. How are aerobic and anaerobic systems involved in nutrient cycling? _____

_____

_____

_____

3. Describe how nutrient cycles can interact in the cycling of matter: _____

_____

_____

_____

4. Compost heaps rely on decomposer organisms to break down plant material into a nutrient-rich, soil-like humus. During the process, the ratio of carbon to nitrogen (C:N) in the heap decreases. Explain why:

_____

_____

_____

139 Producers

Key Question: How do producers (autotrophs) make their own food?

What is a producer?

▶ A **producer** is an organism that can make its own food. Most producers utilize the energy from the sun to do this, but some organisms use chemical energy.

▶ Producers are also called autotrophs, which means self feeding.

▶ Plants, algae, and some bacteria and protists are producers.

▶ Most producers are photoautotrophs, and use the energy in sunlight to make their food. The process by which they do this is called **photosynthesis**.

▶ Some producers are chemoautotrophs and use the chemical energy in inorganic molecules, e.g. hydrogen sulfide, to make their food.

Photosynthesis transforms sunlight energy into chemical energy. The chemical energy is stored as glucose (a sugar), and the energy is released when this undergoes further metabolic processes. The inputs and outputs of photosynthesis are shown on the leaf diagram (right).

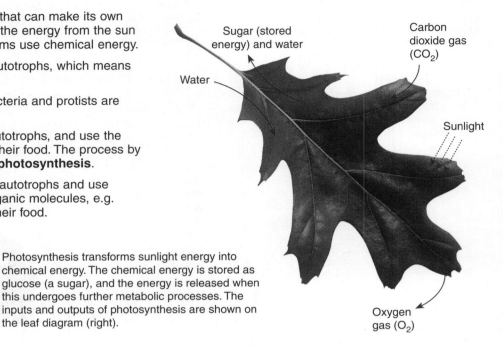

Sugar (stored energy) and water

Water

Carbon dioxide gas (CO_2)

Sunlight

Oxygen gas (O_2)

Photosynthesis by marine algae provides oxygen and absorbs carbon dioxide. Most algae are microscopic but some, like this kelp, are large.

On land, vascular plants (plants with transport tissues) are the main producers of food.

Producers, e.g. grasses, make their own food, and are also the ultimate source of food and energy for consumers, such as these cows.

1. (a) What is a producer? _____

   _____

   (b) Name some organisms that are producers: _____

   _____

2. Where do producers get their energy from? _____

   _____

3. Why are producers so important in an ecosystem? _____

   _____

   _____

   _____

LS2.B EM

©2022 **BIOZONE** International
ISBN: 978-1-98-856692-4
Photocopying Prohibited

140 Consumers

Key Question: How do consumers (heterotrophs) obtain their food?

What is a consumer?

Consumers (heterotrophs) are organisms that cannot make their own food and must get their food by consuming other organisms (by eating or extracellular digestion). Animals, fungi, and some bacteria are consumers. Consumers (herbivores, carnivores, decomposers) are categorized according to where they get their energy from.

Consumers need producers

Consumers rely on producers for survival, even if they do not consume them directly. Herbivores, such as the rabbit below, gain their energy by eating plants. Although higher level consumers, such as the eagle, may feed off herbivores, they still ultimately rely on plants to sustain them. Without the plants, the rabbit would not survive, and the eagle could not eat it.

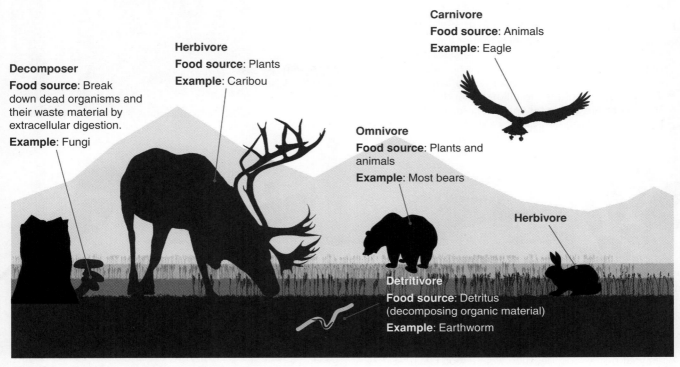

Decomposer
Food source: Break down dead organisms and their waste material by extracellular digestion.
Example: Fungi

Herbivore
Food source: Plants
Example: Caribou

Carnivore
Food source: Animals
Example: Eagle

Omnivore
Food source: Plants and animals
Example: Most bears

Herbivore

Detritivore
Food source: Detritus (decomposing organic material)
Example: Earthworm

Schematic showing the movement of energy and matter from producers to consumers

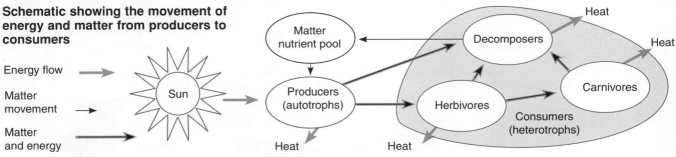

Energy flow →

Matter movement →

Matter and energy →

Sun → Producers (autotrophs) → Matter nutrient pool → Decomposers → Herbivores → Carnivores

Consumers (heterotrophs)

Heat

1. (a) What is a consumer? _____

_____

(b) Consumers are categorized into different groups. What is this categorization based on? _____

_____

2. Where do consumers get their energy from? _____

3. In the schematic system above, how are the movements of matter and energy different? _____

_____

EM   LS2.B

141 Food Chains

Key Question: How do food chains model the feeding relationships between organisms?

Food chains

▸ Organisms in ecosystems interact through their feeding (or trophic) relationships. These interactions can be shown in a **food chain**, which is a simple model to illustrate how energy and matter, in the form of food, pass from one organism to the next. Each organism in the chain is a food source for the next.

Trophic levels

▸ The levels of a food chain are called **trophic** (feeding) **levels**. An organism is assigned to a trophic level based on its position in the food chain. Organisms may occupy different trophic levels in different food chains or during different stages of their life.

▸ Arrows link the organisms in a food chain. The direction of the arrow shows the flow of energy and matter through the trophic levels. At each link, energy is lost from the system as heat. This loss of energy limits how many links can exist. Most food chains begin with producers, which use the energy in sunlight to make their own food. Therefore, sunlight is the ultimate source of energy for life on Earth. Producers are eaten by primary consumers (herbivores). Secondary (and higher level) consumers eat other consumers, as shown below.

Great white heron spears a fish to eat

350z33 CC 3.0

Producers	**Primary consumers** **Herbivores**	**Secondary consumers** **Omnivores & carnivores**	**Tertiary consumers** **Omnivores & carnivores**
Trophic level: 1	Trophic level: 2	Trophic level: 3	Trophic level: 4

Mike baird-WP

1. In your own words, describe what a food chain is: _____

_____

_____

2. (a) A simple food chain for a cropland ecosystem is pictured below. Label the organisms with their trophic level and trophic status, e.g. primary consumer.

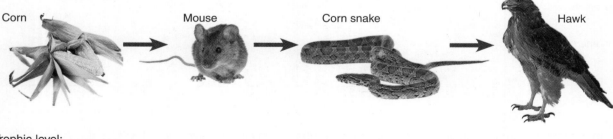

Corn   Mouse   Corn snake   Hawk

Trophic level: _____ _____ _____ _____

Trophic status: _____ _____ _____ _____

(b) What is the ultimate energy source for both the food chains pictured on this page? _____

(c) Why are there rarely more than five or six links in a food chain? _____

_____

 | LS2.B | EM |

©2022 **BIOZONE** International
ISBN: 978-1-98-856692-4
Photocopying Prohibited

142 Food Webs

Key Question: How can we show the complex feeding relationships between all the organisms in a community?

▶ If we show all the connections between all the food chains in an ecosystem, we can create a web of feeding interactions called a **food web**. A food web is a model to show who eats what, in a community.

▶ Food webs are simplified representations of real ecosystems and frequently do not (or cannot) show all the interactions occurring in a real system. The flow of energy through the trophic linkages is in one direction, as opposed to the cyclic flow of matter. The biomass (mass of biological material) in the system represents stored energy and decreases from the base of each food chain to the top because energy is lost to the environment with each transfer. A simplified food web for a lake ecosystem is shown below.

A simple food web for a lake ecosystem

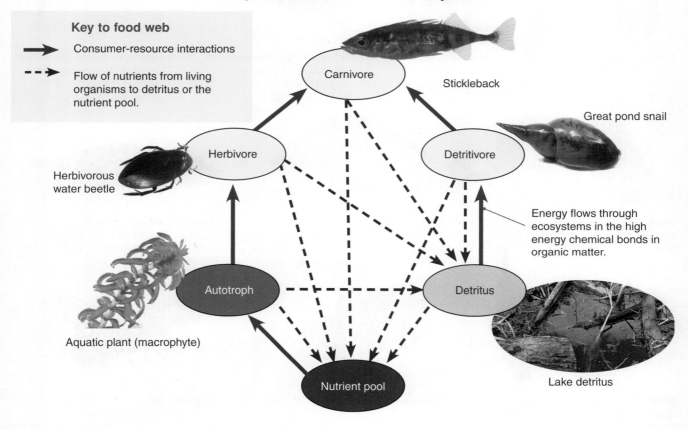

Key to food web

→ Consumer-resource interactions

⇢ Flow of nutrients from living organisms to detritus or the nutrient pool.

Carnivore — Stickleback

Great pond snail

Herbivore

Detritivore

Herbivorous water beetle

Autotroph

Detritus

Energy flows through ecosystems in the high energy chemical bonds in organic matter.

Aquatic plant (macrophyte)

Nutrient pool

Lake detritus

1. What is a food web? _____

_____

2. Why would an ecosystem with only a few different types of organism have a less complex food web than an ecosystem with many different types of organism?

_____

_____

3. Why is it difficult to accurately predict the effect on the lake ecosystem of removing one trophic element (producer or consumer) from this simplified food web?

_____

_____

_____

©2022 **BIOZONE** International
ISBN: 978-1-98-856692-4
Photocopying Prohibited

EM LS2.B

143 Constructing Food Webs

Key Question: How can we use information from food chains to construct a food web?

Knowing what the inhabitants of an ecosystem feed on allows **food chains** to be constructed. Food chains can be used to construct a **food web**. The organisms below are typical of those found in many lakes. For simplicity, only a few organisms are represented here. Real lake communities have hundreds of different species interacting together. Your task is to assemble the organisms below into a food web in a way that shows how they are interconnected by their feeding relationships.

Autotrophic protists (algae)
Chlamydomonas (above left), and some diatoms (above right) are photosynthetic.

Macrophytes
Aquatic green plants are photosynthetic.

Protozoan (e.g. Paramecium)
Diet: Mainly bacteria and microscopic green algae such as *Chlamydomonas*.

Asplanchna (zooplankton)
A large, carnivorous rotifer.
Diet: Protozoa and young zooplankton, e.g. *Daphnia*.

Daphnia (zooplankton)
Small freshwater crustacean.
Diet: Planktonic algae.

Common carp
Diet: Mainly feeds on bottom-living insect larvae and snails, but will also eat some plant material (not algae).

Three-spined stickleback
Common in freshwater ponds and lakes. **Diet**: Small invertebrates such as *Daphnia* and insect larvae.

Diving beetle (adults and larvae)
Diet: Aquatic insect larvae and adult insects. They will also scavenge from detritus. Adults will also take fish fry.

Herbivorous water beetle
Diet: Adults feed on macrophytes. Young beetle larvae are carnivorous, feeding primarily on pond snails.

Dragonfly larva
Large aquatic insect larvae.
Diet: Small invertebrates including *Hydra*, *Daphnia*, insect larvae, and leeches.

Great pond snail
Diet: Omnivorous. Main diet is macrophytes but will eat decaying plant and animal material also.

Leech
Fluid feeding predators.
Diet: Small invertebrates, including rotifers, small pond snails, and worms.

Pike
Diet: Smaller fish and amphibians. They are also opportunistic predators of rodents and small birds.

Mosquito larva
Diet: Planktonic algae.

Hydra
A small, carnivorous cnidarian.
Diet: small *Daphnia* and insect larvae.

Detritus
Decaying organic matter (includes bacterial decomposers).

  LS2.B EM

©2022 **BIOZONE** International
ISBN: 978-1-98-856692-4
Photocopying Prohibited

1. From the information provided for the lake food web components on the previous page, construct twelve different food chains to show the feeding relationships between the organisms. Some food chains may be shorter than others and most species will appear in more than one food chain. An example has been completed for you.

Example 1: Macrophyte ——————▶ Herbivorous water beetle ——————▶ Carp ——————▶ Pike

(a) _____

(b) _____

(c) _____

(d) _____

(e) _____

(f) _____

(g) _____

(h) _____

(i) _____

(j) _____

(k) _____

(l) _____

2. Use the food chains you created above to help you to draw up a food web for this community in the box below. Use the information supplied on the previous page to draw arrows showing the flow of energy between species; only energy from (not to) the detritus is required.

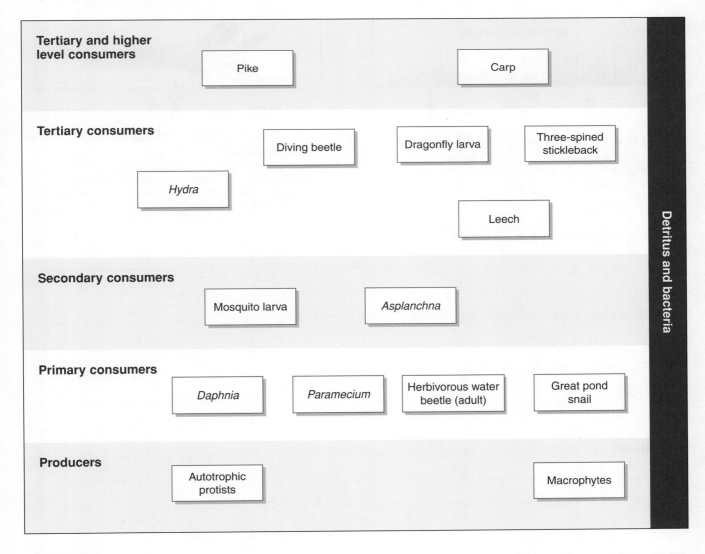

Tertiary and higher level consumers

Pike   Carp

Tertiary consumers

Diving beetle   Dragonfly larva   Three-spined stickleback

Hydra

Leech

Secondary consumers

Mosquito larva   Asplanchna

Primary consumers

Daphnia   Paramecium   Herbivorous water beetle (adult)   Great pond snail

Producers

Autotrophic protists   Macrophytes

Detritus and bacteria

144 Energy Inputs and Outputs

Key Question: What is the difference in energy inputs and outputs in producers and consumers?

▶ The total amount of energy captured by **photosynthesis** is the gross primary production. Net primary production is the amount of energy available to herbivores after **respiration** losses.

▶ The gross primary production (GPP) of any ecosystem will depend on the capacity of the producers to capture light energy and fix carbon in organic compounds.

▶ The net primary production (NPP) is then determined by how much of the GPP goes into plant biomass after the respiratory needs of the producers are met. This will be the amount available to the next **trophic level.**

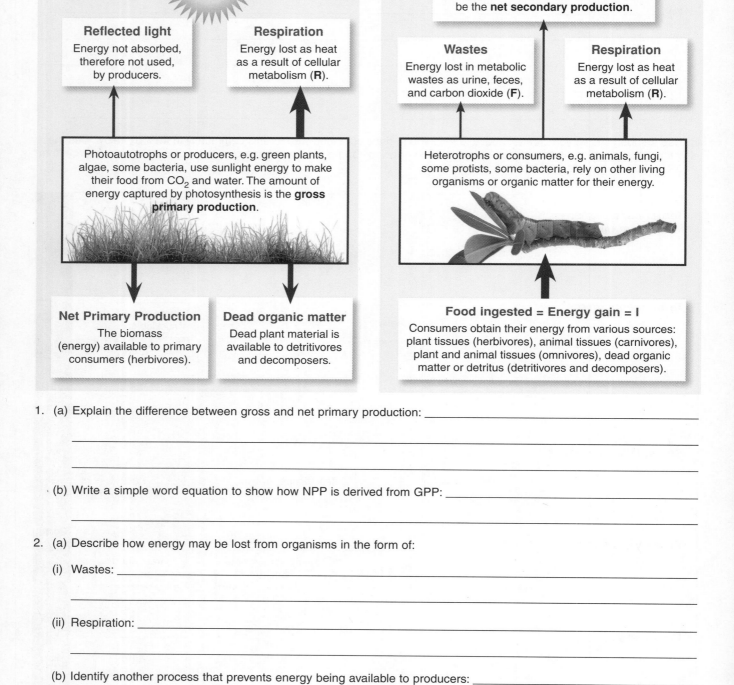

Reflected light
Energy not absorbed, therefore not used, by producers.

Respiration
Energy lost as heat as a result of cellular metabolism (**R**).

Net production
Biomass (energy) available to next trophic level (**N**). For herbivores, this amount will be the **net secondary production**.

Wastes
Energy lost in metabolic wastes as urine, feces, and carbon dioxide (**F**).

Respiration
Energy lost as heat as a result of cellular metabolism (**R**).

Photoautotrophs or producers, e.g. green plants, algae, some bacteria, use sunlight energy to make their food from CO_2 and water. The amount of energy captured by photosynthesis is the **gross primary production**.

Heterotrophs or consumers, e.g. animals, fungi, some protists, some bacteria, rely on other living organisms or organic matter for their energy.

Net Primary Production
The biomass (energy) available to primary consumers (herbivores).

Dead organic matter
Dead plant material is available to detritivores and decomposers.

Food ingested = Energy gain = I
Consumers obtain their energy from various sources: plant tissues (herbivores), animal tissues (carnivores), plant and animal tissues (omnivores), dead organic matter or detritus (detritivores and decomposers).

1. (a) Explain the difference between gross and net primary production: _____

_____

_____

(b) Write a simple word equation to show how NPP is derived from GPP: _____

_____

2. (a) Describe how energy may be lost from organisms in the form of:

(i) Wastes: _____

_____

(ii) Respiration: _____

_____

(b) Identify another process that prevents energy being available to producers: _____

_____

©2022 **BIOZONE** International
ISBN: 978-1-98-856692-4
Photocopying Prohibited

145 Energy Flow in Ecosystems

Key Question: How does energy flow through an ecosystem?

Conservation of energy and trophic efficiency

▸ Energy flows through an ecosystem from one **trophic level** to the next. Only 5-20% of energy is transferred from one trophic level to the next.

▸ The Law of Conservation of Energy states that energy cannot be created or destroyed, only transformed from one form, e.g. light energy, to another, e.g. chemical energy, in the bonds of molecules.

▸ Each time energy is transferred (as food) from one trophic level to the next, some energy is given out as heat, usually during **cellular respiration**. This means the amount of energy available to the next trophic level is less than at the previous level.

▸ Potentially, we can account for the transfer of energy from its input, as solar radiation, to its release, as heat from organisms because energy is conserved. The percentage of energy transferred from one trophic level to the next is the trophic efficiency. It varies between 5% and 20% and measures the efficiency of energy transfer. An average figure of 10% trophic efficiency is often used. This is called the ten percent rule.

Calculating available energy

The energy available to each trophic level will equal the amount entering that trophic level, minus total losses from that level (energy lost as heat + energy lost to detritus).

Heat energy is lost from the ecosystem to the atmosphere. Other losses become part of the detritus and may be utilized by other organisms in the ecosystem.

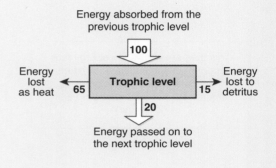

Energy absorbed from the previous trophic level

100

Energy lost as heat ← 65 | **Trophic level** | 15 → Energy lost to detritus

20

Energy passed on to the next trophic level

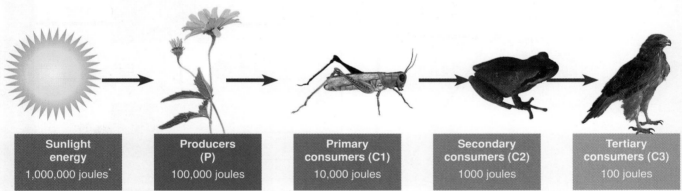

Sunlight energy	Producers (P)	Primary consumers (C1)	Secondary consumers (C2)	Tertiary consumers (C3)
1,000,000 joules*	100,000 joules	10,000 joules	1000 joules	100 joules

*Note: joules are units of energy

1. Why is the energy available to a particular trophic level less than the energy in the previous trophic level? _____

_____

_____

_____

2. Why must ecosystems receive a continuous supply of energy from the Sun? _____

_____

_____

3. (a) What is trophic efficiency? _____

_____

   (b) In general, how much energy is transferred between trophic levels? _____

©2022 **BIOZONE** International
ISBN: 978-1-98-856692-4
Photocopying Prohibited

EM SSM LS2.B

Quantifying energy transfer and trophic efficiency in an ecosystem

4. Identify the process occurring at each of points 1-4 on the diagram.

1: _____

2: _____

3: _____

4: _____

5. Calculate the amount of energy transferred to each trophic level. Write your answers in the spaces labeled (a)-(d):

Sunlight falling on plant surfaces
7,000,000

Light absorbed by plants
1,700,000

1
Producers
87,400

22,950   50,450

a   E = _____

This diagram shows the energy flow through a hypothetical ecosystem. Numbers represent kilojoules of energy per square meter per year (kJ/m²/yr)

4600   **Primary (1°) consumers**   7800

b _____ **2** E = _____

4

Energy imported into the system.   2000   **Detritus**   180   **Secondary (2°) consumers**   1330

Energy exported out of the system.   10,465

c _____ E = _____

19,300

100

Tertiary (3°) consumers   **3** 55

d _____

Decomposers and detritivores feeding on each other.   **Decomposers and detritivores**   19,200

Energy lost to the system as heat

6. (a) Calculate the percentage of light energy that is absorbed when it falls on the plants:

   Light absorbed by plants ÷ sunlight falling on plant surfaces x 100: _____

   (b) Calculate the percentage of absorbed light energy that is converted (fixed) into producer energy:

   Producers ÷ light energy x 100: _____

   (c) What percentage of light energy is absorbed but not fixed? _____

   (d) Account for the difference between the amount of energy absorbed and the amount fixed by producers: _____

   _____

   _____

7. Calculate the percentage efficiency of transfer between each trophic level. Write each figure on the diagram in the spaces provided (E =). Which transfer is the most efficient?

   _____

©2022 **BIOZONE** International
ISBN: 978-1-98-856692-4
Photocopying Prohibited

146 Ecological Pyramids

Key Question: How can the number of organisms, amount of energy, or amount of biomass at each trophic level be represented in an ecosystem?

The energy, biomass, or numbers of organisms at each trophic level in any ecosystem can be represented by an **ecological pyramid**. The first trophic level is placed at the bottom of the pyramid and subsequent trophic levels are stacked on top in their "feeding sequence". Ecological pyramids provide a convenient model to illustrate the relationship between different trophic levels in an ecosystem.

▶ Pyramid of numbers shows the numbers of individual organisms at each trophic level.

▶ Pyramid of biomass measures the mass of the biological material at each trophic level.

▶ Pyramid of energy shows the energy contained within each trophic level. Pyramids of energy and biomass are usually quite similar in appearance.

▶ This generalized ecological pyramid (right) shows a conventional pyramid shape, with a large number of producers at the base, and decreasing numbers of consumers at each successive trophic level.

▶ Ecological pyramids for this plankton-based ecosystem have a similar appearance, regardless of whether we construct them using energy, or biomass, or numbers of organisms.

▶ Units refer to biomass or energy. The images provide a visual representation of the organisms present.

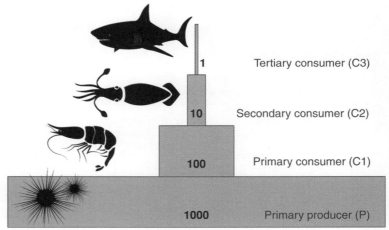

Not to scale

1. (a) What major group is missing from the pyramid above? _____

   (b) Explain the significance of this group in a food web: _____

   _____

   _____

There are benefits and disadvantages to each type of pyramid.

▶ A pyramid of numbers provides information about the number of organisms at each level, but it does not account for their size, which can vary greatly. For example, one large producer might support many small consumers.

▶ A pyramid of biomass is often more useful because it accounts for the amount of biological material at each level. The number of organisms at each level is multiplied by their mass to produce biomass.

▶ While number and biomass pyramids provide information about an ecosystem's structure, a pyramid of energy provides information about function, i.e. how much energy is fixed, lost, and available to the next trophic level.

C3	Weasels
C2	Birds
C1	Insects
P	Trees

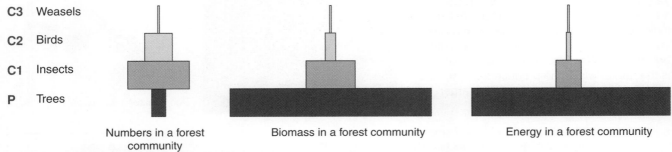

Numbers in a forest community   Biomass in a forest community   Energy in a forest community

2. What is the advantage of using a biomass or energy pyramid, rather than a pyramid of numbers, to express the relationship between different trophic levels?

   _____

   _____

   _____

©2022 **BIOZONE** International
ISBN: 978-1-98-856692-4
Photocopying Prohibited

EM SSM LS2.B

A pyramid of numbers can sometimes be inverted

In some ecosystems, e.g. a forest ecosystem (right), a few large producers can support all the organisms at the higher trophic levels. This is due to the large size of the producers, e.g. a large tree can support many individual consumer organisms. A pyramid of energy for this system would be a conventional pyramid shape because the energy of the producers is enough to support the consumer levels.

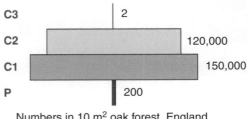

C3 — 2
C2 — 120,000
C1 — 150,000
P — 200

Numbers in 10 m² oak forest, England

3. The forest community above has relatively few producers. How can it support a large number of consumers?

_____

_____

_____

_____

4. The table (below) shows the number of organisms at each trophic level of a grassland community.

   (a) Draw the pyramid of numbers for this data in the space below:

Numbers in a grassland community	
Trophic level	**Number of organisms**
Producer	1,500,000
Primary consumer	200,000
Secondary consumer	90,000
Tertiary consumer	1

   (b) Do you think the pyramid of energy would be similar or different? Explain your answer: _____

   _____

   _____

   _____

5. Would a pyramid of energy ever have an inverted (upturned) shape? Explain your reasoning: _____

   _____

   _____

An unusual biomass pyramid

You can see from the previous examples that biomass and energy pyramids usually look very similar. However, in some ecosystems, a single (one-time) measure of biomass may indicate that a smaller, producer biomass is supporting a larger, consumer biomass (figure below right). What this pyramid does not show is the rate at which the producers (algae) are reproducing in order to support the larger biomass of consumers.

Phytoplankton (algae)   NOAA   Zooplankton (copepods)

6. Give a possible explanation of how a small biomass of producers (algae) can support a larger biomass of consumers (zooplankton):

   _____

   _____

   _____

   _____

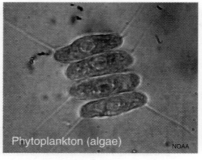
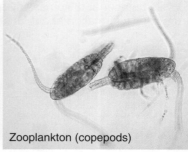

Zooplankton and bottom fauna — 21 g/m²

Algae — 4 g/m²

Biomass

147 Investigating Ecological Pyramids

Key Question: What patterns do we see in ecological pyramids of real-world examples?

Investigation 7.1   Exploring biomass pyramids

See appendix for equipment list.

1. You can work individually or in pairs for this investigation. It makes use of HHMI's online interactive module "Exploring **Biomass Pyramids**". The module is based on real research from an aquatic ecosystem in Panama (Mary Power, 1984). The work examined the ecology of armored catfish (Ancistris sp.) in the Rio Frijoles. These small fish browse algae growing on the substrate. In this investigation, you will collect and analyze data from a virtual river to construct pyramids of energy and biomass. The investigation includes embedded questions, which you will answer in order to proceed.

Aquarium specimen of armored catfish (*Ancistris* species) showing suckered mouth

2. Access the interactive module via BIOZONE's Resource Hub or by typing www.biointeractive.org/classroom-resources/exploring-biomass-pyramids.

3. Launch the interactive from the button on the left hand corner of the screen. Read through the introduction, then click the LAUNCH FIELD STUDY button.

4. The next screen will invite you to explore the pools of the Rio Frijoles. Once you have done that, you can commit to a pool using COMMIT TO POOL button at the bottom of the screen.

5. Follow the on-screen instructions to make a prediction about the shape of the biomass pyramid for this ecosystem. Once you have done this, move on to sample the algal community and quantify its biomass, and then count the catfish and quantify their biomass.

1. Do your calculations from the investigation support your original prediction? Explain: _____

_____

_____

2. Continue with the interactive to run the trophic simulator and examine the productivity of algae over a longer period of time. What does the pyramid of biomass look like now?

_____

_____

3. You will be asked to summarize your findings. Paraphrase your summary below: _____

_____

_____

_____

_____

_____

4. If you wish, continue the interactive session to explore how algal productivity is affected by the amount of sunlight reaching the pond and how this affects the number of consumers that can be supported. At the end of the interactive session, you can generate a report. Attach your report to this page.

EM   LS2.B

148 Cycles of Matter

Key Question: How does matter cycle through the biotic and abiotic compartments of Earth's ecosystems?

Nutrients cycle through ecosystems

▶ Nutrient cycles move and transfer chemical elements, e.g. carbon, hydrogen, nitrogen, and oxygen, through an ecosystem. Because these elements are part of many essential nutrients, their cycling is called a nutrient cycle, or a biogeochemical cycle. The term biogeochemical means that biological, geological, and chemical processes are involved in nutrient cycling.

▶ In a nutrient cycle, the nutrient passes through the biotic (living) and abiotic (physical) components of an ecosystem (see diagram below). Recall that energy drives the cycling of matter within and between systems. Matter is conserved throughout all these transformations, although it may pass from one ecosystem to another.

Processes in a generalized biogeochemical cycle

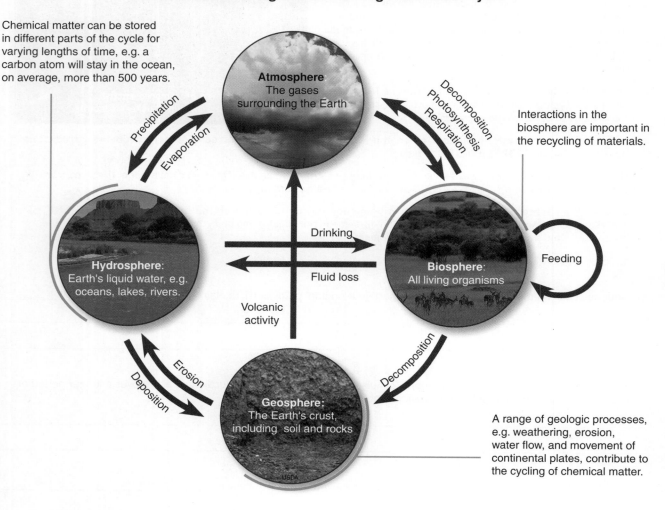

Chemical matter can be stored in different parts of the cycle for varying lengths of time, e.g. a carbon atom will stay in the ocean, on average, more than 500 years.

Atmosphere: The gases surrounding the Earth

Precipitation

Evaporation

Decomposition
Photosynthesis
Respiration

Interactions in the biosphere are important in the recycling of materials.

Drinking

Fluid loss

Feeding

Hydrosphere: Earth's liquid water, e.g. oceans, lakes, rivers.

Biosphere: All living organisms

Volcanic activity

Erosion

Deposition

Decomposition

Geosphere: The Earth's crust, including soil and rocks

A range of geologic processes, e.g. weathering, erosion, water flow, and movement of continental plates, contribute to the cycling of chemical matter.

1. What is a nutrient cycle? _____

_____

_____

2. Why do you think it is important that matter is cycled through an ecosystem? _____

_____

_____

_____

©2022 **BIOZONE** International
ISBN: 978-1-98-856692-4
Photocopying Prohibited

149 The Hydrologic Cycle

Key Question: What processes cycle water around the biosphere, atmosphere, hydrosphere, and geosphere?

▸ The **hydrologic cycle** results from the cycling of water from the oceans to the land and back. About 97% of the water on Earth is stored in the oceans, which contain more than 1.3 billion cubic kilometers of water. Less than 1% of Earth's water is freely available fresh water, in lakes and streams.

▸ Water evaporates from water bodies into the atmosphere and falls as precipitation, e.g. rain. Precipitation onto land is transported back to the oceans by rivers and streams or returned to the atmosphere by evaporation or transpiration (evaporation from plant surfaces).

▸ Water can cycle very quickly if it remains near the Earth's surface, but in some circumstances it can remain locked away for hundreds or even thousands of years, e.g. in deep ice layers at the poles or in groundwater.

▸ Humans intervene in the water cycle by using water for their own needs. Irrigation from rivers and lakes changes evaporation patterns, lowers lake levels, and reduces river flows.

Water is the only substance on Earth that can be found naturally as a solid, liquid, or gas. It has the unique property of being less dense as a solid than a liquid, causing water to freeze from the top down and therefore float, and it has an unexpectedly high boiling point compared to other, similar molecules.

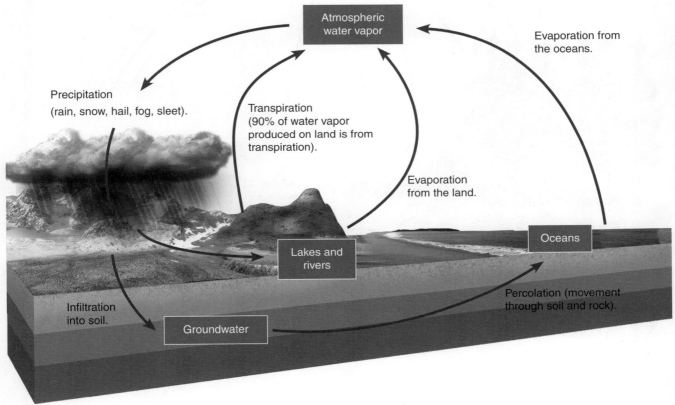

1. What is the main storage reservoir for water on Earth? _____

2. Explain the two processes by which water moves from the land or oceans to the atmosphere: _____
   _____
   _____

3. Identify the feature of water that allows it to cycle and be available for photosynthesis and respiration: _____
   _____

150 The Carbon Cycle

Key Question: How does carbon cycle between the atmosphere, biosphere, geosphere, and hydrosphere?

▸ All life is carbon-based. Carbon cycles between the atmosphere, biosphere, geosphere, and hydrosphere. Photosynthesis and respiration are central to this.

▸ Carbon is the essential element of life. Its unique properties allow it to form an almost infinite number of different molecules. In living systems, the most important of these are carbohydrates, fats, nucleic acids, and proteins.

▸ Carbon in the atmosphere is found as carbon dioxide (CO_2). In rocks, it is most commonly found as either coal (mostly carbon) or limestone (calcium carbonate).

▸ The most important processes in the **carbon cycle** are photosynthesis and respiration.

▸ **Photosynthesis** removes carbon from the atmosphere and converts it to organic molecules. This organic carbon may eventually be returned to the atmosphere through respiration.

▸ Carbon cycles at different rates depending on where it is. On average, carbon remains in the atmosphere as CO_2 for about 5 years; in plants and animals for about 10 years; and in oceans for about 400 years. Carbon can remain in rocks, e.g. coal, for millions of years.

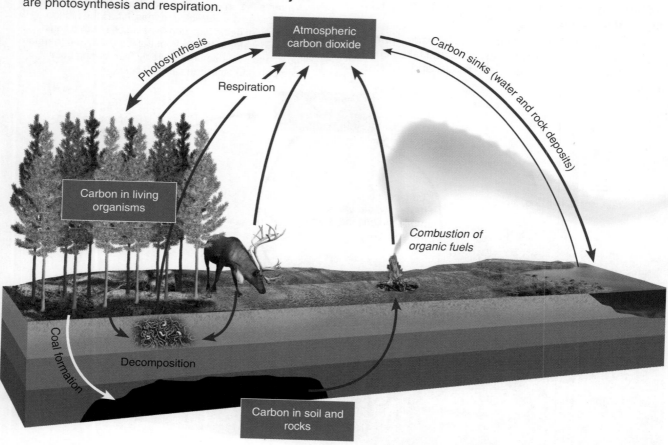

1. (a) In what form is carbon found in the atmosphere? _____

   (b) In what three important molecules is carbon found in living systems? _____

   (c) In what two forms is carbon found in rocks? _____

2. (a) Name two processes that remove carbon from the atmosphere: _____

   (b) Name two processes that add carbon to the atmosphere: _____

3. What is the effect of deforestation, and burning of coal and oil on carbon cycling? _____

   _____

   _____

   _____

   LS2.B PS3.D SSM

©2022 **BIOZONE** International
ISBN: 978-1-98-856692-4
Photocopying Prohibited

151 Modeling the Carbon Cycle

Key Question: How can a simple model be used to represent the carbon cycle?

Investigation 7.2   A model of the carbon cycle

See appendix for equipment list.

> ⚠ **Living organisms should be handled with care and respect.**

1. In this investigation, you will make a simple ecobottle to model the **carbon cycle** in a small closed ecosystem. Your group will be provided with the following equipment: A large, clear soda bottle with a lid; filtered pond water; aquarium gravel; a source of detritus, e.g. dead leaves; aquatic plant (such as Cabomba); small pond snails.

2. Use the equipment to set up a bottle ecosystem. You will need to think about how long you wait before you close the system off, how much air gap you will have, how much organic material you will add, and where you will put your ecosystem (light/dark).

3. Draw a picture of your bottle ecosystem or take a photograph and attach (right). Label the picture to include important design features.

4. Leave your bottle ecosystem for a week. Observe it carefully at various times during the week. After a week, note down any changes since you set it up.

5. Return any living organisms back to the aquarium and dispose of any waste materials.

1. (a) What produces the O_2 in your system?

   _____

   (b) What produces the CO_2 in your system?

   _____

2. The pond water contains small microorganisms. What is their role in this system?

   _____

   _____

3. Was your system stable? Explain why (or why not):

   _____

   _____

   _____

   _____

4. In the space provided, draw a simple diagram to show how carbon cycles between the aquatic plant and the animals:

 EM   SSM   CE   PS3.D   LS2.B

152 The Oxygen Cycle

The importance of the oxygen cycle

▸ The oxygen cycle describes the movement of oxygen (O_2) through an ecosystem. The oxygen cycle is closely linked to the carbon cycle. The **oxygen cycle** describes the movement of oxygen between the biotic and abiotic components of ecosystems. **Photosynthesis** is the main source of oxygen in ecosystems.

▸ Oxygen is involved to some degree in all the other biogeochemical cycles, but is closely linked to the carbon cycle in particular. This is because most producers utilize carbon dioxide in photosynthesis, and produce oxygen as a waste product. The oxygen is used in **cellular respiration** and carbon dioxide is produced as a waste product. This is the oxygen-carbon dioxide cycle, simplified in the diagram (right).

The link between the oxygen and carbon cycles

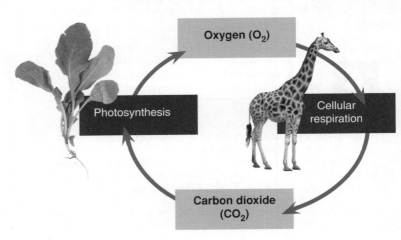

Oxygen (O_2)

Photosynthesis

Cellular respiration

Carbon dioxide (CO_2)

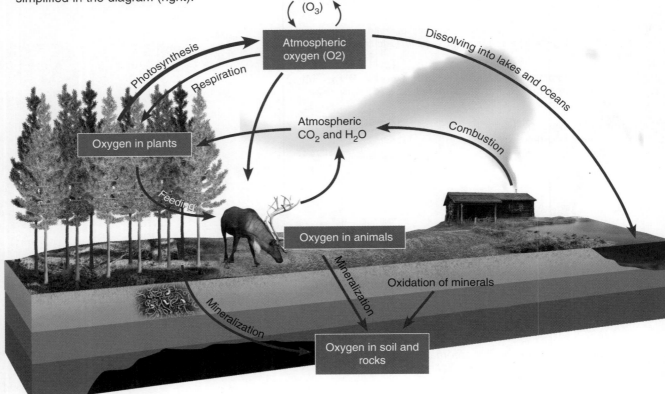

Ozone (O_3)

Atmospheric oxygen (O2)

Photosynthesis

Respiration

Dissolving into lakes and oceans

Oxygen in plants

Atmospheric CO_2 and H_2O

Combustion

Feeding

Oxygen in animals

Mineralization

Oxidation of minerals

Mineralization

Mineralization

Oxygen in soil and rocks

1. What is the main source of oxygen for the oxygen cycle? _____

_____

2. Why are the oxygen cycle and carbon cycle so interdependent? _____

_____

_____

_____

_____

   LS2.B SSM EM

©2022 **BIOZONE** International
ISBN: 978-1-98-856692-4
Photocopying Prohibited

153 Role of Photosynthesis in Carbon Cycling

Key Question: What role do the processes of photosynthesis and respiration play in carbon cycling?

Photosynthesis and carbon

▶ **Photosynthesis** removes carbon from the atmosphere and adds it to the biosphere. Photosynthesis removes carbon from the atmosphere by fixing the carbon in CO_2 into carbohydrate molecules. Plants use the carbohydrates, e.g. glucose, to build structures such as wood.

▶ Some carbon may be returned to the atmosphere during **respiration,** either from the plant or from animals. If the amount or rate of carbon fixation is greater than that released during respiration, then carbon will build up in the biosphere and be reduced in the atmosphere (diagram, right).

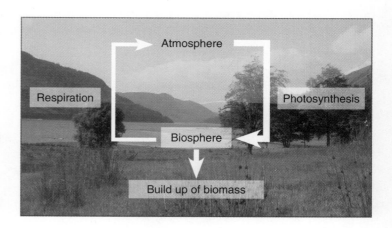

Respiration and carbon

▶ Respiration removes carbon from the biosphere and adds it the atmosphere. Cellular respiration releases carbon into the atmosphere as carbon dioxide as a result of the breakdown of glucose.

▶ If the rate of carbon release is greater than that fixed by photosynthesis then, over time, carbon may accumulate in the atmosphere (diagram bottom right). Before the Industrial Revolution, many thousands of gigatonnes (Gt) of carbon were contained in the biosphere of the Earth's crust, e.g. as coal.

▶ Deforestation and the burning of fossil fuels have increased the amount of carbon in the atmosphere.

Carbon cycling simulation

Plants move about 120 Gt of carbon from the atmosphere to the biosphere a year. Respiration accounts for about 60 Gt of carbon a year. A simulation was carried out to study the effect of varying the rates of respiration and photosynthesis on carbon deposition in the biosphere or atmosphere. To keep the simulation simple, only the effects to the atmosphere and biosphere were simulated. Effects such as ocean deposition and deforestation were not studied. The results are shown in the tables (right and below).

Table 1: Rate of photosynthesis equals the rate of respiration.

Years	Gt carbon in biosphere	Gt carbon in atmosphere
0	610	600
20	608	600
40	608	600
60	609	598
80	612	598
100	610	596

Table 2: Rate of photosynthesis increases by 1 Gt per year.

Years	Gt carbon in biosphere	Gt carbon in atmosphere
0	610	600
20	632	580
40	651	558
60	671	538
80	691	518
100	710	498

Table 3: Rate of cellular respiration increases by 1 Gt per year.

Years	Gt carbon in biosphere	Gt carbon in atmosphere
0	610	600
20	590	619
40	570	641
60	548	664
80	528	686
100	509	703

SSM PS3.D LS2.B

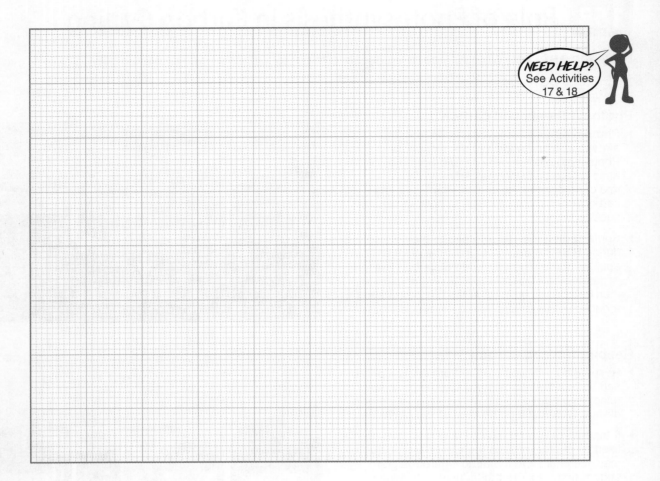

1. Plot the data for tables 1, 2, and 3 on the grid provided (above). Include a key, and appropriate titles and axes.

2. (a) What is the effect of increasing the rate of photosynthesis on atmospheric carbon?

   _____

   (b) i. What is the effect of increasing the rate of photosynthesis on biospheric carbon? _____

   _____

   ii. How does this effect occur? _____

   _____

   _____

3. What is the effect of increasing the rate of cellular respiration on atmospheric and biospheric carbon? _____

   _____

   _____

4. In the real world, respiration is not necessarily increasing in comparison to photosynthesis, but many human activities cause the same effect.

   (a) Name two human activities that have the same effect on atmospheric carbon as increasing the rate of cellular respiration:

   _____

   (b) What effect does this extra atmospheric carbon have on the global climate? _____

   _____

©2022 **BIOZONE** International
ISBN: 978-1-98-856692-4
Photocopying Prohibited

154 The Nitrogen Cycle

Key Question: How is nitrogen conserved as it moves through an ecosystem?

▸ Nitrogen is essential for building proteins. Nitrogen gas is converted to nitrates, which are taken up by plants. Animals gain nitrogen by feeding off plants or animals. Nearly eighty percent of the Earth's atmosphere is made of nitrogen gas. As a gas, nitrogen is very stable and unreactive, effectively having no interaction with living systems. However, nitrogen is extremely important in the formation of amino acids, which are the building blocks of proteins.

▸ Nitrogen may enter the biosphere during lightning storms. Lightning produces extremely high temperatures in the air (around 30,000°C). At such high temperatures, nitrogen reacts with oxygen in the air to form ammonia and nitrates which dissolve in water and are washed into the soil.

▸ Some bacteria can fix nitrogen directly from the air. Some of these bacteria are associated with plants (especially legumes) and produce ammonia (NH_3). This can be converted to nitrates (NO_3^-) by other bacteria. Other bacteria produce nitrites (NO_2^-).

▸ Nitrates are absorbed and used by plants to make amino acids. Animals gain their nitrogen by feeding on plants (or on herbivores).

▸ Nitrogen is returned to the atmosphere by denitrifying bacteria which convert nitrates back into nitrogen gas.

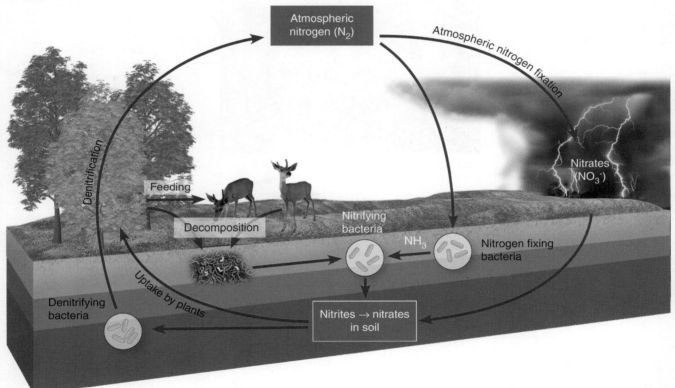

1. Name two processes that fix atmospheric nitrogen: _____

   _____

2. What process returns nitrogen to the atmosphere? _____

3. What essential organic molecule does nitrogen help form? _____

4. Why do farmers often plant legumes between cropping seasons, then plow them into the soil rather than harvest them?

   _____

   _____

   _____

5. Where do animals get their nitrogen from? _____

   _____

©2022 **BIOZONE** International
ISBN: 978-1-98-856692-4
Photocopying Prohibited

155 Review Your Understanding

Key Question: How did energy and matter move through ecosystems when dinosaurs were the dominant species?

1. Dinosaurs were heterotrophs that relied on other living organisms or organic particulate matter for their energy, including herbivores and carnivores. Study the diagram below, on energy flow relating to **heterotrophs**. Explain how the activities of autotrophs (plants) and heterotrophs enable the flow of energy in an ecosystem:

_____

_____

_____

_____

_____

Respiration
Heat given off from metabolic activity.

Wastes
Metabolic waste products are released, e.g. as urine, feces, carbon dioxide.

Dead tissue
Available to detritivores and decomposers.

Heterotrophs

Recreation of a young carnivorous Utahraptor chasing a reptile.

Growth and new offspring
New offspring as well as growth and weight gain.

Eaten by carnivores
Some tissue eaten by carnivores and omnivores.

Food

2. Describe how energy may be lost from dinosaurs in the form of:

   (a) Wastes: _____

   (b) Respiration: _____

3. Explain why so little energy is available for dinosaur growth and reproduction, regardless of trophic group: _____

_____

_____

_____

4. Plants require a number of elements, including oxygen, carbon, hydrogen, and nitrogen. Complete the chart below to list the form of the element that the plant uses (molecule/compound); the original location of the element, the part of the plant where the element enters, and the plant processes that use each element:

Element	Molecules entering plant	Origin of molecule	Part of plant where molecule enters	Main plant processes using each molecule
Oxygen				
Carbon				
Hydrogen				
Nitrogen				Incorporated into amino acids to build structural protein in plant tissue and chlorophyll. Used in many processes, including photosynthesis.

©2022 **BIOZONE** International
ISBN: 978-1-98-856692-4

156 Summing Up

The gross primary production (GPP) of any ecosystem will be determined by the efficiency with which solar energy is captured by photosynthesis. The efficiency of subsequent energy transfers will determine the amount of energy available to consumers. These energy transfers can be quantified using measurements of dry mass.

Production vs productivity: What's the difference?

Strictly speaking, the primary production of an ecosystem is distinct from its productivity, which is the amount of production per unit time (a rate). However, because values for production (accumulated biomass) are usually given for a certain period of time in order to be meaningful, the two terms are often used interchangeably.

Corn field

Mature pasture

In this activity, you will calculate energy and biomass transfers in real and experimental systems.

1. The energy budgets of two agricultural systems (4000 m² area) were measured over a growing season of 100 days. The results are tabulated (right).

   (a) For each system, calculate the percentage efficiency of energy utilization (how much incident solar radiation is captured by photosynthesis):

   Corn: _____

   Mature pasture: _____

   (b) For each system, calculate the percentage losses to respiration:

   Corn: _____

   Mature pasture: _____

   (c) For each system, calculate the percentage efficiency of NPP:

   Corn: _____

   Mature pasture: _____

   (d) Which system has the greatest efficiency of energy transfer to biomass? _____

	Corn field	Mature pasture
	kJ x 10⁶	kJ x 10⁶
Incident solar radiation	8548	1971
Plant utilization		
Net primary production (NPP)	105.8	20.7
Respiration (R)	32.2	3.7
Gross primary production (GPP)	138.0	24.4

Estimating NPP in *Brassica rapa*

Background

Brassica rapa (right) is a fast growing brassica species that can complete its life cycle in as little as 40 days, if growth conditions are favorable. A class of students wished to estimate the gross and net primary productivity of a crop of these plants using wet and dry mass measurements made at three intervals, over 21 days.

The method

▸ Seven groups of three students each grew 60 *B. rapa* plants in plant trays under controlled conditions. On day 7, each group made a random selection of 10 plants and removed them, with roots intact. The 10 plants were washed, blotted dry, and then weighed collectively (giving wet mass).

▸ The 10 plants were placed in a ceramic drying bowl and placed in a drying oven at 200°C for 24 hours, then weighed (giving dry mass).

▸ On day 14 and again on day 21, the procedure was repeated with a further 10 plants (randomly selected).

▸ The full results for group 1 are presented in Table 1 on the next page. You will complete the calculation columns.

EM LS2.B

Table 1: Group 1's results for growth of 10 *B. rapa* plants over 21 days

Age in days	Wet mass of 10 plants (g)	Dry mass of 10 plants (g)	Percent biomass	Energy in 10 plants (kJ)	Energy per plant (kJ)	NPP (kJ/plant/day)
7	19.6	4.2				
14	38.4	9.3				
21	55.2	15.5				

2. Calculate percent biomass using the equation: % biomass = dry mass ÷ wet mass x 100. Enter the values in Table 1.

3. Each gram of dry biomass is equivalent to 18.2 kJ of energy. Calculate the amount of energy per 10 plants and per plant for plants at 7, 14, and 21 days. Enter the values in Table 1.

4. Calculate the Net Primary Productivity per plant, i.e. the amount of energy stored as biomass per day (kJ/plant/d). Enter the values in Table 1. We are using per plant in this exercise as we do not have a unit area of harvest.

5. The other 6 groups of students completed the same procedure and, at the end of the 21 days, the groups compared their results for NPP. The results are presented in Table 2, right.

   Transfer group 1's NPP results from Table 1 to complete the table of results and calculate the mean NPP for *B. rapa*.

Table 2: Class results for NPP of *B. rapa* over 21 days

Time in days (d)	Group NPP (kJ/plant/day)							
	1	2	3	4	5	6	7	Mean NPP
7		1.05	1.05	1.13	1.09	1.13	1.09	
14		1.17	1.21	1.25	1.21	1.25	1.17	
21		1.30	1.34	1.30	1.34	1.38	1.34	

6. On the grid, plot the class mean NPP vs time.

7. (a) What is happening to the NPP over time?

   _____

   _____

   (b) Explain why this is happening: _____

   _____

   _____

   _____

   _____

   _____

   _____

8. What would you need to know to determine the gross primary productivity of *B. rapa*?

   _____

   _____

   _____

9. Net production in consumers (N), or secondary production, can be expressed as N = I - (F+R). Red meat contains approximately 700 kJ per 100 grams. If N = 20% of the energy gain (I), how much energy is lost as F and R?

   _____

10. Devise a methodology and calculations to determine the net secondary production and respiratory losses of 12 day old cabbage white caterpillars feeding on brussels sprouts for 3 days. Begin with 10 caterpillars and ~30 g brussels sprouts. How would you calculate the efficiency of energy transfer from producers to consumers? What assumptions are you making in your calculations?
    Staple your methodology to this page. You will need to know:

    Energy value of plant material: dry mass x 18.2 kJ.
    Energy value of animal material: dry mass x 23.0 kJ
    Energy value of egested waste (frass): mass x 19.87 kJ

NEED HELP?
See Activity
144

©2022 **BIOZONE** International
ISBN: 978-1-98-856692-4
Photocopying Prohibited

CHAPTER 8

The Dynamic Ecosystem

Activity number

Anchoring Phenomenon

A mammoth task: How could bringing back the mammoth help restore a lost ecosystem?   157 171

What is a dynamic ecosystem?

☐ 1 Distinguish between small or large scale events that may permanently affect an ecosystem. Describe the relationship between ecosystem stability, resilience, and biodiversity.   158

☐ 2 Analyze data from coral ecosystems to explain factors affecting resilience. Discuss the effect of overharvesting on the resilience of an ecosystem. Use a model of spruce budworm and balsam fir population to discuss ecosystem resilience.   159 160

☐ 3 Using evidence from a number of examples, define keystone species and explain their significance on the stability of an ecosystem.   161

☐ 4 Discuss the permanent impact of extreme changes on an ecosystem. Use information from a model to evaluate ecosystem change caused by the eruption of Mount St. Helens.   162

What effects do human impacts have on the ecosystem?

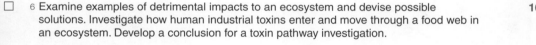

☐ 5 Analyze evidence, in the form of data and case studies, to make claims about the cause of climate change. Use evidence of climate change impacts from a Florida Everglades case study to investigate current and possible future effects to the ecosystem and its biodiversity.   163

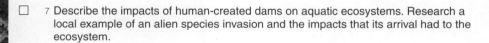

☐ 6 Examine examples of detrimental impacts to an ecosystem and devise possible solutions. Investigate how human industrial toxins enter and move through a food web in an ecosystem. Develop a conclusion for a toxin pathway investigation.   164

☐ 7 Describe the impacts of human-created dams on aquatic ecosystems. Research a local example of an alien species invasion and the impacts that its arrival had to the ecosystem.   165 166

☐ 8 Analyze a range of information types to identify how fish stocks become over-exploited in an ecosystem. Use a model to investigate how humans can impact fish stocks due to fishing activity. Evaluate the effectiveness of fish farming as a solution for overfishing.   167 168

☐ 9 Understand the effects of deforestation due to human activity. Describe the conflicting aims of human requirements and conservation, using a case study of the spotted owl in the Pacific Northwest.   169

☐ 10 Model a solution to a hypothetical case study in a bird breeding reserve, considering economic, cultural, and environmental values.   170

☐ 11 Consider the consequences of reintroducing mammoths into an ecosystem.   171

☐ 12 Consider and respond to a claim about the future impacts of climate change on bird diversity.   172

157 A Mammoth Task

Key Question: How could bringing back the mammoth help restore a lost ecosystem?

Mammoth ecosystems

▶ Mammoths belonged to the same family as modern Asian and African elephants and lived from 300,000 to 10,000 years ago. Alongside other large, grazing herbivores, they occupied an ecosystem of treeless grasslands, often covered in ice sheets through winter. As a keystone species, their actions of scraping off snow with their tusks, grazing, and trampling the grassland maintained the landscape, keeping the ground compacted and frozen.

▶ Evidence suggests humans may have contributed to the extinction of large grazers, including mammoths, in these frozen lands. This resulted in a change in vegetation. Without the trampling effect of the grazers, the ground grew softer and small shrubs and trees grew. It began to thaw, melting the permafrost cover and fundamentally changing the ecosystem.

1. What was the effect of large grazers on the tundra?

   _____

   _____

   _____

2. Frozen tundra prevented millions of tonnes of organic material being recycled back into the environment. What would be the effect of the tundra melting? How might this affect things like climate change?

   _____

   _____

   _____

Frozen tundra, much as it would have appeared for mammoths   Modern tundra

3. A conservation group is attempting to recreate the mammoth's ecosystem, called Pleistocene Park in Siberia, Russia. Over 100 species of animal have been brought in, including bison, reindeer and moose. There is also future hope that eventually, they may be able to bring in a genetically modified "mammoth".

   (a) Why do you think other species are required to be introduced alongside the mammoth?

   _____

   _____

   _____

   (b) How might bringing back the mammoth affect the modern tundra? What effect might this have on climate change?

   _____

   _____

   _____

   _____

©2022 **BIOZONE** International
ISBN: 978-1-98-856692-4
Photocopying Prohibited

158 Ecosystem Dynamics

Key Question: How do ecosystems respond to short-term and cyclical changes, but remain relatively stable in the long term?

What is an ecosystem?

▶ An **ecosystem** consists of a community of organisms and their physical environment. For example, a forest ecosystem consists of all the organisms within the defined area of the forest, along with the physical factors in the forest such as the temperature and the amount of wind or rain.

The dynamic ecosystem

▶ Ecosystems are dynamic in that they are constantly changing. Many ecosystem components, including the seasons, predator-prey cycles, and disease cycles, are cyclical. Some cycles may be short term, such as the change of seasons, or long term, such as the growth and retreat of deserts.

▶ Although ecosystems may change constantly over the short term, they may be relatively static over longer periods. For example, some tropical areas have wet and dry seasons, but over hundreds of years the ecosystem as a whole remains unchanged.

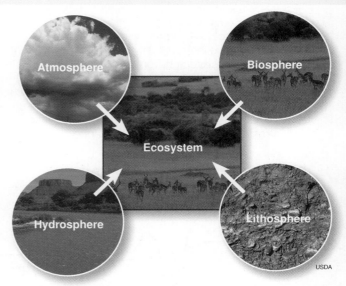

Ecosystems are a result of the interactions between biological (biotic) and physical (abiotic) factors.

An ecosystem may remain stable for many hundreds or thousands of years, provided that the components interacting within it remain stable.

Small scale changes usually have little effect on an ecosystem. Fire or flood may destroy some parts, but enough is left for the ecosystem to return to its original state relatively quickly.

Large scale disturbances such as volcanic eruptions, sea level rise, or large scale open cast mining remove all components of the ecosystem, changing it forever.

1. What is meant by the term dynamic ecosystem? _____

_____

_____

_____

2. (a) Describe two small scale events that an ecosystem may recover from: _____

_____

_____

(b) Describe two large scale events that an ecosystem may not recover from: _____

_____

_____

 SC SPQ LS2.C

Ecosystem stability

▶ Ecosystem stability has various components, including inertia (the ability to resist disturbance) and resilience (the ability to recover from external disturbances).

▶ Ecosystem stability is closely linked to biodiversity, with more diverse systems being more stable. Researchers hypothesize that this is because greater diversity results in a greater number of biotic interactions, and few, if any, vacant niches. The system is protected from change because it is resistant to invasions and there are enough species present to protect ecosystem functions if a species is lost. This hypothesis is supported by experimental evidence but there is uncertainty over what level of biodiversity provides stability or what factors will stress a system beyond its tolerance.

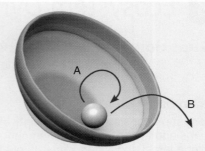

The stability of an ecosystem can be illustrated by a ball in a tilted bowl. Given a slight disturbance the ball will eventually return to its original state (**line A**). However, given a large disturbance the ball will roll out of the bowl and the original state with never be restored (**line B**).

Response to environmental change

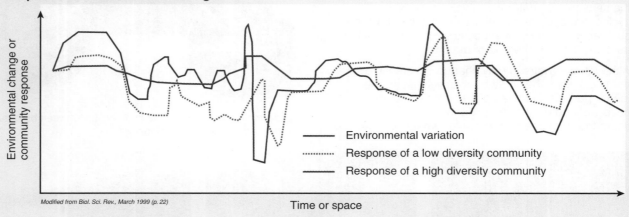

Modified from Biol. Sci. Rev., March 1999 (p. 22)

Legend:
— Environmental variation
·········· Response of a low diversity community
— Response of a high diversity community

Time or space

▶ In models of ecosystem function, higher species diversity increases the stability of ecosystem functions, such as productivity and nutrient cycling. In the graph above, note how the low diversity system varies more consistently with the environmental variation, whereas changes in the high diversity system are more gradual.

▶ In any one ecosystem, some species have a disproportionate effect on ecosystem stability due to their key role in some ecosystem function, e.g. nutrient recycling. These species are called **keystone** (key) **species**.

3. Why is ecosystem stability higher in ecosystems with high biodiversity than in ones with low biodiversity?

_____

_____

4. The effect of changes in ecosystems can be difficult to measure in the field, so researchers often build small scale simulations of an ecosystem. The graph right shows the effect of adding nutrients to a marine ecosystem (as in nutrient runoff from land into the sea). Algal growth-promoting medium was added at 2, 10, or 20% to seawater, together with 0.1 mL of an algal mix. Two days after adding the growth medium and algae, six copepods were added to each chamber. The chambers were sealed and the population size in each chamber was measured over time.

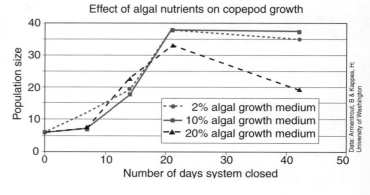

Effect of algal nutrients on copepod growth

Population size

Number of days system closed

Legend:
●--- 2% algal growth medium
■— 10% algal growth medium
▲-- 20% algal growth medium

Data: Armentrout, B & Kappes, H; University of Washington

(a) Which chamber had the greatest environmental disturbance?

_____

(b) Which chamber(s) were able to withstand the environmental disturbance? _____

(c) What does this tell us about the stability and resilience of the system being studied? _____

_____

_____

©2022 **BIOZONE** International
ISBN: 978-1-98-856692-4
Photocopying Prohibited

159 The Resilient Ecosystem

Key Question: How is the resilience of a ecosystem affected by its biodiversity, health, and the frequency with which it is disturbed?

Factors affecting ecosystem resilience

Resilience is the ability of the **ecosystem** to recover after disturbance and is affected by three important factors: diversity, ecosystem health, and frequency of disturbance. Some ecosystems are naturally more resilient than others.

Ecosystem biodiversity

The greater the diversity of an ecosystem, the greater the chance that all the roles (niches) in an ecosystem will be occupied, making it harder for invasive species to establish and easier for the ecosystem to recover after a disturbance.

Ecosystem health

Intact ecosystems are more likely to be resilient than ecosystems suffering from species loss or disease.

Disturbance frequency

Single disturbances to an ecosystem can be survived, but frequent disturbances make it more difficult for an ecosystem to recover. Some ecosystems depend on frequent natural disturbances for their maintenance, e.g. grasslands rely on natural fires to prevent shrubs and trees from establishing. The various grass species in this ecosystem have evolved to survive frequent fires.

A study of coral and algae cover at two locations in Australia's Great Barrier Reef (right) showed how ecosystems recover after a disturbance. At Low Isles, frequent disturbances, e.g. from cyclones, made it difficult for corals to reestablish, while at Middle Reef, infrequent disturbances made it possible to coral to reestablish its dominant position in the ecosystem.

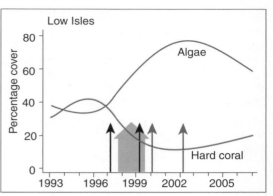

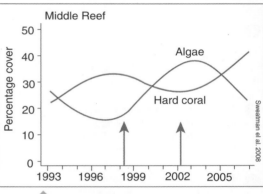

 Crown of thorns starfish outbreak

 Cyclones   Bleaching event

Resilience and harvesting

It is important to consider the resilience of an ecosystem when harvesting resources from it. For example, logging and fishing remove organisms from an ecosystem for human use. Most ecosystems will be resilient enough to withstand the removal of a certain number of individuals. However, excessive removal may go beyond the ecosystem's ability to recover. Examples include the **overfishing** of North Sea cod and deforestation in the Amazon basin.

1. Define ecosystem resilience: _____

_____

2. Why did the coral at Middle Reef remain abundant from 1993 to 2005 while the coral at Low Isles did not?

_____

_____

_____

3. How might over-harvesting affect an ecosystem's resilience? _____

_____

_____

_____

160 A Case Study in Ecosystem Resilience

Key Question: How are resilient ecosystems able to recover from moderate fluctuations?

Spruce budworm and balsam fir

A case study of **ecosystem resilience** is provided by the spruce-fir forest community in northern North America. Organisms in the community include the spruce budworm, and balsam fir, spruce, and birch trees. Despite its name, the spruce budworm is a greater pest to balsam fir than spruce. The community fluctuates between two extremes:

▶ During budworm outbreaks, the environment favors the spruce and birch species.

▶ Between spruce budworm outbreaks, the environment favors the balsam fir.

Balsam fir

Spruce budworm

1 Under certain environmental conditions, the spruce budworm population grows so rapidly it overwhelms the ability of predators and parasites to control it.

2 The budworm feeds on balsam fir (despite their name), killing many trees. The spruce and birch trees are left as the major species.

3 The population of budworm eventually collapses because of a lack of food.

4 Balsam fir saplings grow back in thick stands, eventually out-competing the spruce and birch. Evidence suggests these cycles have been occurring for possibly thousands of years.

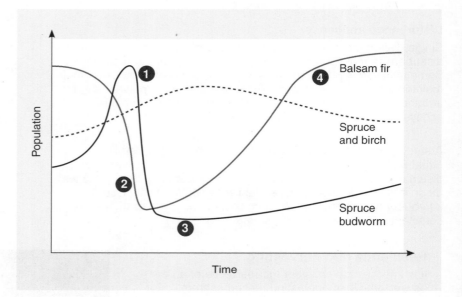

1. (a) Why could predators not control the budworm population? _____

   _____

   _____

   _____

   (b) What was the cause of the budworm population collapse after its initial rise? _____

   _____

2. Under what conditions does the balsam fir out-compete the spruce and birch? _____

   _____

3. In what way is the system resilient in the long term? _____

   _____

   _____

   _____

   _____

   _____

   LS2.C   SC

©2022 **BIOZONE** International
ISBN: 978-1-98-856692-4
Photocopying Prohibited

161 Keystone Species

Key Question: What are keystone species?

Keystone species

▶ **Keystone species** play a crucial role in **ecosystems**. Their actions are key to maintaining the dynamic equilibrium of an ecosystem. Some species have a disproportionate effect on the stability of an ecosystem. These species are called keystone species (or key species). The term keystone species comes from the analogy of the keystone in a true arch. If the keystone is removed, the arch collapses.

▶ The role of the keystone species varies from ecosystem to ecosystem but if they are lost, the ecosystem can rapidly change, or collapse completely. The pivotal role of keystone species is a result of their influence in some aspect of ecosystem functioning, e.g. as predators, prey, or processors of biological material.

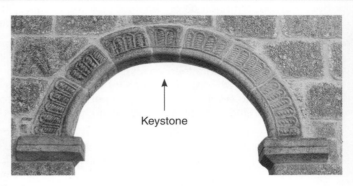

Keystone

An archway is supported by a series of stones, the central one being the keystone. Although this stone is under less pressure than any other stone in the arch, the arch collapses if it is removed.

Keystone species make a difference

▶ The idea of the keystone species was first hypothesized in 1969 by US scientist, Robert Paine. He studied an area of rocky seashore, noting that diversity seemed to correlate with the number of predators present. To test this, he removed the starfish from an 8 m by 2 m area of seashore. Initially, the barnacle population increased rapidly before collapsing and being replaced by mussels and gooseneck barnacles. Eventually, the mussels crowded out the gooseneck barnacles and the algae that covered the rocks. Limpets that fed on the algae were lost. The number of species in the study area dropped from 15 to 8.

Ochre starfish - Paine removed these in his famous study.

Elephants play a key role in maintaining the savannas by pulling down even very large trees for food. This activity maintains the grasslands.

The burrowing of prairie dogs increases soil fertility and channels water into underground stores. Their continuous grazing promotes grass growth and diversity.

Mountain lions prey on a wide range of herbivores and to some extent dictate their home ranges and the distribution of scavenger species.

1. Define the term keystone species: _____

_____

_____

2. Prairie dog colonies are often destroyed by ranchers who believe they compete with cattle for food. How might this lead to the collapse of prairie ecosystems?

_____

_____

_____

_____

SC LS2.C

Sea otters as keystone species

▶ Sometimes, the significant keystone effects of a species becomes evident when a species declines rapidly to the point of near extinction. This is illustrated by the sea otter example described below.

▶ Sea otters live along the Northern Pacific coast of North America and have been hunted for hundreds of years for their fur. Commercial hunting didn't fully begin until about the mid 1700s, when large numbers were killed and their fur sold to overseas markets.

▶ The drop in sea otter numbers had a significant effect on the local marine environment. Sea otters feed on shellfish, particularly sea urchins. Sea urchins eat kelp, which provides habitat for many marine creatures. Without the sea otters to control the sea urchin population, sea urchin numbers increased and the kelp forests were severely reduced.

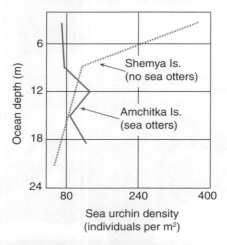

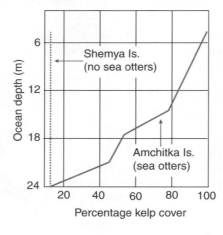

Sea otters are critical to ecosystem function. When their numbers were significantly reduced by the fur trade, sea urchin populations exploded and the kelp forests, on which many species depend, were destroyed.

The effect can be seen on Shemya and Amchitka Islands. Where sea otters are absent, large numbers of sea urchins are found, and kelp are almost absent.

Kelp are large seaweeds belonging to the brown algae. There are many forms and species of kelp. Giant kelp can grow to 45 m long.

Similar to how forests on land provide diverse habitats for terrestrial species, kelp provides habitat, food, and shelter for a variety of marine animals.

Sea urchins kill kelp by eating the holdfast that secures the kelp to the seabed. Unchecked urchin populations can quickly turn a kelp forest into an "urchin barren".

3. What is the importance of kelp in the ecosystem? _____

_____

4. (a) What effect do sea otters have on sea urchin numbers? _____

_____

(b) What effect do sea urchins have on kelp cover? _____

_____

(c) What evidence is there that the sea otter is a keystone species in these Northern Pacific coastal ecosystems?

_____

_____

_____

5. From the new evidence above, if your definition of a keystone species has changed go back to Q1. and edit it.

©2022 **BIOZONE** International
ISBN: 978-1-98-856692-4
Photocopying Prohibited

162 Ecosystem Changes

Key Question: Can there be such severe disturbances to ecosystems that they never return to their original state?

▸ **Ecosystems** are dynamic, constantly fluctuating between particular conditions. However, large scale changes can occasionally occur to completely change the ecosystem. These include **climate change**, volcanic eruptions, or large scale fires.

Human influenced changes

▸ Dolly Sods is a rocky high plateau area in the Allegheny Mountains of eastern West Virginia, USA. Originally, the area was covered with spruce, hemlock, and black cherry. During the 1880s, logging began in the area and virtually all of the commercially viable trees were cut down. The logging caused the underlying humus and peat to dry out. Sparks from locomotives and campfires frequently set fire to this dry peat, producing fires that destroyed almost all the remaining forest. In some areas, the fires were so intense that they burnt everything right down to the bedrock, destroying seed banks. One fire, during the 1930s, destroyed over 100 km² of forest. The forests have never recovered. What was once a forested landscape is now mostly open meadow.

The original dense forests of Dolly Sods included spruce, hemlock, white oak and black cherry. These species have been replaced mostly by maple, birch, beech, and low growing scrub.

1. (a) Identify the large scale ecosystem change that occurred at Dolly Sods: _____

_____

(b) What caused this change in the ecosystem? _____

_____

(c) Explain why the ecosystem has not been able to recover quickly after the change: _____

_____

_____

Natural changes

▸ Volcanic eruptions can cause extreme and sudden changes to local, or even global, ecosystems. The eruption of Mount St. Helens in 1980 provides a good example of how the natural event of a volcanic eruption can cause extreme and long lasting changes to an ecosystem.

Mount St. Helens, one day before the eruption.

Mount St. Helens, post-eruption.

Before the 1980 eruption, Mount St. Helens had an almost perfect and classic conical structure. The "old growth" forests surrounding it were predominantly conifer, including Douglas-fir, western red cedar, and western white pine. Around 35 mammal species and numerous bird species occupied the forests, with amphibians and fish living in the surrounding lakes and streams.

Destruction from the heat, ash, and mudslides was patchwork and random to the ecosystem. Surviving plants, animals, and fungi acted as source populations to "reseed" the ecosystem. The volcanic ash provided plentiful nutrients for lakes and the biodiversity of many lakes multiplied by around 4 x. However, around 230 square kilometers of forest habitats were lost.

SC LS2.C

Schematic of eruption and recovery

The eruption covered about 600 km² (dark blue) in ash (up to 180 m deep in some areas) and blasted flat 370 km² of forest. The schematic shows the general area of bare land per decade since the eruption.

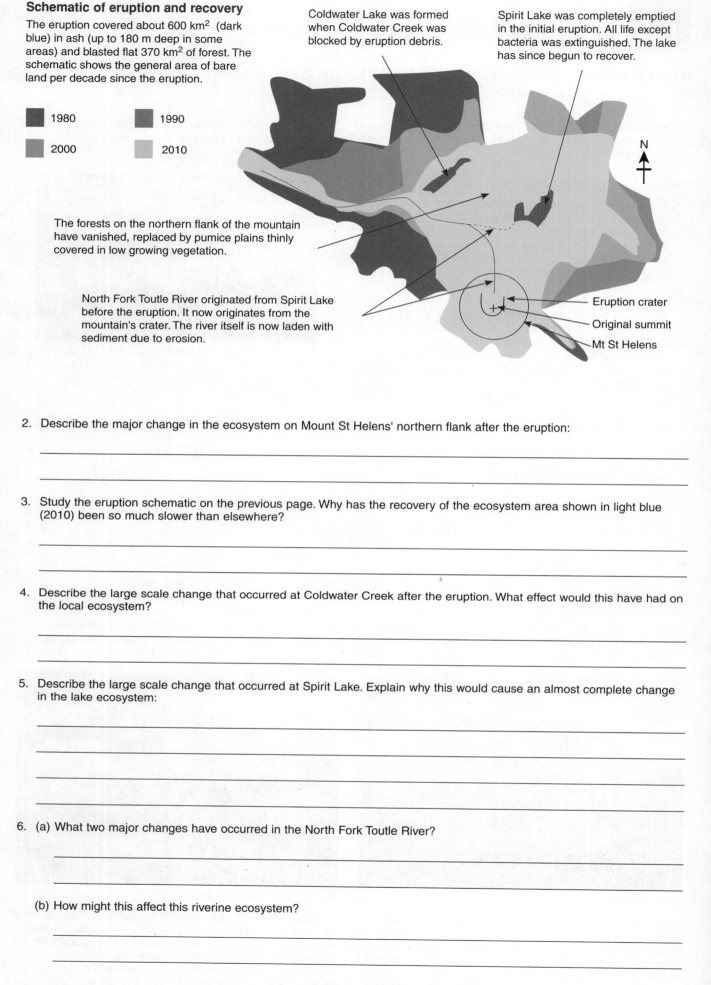

■ 1980   ■ 1990
■ 2000   ■ 2010

Coldwater Lake was formed when Coldwater Creek was blocked by eruption debris.

Spirit Lake was completely emptied in the initial eruption. All life except bacteria was extinguished. The lake has since begun to recover.

The forests on the northern flank of the mountain have vanished, replaced by pumice plains thinly covered in low growing vegetation.

North Fork Toutle River originated from Spirit Lake before the eruption. It now originates from the mountain's crater. The river itself is now laden with sediment due to erosion.

Eruption crater
Original summit
Mt St Helens

2. Describe the major change in the ecosystem on Mount St Helens' northern flank after the eruption:

_____

_____

3. Study the eruption schematic on the previous page. Why has the recovery of the ecosystem area shown in light blue (2010) been so much slower than elsewhere?

_____

_____

4. Describe the large scale change that occurred at Coldwater Creek after the eruption. What effect would this have had on the local ecosystem?

_____

_____

5. Describe the large scale change that occurred at Spirit Lake. Explain why this would cause an almost complete change in the lake ecosystem:

_____

_____

_____

_____

6. (a) What two major changes have occurred in the North Fork Toutle River?

_____

_____

(b) How might this affect this riverine ecosystem?

_____

_____

_____

163 Climate Change and Ecosystem Change

Key Question: What effects will the long term warming of the Earth's atmosphere have on sea levels and land temperatures?

The Earth's climate is changing rapidly

▸ There is evidence that the Earth's atmosphere is experiencing a period of accelerated warming. Fluctuations in the Earth's surface temperature as a result of climate shifts are normal, and the current period of warming climate is partly explained by warming after the end of the last glacial that finished 12,000 years ago.

▸ Since the mid 20th century, the Earth's surface temperature has been increasing rapidly. This phenomenon is a consequence of climate change and scientists attribute it to the increase in atmospheric levels of CO_2 and other greenhouse gases emitted into the atmosphere as a result of human activity.

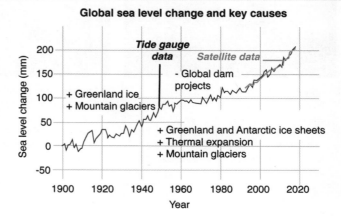

Global sea level change and key causes

▸ The historically warmest recorded years have occurred after 2015 and, since the 1980s, each consecutive decade has been warmer than the previous one. The average global temperature reached 1.11°C above pre-industrial levels in 2021, and is quickly climbing towards the 1.5°C limit that the Paris Agreement acknowledged would prevent the worst climate change impacts. Most countries are now working towards reducing emissions.

Potential effects of climate change

Hurricane Katrina damage, Mississippi

Rising sea levels: It is predicted that sea levels will rise between 300 - 600 mm above their current mean level by 2100. The rise is due to both thermal expansion (ocean water takes up more space when it is warmer) and because of melting of glaciers and ice shelves. Rising sea level will inundate coastal and low lying ecosystems and increase erosion.

The ice-albedo effect refers to the ability of ice to reflect sunlight. Cooling tends to increase ice cover, so more sunlight is reflected from the surface of the ice. Warming reduces ice cover and more solar energy is absorbed, resulting in more warming. Ice has a stabilizing effect on global climate, reflecting nearly all the sun's energy that hits it.

Weather patterns: Climate change may cause regional changes in weather patterns. High intensity hurricanes occur more frequently now than in the past, driven by higher ocean surface temperatures. Storm surge may become more frequent. A storm surge is a rise in sea level occurring during an intense storm. Strong storm winds push the water onshore, causing flooding.

1. What is **climate change**? _____

   _____

2. What is the major cause of **anthropogenic climate change**? _____

   _____

3. How does climate change contribute to rising sea levels? _____

   _____

4. Explain how the level of ice cover can affect global climate: _____

   _____

   _____

   _____

SC   LS2.C

Case study: Climate change and sea level rise in the Florida Everglades

Low lying wetlands such as the Everglades risk being inundated by seawater from sea level rise. The Everglades (above) contains a variety of ecosystems, including mangrove forests, sawgrass marshes, cypress swamps, and pine lands.

Studies have found that the small fish that make up the foraging base of many coastal species do better under less saline conditions. As the saltwater-freshwater interface moves inland, the production of the Everglades will decline due to lower food sources.

Peat collapse is another effect of saltwater intrusion. The intrusion of saltwater removes the peat soil. Peat bogs support a variety of freshwater species and the plants and ecosystem they support helps filter the fresh water humans depend on.

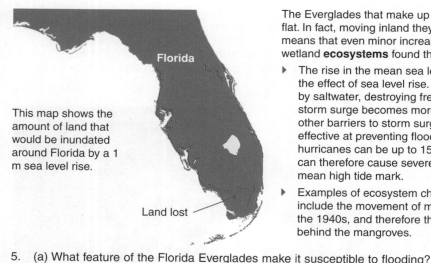

This map shows the amount of land that would be inundated around Florida by a 1 m sea level rise.

The Everglades that make up a large part of southern Florida are extremely flat. In fact, moving inland they increase just 5 cm in altitude per kilometer. This means that even minor increases in sea level will have major effects on the wetland **ecosystems** found there.

▸ The rise in the mean sea level and inundation of the land is only part of the effect of sea level rise. As the sea level rises, ground water is intruded by saltwater, destroying freshwater habitat from underneath. In addition, storm surge becomes more important. As outlying islands, reefs, and other barriers to storm surges become closer to sea level, they are less effective at preventing flooding during storms. Storm surges during severe hurricanes can be up to 15 m above the high tide level. The storm surge can therefore cause severe erosion and damage to the land far beyond the mean high tide mark.

▸ Examples of ecosystem change in the Everglades due to sea level change include the movement of mangrove forests approximately 3 km inland since the 1940s, and therefore the equivalent reduction in freshwater ecosystems behind the mangroves.

5. (a) What feature of the Florida Everglades make it susceptible to flooding? _____

_____

(b) Describe the effects of sea level rise on the ecosystems of the Everglades: _____

_____

_____

_____

_____

6. In the US, the American crocodile is only found in southern Florida (below). They can tolerate high temperatures and brackish water (water that is a mix of sea water and fresh water). The American crocodile is less tolerant of cold than the American alligator. The alligator is more widespread and is found throughout the southeastern US and coexists with the American crocodile in southern Florida.

(a) What could happen to the American crocodile range in Florida if sea levels rise?

_____

_____

(b) How might climate change affect its range overlap with the American alligator?

_____

_____

_____

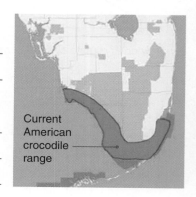

Current American crocodile range

164 Human Impact on Ecosystems

Key Question: What impact does human activity, either deliberate or accidental, have on ecosystems?

▸ Human activity can have major effects on **ecosystems**. These impacts can be so large that they are irreversible, and include pollution of the air, water, or soil of ecosystems; habitat destruction, including deforestation; and removal of important species and introduction of invasive species. There are few places remaining on Earth that are not directly affected by human activities. For most of human civilization, these impacts on the environment have had detrimental effects. We are only now, slowly, beginning to minimize the environmental impact of human activities.

Detrimental impacts of human activities on ecosystems

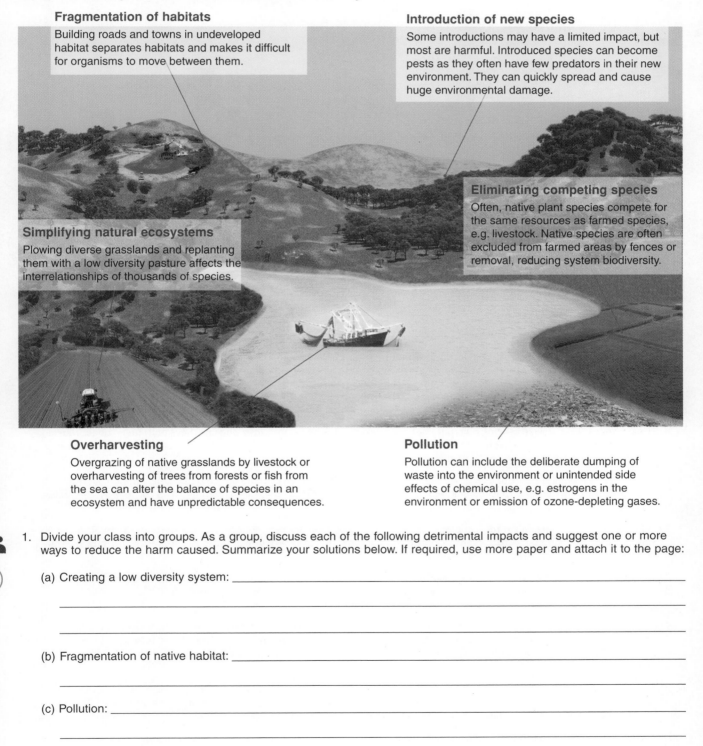

Fragmentation of habitats
Building roads and towns in undeveloped habitat separates habitats and makes it difficult for organisms to move between them.

Introduction of new species
Some introductions may have a limited impact, but most are harmful. Introduced species can become pests as they often have few predators in their new environment. They can quickly spread and cause huge environmental damage.

Eliminating competing species
Often, native plant species compete for the same resources as farmed species, e.g. livestock. Native species are often excluded from farmed areas by fences or removal, reducing system biodiversity.

Simplifying natural ecosystems
Plowing diverse grasslands and replanting them with a low diversity pasture affects the interrelationships of thousands of species.

Overharvesting
Overgrazing of native grasslands by livestock or overharvesting of trees from forests or fish from the sea can alter the balance of species in an ecosystem and have unpredictable consequences.

Pollution
Pollution can include the deliberate dumping of waste into the environment or unintended side effects of chemical use, e.g. estrogens in the environment or emission of ozone-depleting gases.

1. Divide your class into groups. As a group, discuss each of the following detrimental impacts and suggest one or more ways to reduce the harm caused. Summarize your solutions below. If required, use more paper and attach it to the page:

(a) Creating a low diversity system: _____

_____

_____

(b) Fragmentation of native habitat: _____

_____

(c) Pollution: _____

_____

_____

©2022 **BIOZONE** International
ISBN: 978-1-98-856692-4
Photocopying Prohibited

How do toxins move through food chains?

▸ Recall how organisms in an ecosystem are connected through their feeding relationships. Each organism in a food chain is a food source for the next. Many food chains interconnect in food webs, where organisms may feed at several different trophic levels.

▸ What happens when a toxin is released into the environment as a result of human activity? Where does the toxin go and how does it move through ecosystems?

▸ Toxins, such as mercury, pesticides, and DDT, are called persistent. This means they resist degradation from abiotic and biotic factors. They are not excreted (or excreted very slowly) so they accumulate in the tissues of the organisms that eat them.

▸ In this investigation, you will simulate the movement of a persistent toxin such as DDT through a lake food chain and observe what happens to the concentration of the toxin in higher order consumers.

Investigation 8.1 Pathways for toxins in food webs

See appendix for equipment list.

1. Study the image of a lake food chain. DDT was sprayed on the surrounding wetland to control mosquitoes but much of it washed into the lake, where its breakdown product, DDE, was taken up by phytoplankton (algae).

2. What happens to the concentration of DDE in the organisms in the food chain?

_____

In this investigation you will find out for yourselves how this happens.

3. Each group will have a tape and marker, a bag of small colored beads, 10 small test tubes, 3 medium test tubes, 1 large test tube, a 50 mL beaker, and a 100 mL beaker.

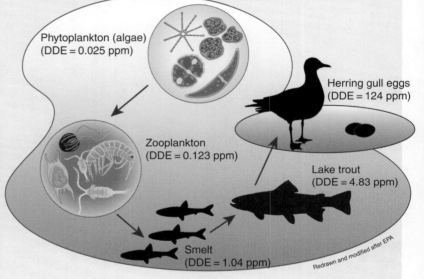

Phytoplankton (algae)
(DDE = 0.025 ppm)

Herring gull eggs
(DDE = 124 ppm)

Zooplankton
(DDE = 0.123 ppm)

Lake trout
(DDE = 4.83 ppm)

Smelt
(DDE = 1.04 ppm)

Redrawn and modified after EPA

4. Use the tape and marker to label the test tubes as follows:
   – small test tubes with the first organism in the food chain (algae)
   – medium test tubes with the second organism in the food chain (zooplankton)
   – large test tube with the third organism in the food chain (smelt)
   – 50 mL beaker with the fourth organism in the food chain (trout)
   – 100 mL beaker with the final organism in the food chain (gull)

5. Use a measuring device (syringe or pipette) to measure 0.5 mL water into one of the small test tubes. Mark the level and tip out the water. This tube will be your reference tube for phytoplankton so write REF on it. Mark the other 9 tubes with the same 0.5 mL level.

6. Place the 10 small test tubes in a rack and sprinkle beads over the tubes to simulate DDE uptake by the phytoplankton organisms (phytoplankters). Keep sprinkling until all the tubes contain ~0.5 mL beads. Set aside your reference tube in the test tube rack.

7. Each zooplankton organism (or zooplankter) eats three phytoplankters. Empty the contents of 3 small tubes into one of the medium-sized tubes. Repeat for each of the other two medium-sized tubes. Mark one of the tubes with the bead level (1.5 mL) and add REF.

8. The smelt eats three zooplankters. Empty the contents of the 3 medium-sized test tubes into the large test tube. Now add 1.5 mL beads to the (now empty) medium-sized test tube you marked REF. Set it aside in the test-tube rack. This is your reference tube for zooplankton.

9. The trout eats three smelt. Mark the large test tube at the level of the beads and then empty the beads into the 50 mL beaker. Refill the large test tube to the line and again, empty the beads into the 50 mL beaker. Repeat once more. Next add beads to the large test tube up to the line and label it REF. Set it aside. This is your reference tube for the smelt.

10. The gull eats two trout. Mark the level of the beads in the 50 mL beaker and empty them into the 100 mL beaker. Refill the 50 mL beaker to the line and again empty it into the 100 mL beaker. Finally, refill the 50 mL beaker to the line and label it REF. Set it aside. This is your reference for the trout.

11. Label your 100 mL beaker REF-GULL. Compare the contents of the 5 organisms (REF tubes).

©2022 **BIOZONE** International
ISBN: 978-1-98-856692-4
Photocopying Prohibited

12. Draw a diagram or flow chart in the space provided to show what you did in your investigation, and how the volume of beads (toxins) in the food chain changed with successive levels in the food chain.

2. (a) What do the beads represent?: _____

(b) What happened to the volume of beads in the 5 containers ("organisms"):_____

_____

_____

_____

(c) Explain why this happened:_____

_____

_____

_____

3. Will all the organisms at one level in the food chain (trophic level) have the same "toxin" level? Explain your answer:

_____

_____

_____

_____

4. This investigation demonstrates the phenomenon of biomagnification. In your group, discuss how could you make it more realistic and write your ideas below:

_____

_____

_____

_____

_____

165 The Effects of Damming

Key Question: What effects do damming have on ecosystems?

The increased importance of hydroelectricity and the production of water reservoirs has resulted in the damming of many rivers. Rivers are dammed for a variety of reasons, including production of electricity, flood control, and to provide water for irrigation. Once seen as an environmentally friendly way to produce renewable electricity, the enormous damage dams can do to **ecosystems,** both upstream and downstream of the dam, is often realized too late. Most of the world's major rivers are now dammed; many more than once.

Impacts of dams

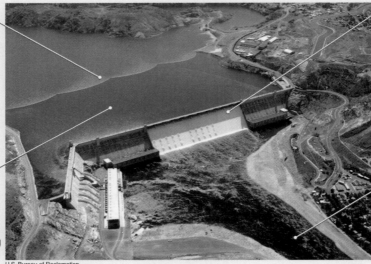

Reservoir
Displaces communities, inundates and fragments ecosystems.

Dam
Fish migration is blocked, separating spawning waters from rearing waters. Sediment accumulates in the reservoir, reducing fertility of downstream soils and leading to erosion of river deltas, which are deprived of their sediment supply.

Water quality
Water quality can be severely reduced for many years after filling. Aquatic communities in free flowing water are destroyed.

Downstream impact
Disrupted water flow and low water quality near the dam reduces biodiversity.

U.S. Bureau of Reclamation

Order of environmental impact of dams

▶ First order impacts occur immediately and are abiotic (physical). They include barrier effects (blocking water flow), and effects to water quality and sediment load.

▶ Second order effects occur soon after the establishment of the dam. These include effects related to the change in the physical and biotic environments. Rates of primary production change and the morphology of the ecosystem changes, e.g. river valley flooding and erosion.

▶ Third order impacts occur after the change in ecosystem and relate to the establishment of new ecosystem: from a flowing water system to a still water system.

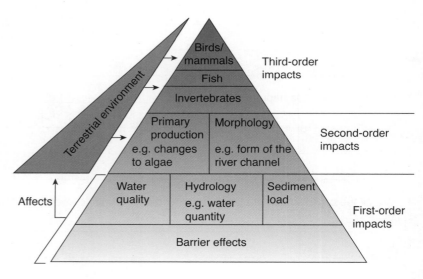

1. Identify three reasons for damming rivers: _____

_____

2. Describe how dams can affect aquatic ecosystems: _____

_____

_____

_____

_____

_____

©2022 **BIOZONE** International
ISBN: 978-1-98-856692-4
Photocopying Prohibited

The Yangtze River

Ship lock

Dam

1987   2006   Pollution behind the dam   Siberian Crane

The Three Gorges Dam (above) on the Yangtze river, China, is 2.3 km wide and 101 m high, with a reservoir 660 km long. The construction of the Three Gorges Dam caused the river water level to rise by 100 m. This rise in water level flooded important wetlands where wading birds, including the Siberian crane, over-wintered. It also flooded 13 cities and hundreds of towns. The waste left behind in the flooded cities and towns continues to pollute the reservoir's waters. Dams reduce flood damage by regulating water flow downstream. However, this also prevents deposition of fertile silts below the dam and increases downstream erosion.

Colorado River

 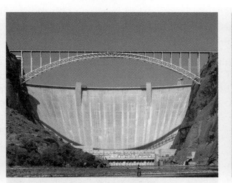 

Glen Canyon Dam

Hoover Dam

A number of dams are found along the Colorado River, which runs from Colorado through to Mexico. The two largest hydroelectric dams on the river are the Glen Canyon Dam and the Hoover Dam. Both dams control water flow through the Colorado River and were controversial even before their construction started.

The construction of Glen Canyon Dam and the Hoover Dam effectively ended the annual flooding of the Colorado River and allowed invasive plants to establish in riparian (riverside) zones. The reduced flow rates of the river have also reduced the populations of many fish species downstream.

The once vast Colorado River delta has been reduced to just 5% of its original size due to the construction of various dams. The river itself no longer reaches the sea. The photo above shows the river as it is, just 3 km below Morelos Dam, an irrigation diversion dam that takes almost all of the remaining river water.

3. Using the pyramid diagram, explain why the environmental impacts of dams occur in a specific order: _____

_____

_____

_____

_____

4. Describe the effects of the Three Gorges Dam on the ecosystems of the Yangtze River: _____

_____

_____

_____

_____

5. Describe the effects of the dams on the Colorado River on the river's ecosystems: _____

_____

_____

_____

166 The Impact of Alien Species

Key Question: What impact do alien species have on ecosystems into which they are brought, either as introduced or invasive species?

▸ Introduced species often have no natural predators or diseases in their new environment. They can proliferate without control and go on to cause serious environmental damage. They have evolved at one place in the world and been transported by humans, either intentionally or inadvertently, to another region. Some of these introductions are directly beneficial to humans and controlled in their impact, e.g. introduced agricultural plants and animals and Japanese clams and oysters (the mainstays of global shellfish industries).

▸ **Invasive species** are introduced species that have a negative effect on the **ecosystems** into which they have been imported. Originally, these species may have been brought by humans as pets, food, ornamental specimens, or decoration, but have escaped into the wild.

▸ In their new environment, they usually lack natural predators, diseases, or other controls to limit their growth.

▸ Other species may have been accidentally transported in cargo shipments or in the ballast water of ships. Some have been deliberately introduced to control another pest species and have themselves become a problem.

▸ Some of the most destructive of all alien species are aggressive plants, e.g. mile-a-minute weed, a perennial vine from Central and South America; Miconia, a South American tree invading Hawaii and Tahiti; and Caulerpa seaweed, the aquarium strain now found in the Mediterranean. Two introductions, one unintentional and the other deliberate, are described below.

Mile-a-minute weed covering woodland in Maryland, USA

Kudzu: a deliberate introduction

Kudzu (*Pueraria lobata*) is a climbing vine native to south-east Asia. It was first introduced to the United States in 1876 and promoted as a forage crop in 1908. Kudzu was widely planted during the Dust Bowl Era to try to conserve soil. In 1940, the government paid farmers almost $20 (about $345 in today's money) a hectare to plant it. However by 1953 the payments had stopped as kudzu escaped farms and invaded woodlands. By 1970, it was declared a weed and in 1997 was placed on the noxious weeds list. Half a billion dollars a year is spent trying to control it. Today, kudzu is estimated to cover 3 million ha of land in the southeastern US.

Red fire ant: an accidental introduction.

Red fire ants (*Solenopsis invicta*) were accidentally introduced into the southeastern states of the United States from South America in the 1920s and have spread north each year. Red fire ants are now resident in 17 US states where they displace populations of native insects and ground-nesting wildlife. They also damage crops and are very aggressive, inflicting a nasty sting. The USDA estimates damage and control costs for red fire ants at more than $6 billion a year.

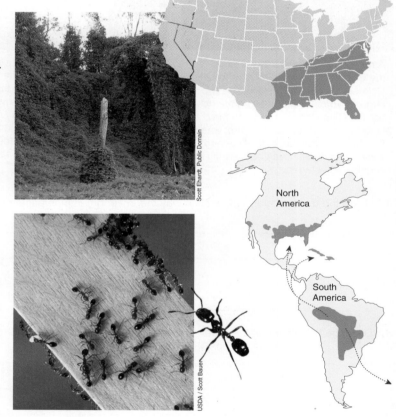

North America

South America

1. Explain why many alien species become invasive when introduced to a new area: _____

_____

_____

_____

2. Research an alien pest species in your area and discuss the impact it has had on a local ecosystem. Write a short summary of your findings and attach it to this page.

    LS2.C LS4.D SC

©2022 **BIOZONE** International
ISBN: 978-1-98-856692-4
Photocopying Prohibited

167 Human Impacts on Fish Stocks

Key Question: What is the impact of unsustainable fishing on fish stocks?

▶ Fishing is an ancient human tradition. It provides food, and is economically, socially, and culturally important. Today, it is a worldwide resource extraction industry. Decades of **overfishing** in all of the world's oceans has pushed commercially important species, such as cod (right), into steep decline. Overfishing has caused the collapse of many fisheries. Unsustainable fishing practices continue throughout the world's oceans.

▶ According to the United Nation's Food and Agriculture Organization (FAO) almost half the ocean's commercially targeted marine fish stocks are either heavily or over-exploited. Without drastic changes to the world's fishing operations, many fish stocks will soon be lost.

Grand Banks fishery

y-axis: Tonnes of cod landed (x 1000), values 0, 300, 600, 900
x-axis: Year, values 1850, 1900, 1950, 2000

Lost fishing gear can entangle all kinds of marine species. This is called **ghost fishing**.

Overfishing has resulted in many fish stocks at historic lows and fishing effort (the effort needed to catch fish) at unprecedented highs.

Huge fishing trawlers are capable of taking enormous amounts of fish. Captures of 400 tonnes at once are common.

The limited selectivity of fishing gear results in millions of marine organisms being discarded for economic, legal, or personal reasons. These organisms are defined as by-catch and include fish, invertebrates, protected marine mammals, sea turtles, and sea birds. Many of the discarded organisms die. Estimates of the worldwide by-catch is approximately 30 million tonnes per year.

Bottom trawls and dredges cause large scale physical damage to the seafloor. Non-commercial, bottom-dwelling species in the path of the net can be uprooted, damaged, or killed. An area of 8 million km² is bottom trawled annually.

Percentage of catch taken

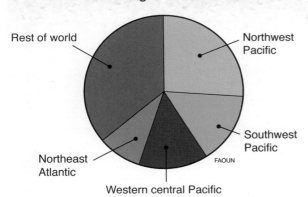

Labels: Rest of world, Northwest Pacific, Southwest Pacific, Western central Pacific, Northeast Atlantic. FAOUN

The single largest fishery is the Northwest Pacific, taking 26% of the total global catch.

Percentage exploitation of fisheries

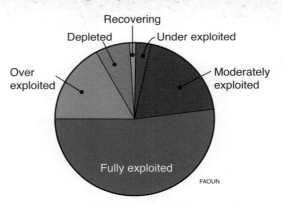

Labels: Recovering, Depleted, Under exploited, Over exploited, Moderately exploited, Fully exploited. FAOUN

52% of the world's fished species are already fully exploited. Any increase in catch from these species would result in over-exploitation. 7% of the fish species are already depleted and 17% are over-exploited.

SC LS4.D LS2.C

Fishing techniques have become so sophisticated, and efforts are on such a large scale, that thousands of tonnes of fish can be caught by one vessel on one fishing cruise. Fishing vessels can reach over 100 m long. Here, 360 tonnes of Chilean jack mackerel are caught in one gigantic net.

Tuna is a popular fish type, commonly found canned in supermarkets and as part of sushi. However, virtually all tuna species are either threatened or vulnerable. Demand for the fish appears insatiable, with a record price of $1.7 million being paid in 2013 for a 221 kg bluefin tuna.

Illegal fishing was a major problem in the 1990s. Thousands of tonnes of catches were being unreported. International efforts have reduced this by an estimated 95%, helping the recovery of some fish stocks. Naval patrol vessels (above) have helped in targeting illegal fishing vessels.

Over-fishing is only one way that humans affect fish populations. The reduction in quality habitats as a result of human activities also has an effect. The bleaching of corals, related to **climate change** and sea water acidification, has the potential to seriously affect fish populations.

It is estimated that 8 million tonnes of plastic finds its way into the sea each year. Plastic can have severe detrimental effects on marine life, especially those that mistake plastic bags for jellyfish or other prey species. Some areas are so polluted it is dangerous to eat fish caught there.

An example of the effect a fishery can have on fish stocks is the Galápagos Island sea cumber population. The commercial fishery began in 1993 and by 2004 the sea cucumber population had dropped by 98%.

1. Why is overfishing of a fish species not sustainable? _____

_____

_____

_____

2. Why is returning by-catch to the sea not always as useful as it might appear? _____

_____

_____

_____

3. What percentage of commercial fish stocks are fully exploited, overexploited, or already depleted? _____

_____

4. Identify three ways in which fish stocks are over-exploited: _____

_____

_____

_____

_____

_____

©2022 **BIOZONE** International
ISBN: 978-1-98-856692-4
Photocopying Prohibited

Investigation 8.2 A model of human impacts on fish stocks

See appendix for equipment list.

1. You will be working in groups of 5-7. Each member of the group will be given 1 small cup and 1 teaspoon.

2. Each group will have 2 bowls, each filled with beads: 60 red beads (representing target fish), 40 yellow beads (representing other fish), and 40 blue beads (representing protected species). One bowl is the ocean bowl, one bowl is the reservoir bowl.

3. For each "fishing season," randomly assign the spatula (representing an ocean trawler) to one student, and a set of chopsticks (representing line and hook traditional fishing) to another. Remaining students use their teaspoon (representing commercial net fishing).

4. Students have 20 seconds altogether as a group to use their "fishing" method to gather as many "target fish" as possible from their ocean bowl.

5. Each student counts their "catch" and records the number, and fishing method, on the data chart below.

Fishing season	Fishing method	Target fish (red)	'Other' fish (yellow)	Protected species (blue)
1				
2				
3				
4				
5				

6. Replenish remaining beads/fish in the ocean bowl from the groups reservoir bowl. Each bead/fish remaining is doubled to account for breeding (add one bead of the correct color for each one left in the ocean bowl).

7. Group members with fewer than 5 target fish (red beads) in their cup sit out the next season, as they have gone "bankrupt".

8. Redistribute spatula and chopsticks to 2 random group members. Start season 2.

9. Repeat for 5 seasons, or until all target fish (red beads) have gone.

10. If a group's "ocean" has run out of "target" fish, they can spread out and "fish" in another group's ocean (bowl) by joining in their next round.

5. How did the type of fishing method affect:

   (a) The total number of fish caught in any season? _____

   _____

   _____

   (b) The proportion of "target fish" caught, compared to other and protected species (bycatch)? _____

   _____

   _____

6. Why do the "oceans" still become depleted of target fish eventually, even when they reproduce each year? _____

   _____

   _____

   _____

7. How did students moving to other oceans to fish affect fish stocks? _____

   _____

   _____

©2022 **BIOZONE** International
ISBN: 978-1-98-856692-4

8. In your original groups, discuss some ideas about sustainable fishing methods and how they might prevent overfishing. Agree to a set of group rules that incorporates the **sustainability** ideas. Repeat the activity. Record new the bead/fish counts in the chart on the previous page, but in a different colored pen or pencil.

   (a) Describe one of your new sustainability rules used in round two, and how it reduced overfishing in the activity:

   _____

   _____

   _____

   (b) What new rule was successful in reducing the bycatch of other fish (yellow) and protected species (blue)?

   _____

   _____

   _____

9. Which of your new rules could be applied to the problem of overfishing in the real world? _____

   _____

   _____

   _____

10. (a) What parts of the fishing investigation activity do you think represented the "real thing" particularly well? Why?

    _____

    _____

    (b) What parts of the fishing investigation activity do you think misrepresented the "real thing"? Why?

    _____

    _____

11. A fishbone chart helps categorise causes of a problem. Write a major problem that is caused by overfishing at the fish head; list four factors that contribute to the problem as headings, and conduct an online search to find some reasons for each factor, listing these underneath. Some reasons may appear in several categories. One heading has been completed for you. Discuss your findings in a group, or as a class.

e.g. Reduction of fish stocks   (a)

Fish being taken before
they can reproduce; too
many fish being taken
every year, reduced
reproduction rates.

PROBLEM

e.g. Collapse of the
Atlantic salmon fishery

(b)

(c)

* Students must remember that correlation does not always equal causation

©2022 **BIOZONE** International
ISBN: 978-1-98-856692-4
Photocopying Prohibited

168 Evaluating a Solution to Overfishing

Key Question: Could fish farming be a solution to overfishing?

▶ The solutions to **overfishing** are complex and although some ideas may at first appear to provide a solution, deeper investigation often finds them to be unsustainable. There is clear evidence that many of the world's most prized fisheries are being overfished. Fisheries such as Atlantic bluefin tuna, North Sea cod, and Atlantic salmon have been overfished precisely because they are prized fisheries. The fish are valuable because people like to eat them and they were once relatively easy to catch.

▶ However, as fish stocks have declined, people have realized that solutions are needed to ensure the fisheries survive, both for the environment and for the economy.

▶ One solution is to reduce the number of fish caught in the wild, while increasing the total number produced, was to farm them.

▶ Aquaculture includes not just fish farming but the farming of any marine species, including mussels and prawns. It is a highly contentious issue, with arguments both for and against it (below).

Fish farming: A solution to the problem.

▶ With fish stocks plummeting and the world's population ever-increasing, the demand for fish has outpaced nature's ability to keep up. Fish farming provides a way to produce fish and other marine products without the need to destroy wild populations.

▶ For example, the Atlantic salmon has been so overfished that wild caught salmon is no longer a viable fishery. Instead, the Atlantic salmon are farmed close to shore in large sea pens.

▶ Around 1.5 million tonnes of salmon are produced from farms a year, compared to less than three thousand tonnes caught in the wild. Returns like these from commercial fish farms help reduce the need to catch wild fish and allow wild fish stocks to recover.

▶ Importantly, fish farming does not have to rely on specific locations. Atlantic salmon, native to the North Atlantic, is farmed in Chile, Canada, Norway, Russia, the UK, and Australia. Similarly, Chinook or King salmon, native to the North Pacific, is farmed in New Zealand and Chile. New Zealand is the world's largest producer of Chinook salmon, producing more farmed salmon than the rest of the world's entire catch.

▶ Fish farming is highly efficient. In the wild, a salmon might need 10 kg of food for every 1 kg of body weight. Farmed fish require just 4 kg of food per kg of body weight. Much of this food includes fish meal.

▶ The increase in fish farming has not caused an increase in the catch of fish for fish meal. Instead, the fish meal required for fish farming has come from fish meal that was once fed to livestock such as pigs and poultry, which now use other feeds types such as grain.

Fish farming: An unsustainable disaster.

▶ Most fish farming is carried out in sea cages rather than containment facilities. This means waste from the fish farm enters the local **ecosystem** directly.

▶ Fish in the pens can be subject to high stocking rates and thus produce a large and concentrated amount of waste, including feces and food waste. In sites without adequate currents, these wastes can build up in the area and not only pollute the water but cause disease in the farmed fish and in local populations of marine animals.

▶ Disease can be a problem in such concentrated cages, quickly transferring between individuals. Various antifouling, antibacterial, and antiparasitic chemicals are used to reduce disease.

▶ In areas without adequate currents, heavy metals may build up in the environment and the fish, creating a hazard for human consumption. Escapes from the cages often occur, especially in high seas. If the fish being farmed are not native, they can compete with native fish. Because farmed fish have lower genetic diversity than wild fish, a large escape could decrease the overall genetic diversity of a wild population.

▶ Feeding the fish requires fish meal and other products in order to produce fish high in omega-3 fats (not made by fish but taken in with their normal diet). 50% of the world's fish oil production is fed to farmed fish.

▶ Many of the farmed fish species are carnivorous and are fed fish meal. However, this itself comes from fishing and a large percentage of the commercial fishing is dedicated to catching bait fish such as sardines specifically for fish meal. This has the potential to drive these fish to extinction.

1. Use the articles above and your own research to evaluate fish farming as a solution to overfishing. Write a short answer to the question "Can fish farming stop overfishing and produce a sustainable resource?" Include reasons and evidence to justify your answer:

_____

_____

_____

_____

169 Deforestation and Species Survival

Key Question: How does deforestation impact species survival?

Deforestation

▶ At the end of the last glacial period, about 10,000 years ago, forests covered an estimated 6 billion hectares, about 45% of the Earth's land surface. Forests currently cover about 4 billion hectares of land (31% of Earth's surface). They include the cooler temperate forests of North and South America, Europe, China, and Australasia, and the tropical forests of equatorial regions. Over the last 5000 years, the loss of forest cover is estimated at 1.8 billion hectares. 5.2 million hectares has been lost in the last 10 years alone. Temperate regions where human civilizations have historically existed the longest (e.g. Europe) have suffered the most but now the vast majority of deforestation is occurring in the tropics. Intensive clearance of forests during settlement of the most recently discovered lands has extensively altered their landscapes (e.g. in New Zealand, 75% of the original forest was lost in a few hundred years).

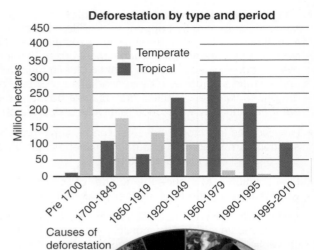

Deforestation by type and period

Legend: Temperate, Tropical. Y-axis: Million hectares (0-450). X-axis periods: Pre 1700, 1700-1849, 1850-1919, 1920-1949, 1950-1979, 1980-1995, 1995-2010.

Causes of deforestation

▶ **Deforestation** is the end result of many interrelated causes, which often center around socioeconomic drivers. In many tropical regions, the vast majority of deforestation is the result of subsistence farming. Poverty and a lack of secure land can be partially solved by clearing small areas of forest and producing family plots. However, huge areas of forests have been cleared for agriculture, including ranching and production of palm oil plantations. These produce revenue for governments through taxes and permits, producing an incentive to clear more forest. Just 14% of deforestation is attributable to commercial logging, although combined with illegal logging it may be much higher.

Causes of deforestation

Logging 14% · Fuel 5% · Subsistence farming 48% · Commercial agriculture 32%

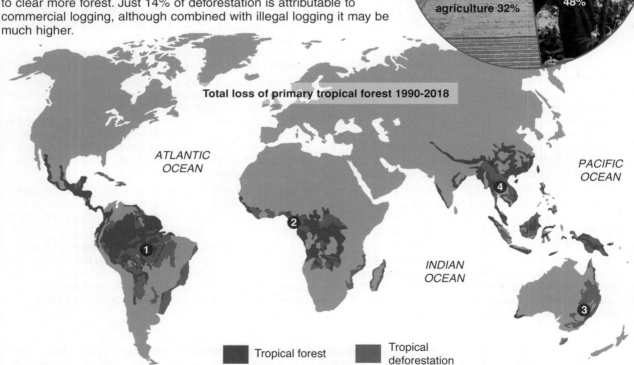

Total loss of primary tropical forest 1990-2018

ATLANTIC OCEAN

PACIFIC OCEAN

INDIAN OCEAN

■ Tropical forest   ■ Tropical deforestation

▶ It is important to distinguish between deforestation involving primary (old growth) forest and deforestation in plantation forests. Plantations are regularly cut down and replaced, and can artificially inflate a country's apparent forest cover or rate of deforestation. The loss of primary forests is far more important as these are refuges of high biodiversity, including for rare species, many of which are endemic to relatively small geographical regions (i.e. they are found nowhere else).

▶ Although temperate deforestation is still a concern, it is in equatorial regions that the pace of deforestation is accelerating (above). This is of global concern as species diversity is highest in the tropics and habitat loss puts a great number of species at risk.

      LS2.C LS4.D SC

©2022 **BIOZONE** International
ISBN: 978-1-98-856692-4
Photocopying Prohibited

The Pacific Northwest and the spotted owl

▶ The forests of the Pacific Northwest of the USA include the primary forests of Oregon and Washington states. They include stands of redwood, Douglas-fir, western red cedar, and shore pine, as well as alder and maple. The region is home to hundreds of species of wildlife, many dependent on the primary (old-growth) forest. One of these, the northern spotted owl, is a keystone species and an important indicator of healthy old growth forest. Listed as threatened under the Endangered Species Act, it was first protected in 1990 after nearly a century of old growth logging. In 1994, in response to the owl's status, the Northwest Forest Plan (NWFP) was implemented to govern land use on federal lands in the Pacific Northwest. In particular, it called for extensive old growth reserve lands and restrictions on old growth logging.

▶ The northern spotted owl ranges over wide areas of the Pacific Northwest. It requires forests with a dense canopy of mature and old-growth trees, abundant logs, standing snags, and live trees with broken tops. Only old growth forests offer these characteristics. When forced to occupy smaller areas of less suitable habitat, the birds are more susceptible to starvation, predation, and competition.

▶ Although the plight of the northern spotted owl triggered the NWFP, it is the fate of the old growth forests, the species they support, and the ecosystem services they provide that are at stake. Despite protections, the number of northern spotted owls is declining at three times the predicted rate and it is increasingly evident that competition from barred owls has a role in this.

US Fish and Wildlife Service

1. Describe the trend in temperate and tropical deforestation over the last 300 years: _____

_____

2. What are some of the causes of deforestation? _____

_____

_____

3. Why is it important to distinguish between total forest loss and loss of primary forests when talking about deforestation?

_____

_____

_____

_____

4. Deforestation in temperate regions has largely stabilized and there has been substantial forest regrowth. However, these second growth forests differ in structure and composition to the forests that were lost. Why might this be of concern?

_____

_____

_____

5. **"The Pacific Northwest old growth forest should be fully protected and not logged."**
Divide your class into two groups, with one group arguing in support of this statement and one group arguing against it. Present group arguments to the class as a whole. Complete your own research, ensuring validity, using the **Resource Hub** for more information. Present your arguments as a report and summarize the main points in the space below:

_____

_____

_____

_____

_____

_____

170 Modeling a Solution

Key Question: What factors should be considered when modeling a conservation solution?

▶ **Conservation** efforts are often a compromise between environmental, economic, and cultural needs. Deciding on a course of action for preserving **biodiversity** is not always simple, and compromises must often be made.

▶ The map below shows a hypothetical area of 9,300 ha (93 km²) in which two separate populations of an endangered bird species exist within a forested area of public land. A proposal to turn part of the area into a wildlife reserve has been put forward by local conservation groups. However, the area is known to have large deposits of economically viable minerals and is frequented by hikers. Hunters also spend time in the area because part of it has an established population of introduced game animals. The proposal would allow a single area of up to 1,500 ha (15 km²) to be reserved exclusively for conservation efforts.

1. Study the map below and draw on to the map where you would place the proposed reserve, taking into account economic, cultural, and environmental values. On a separate sheet, write a report justifying your decision as to where you placed the proposed reserve.

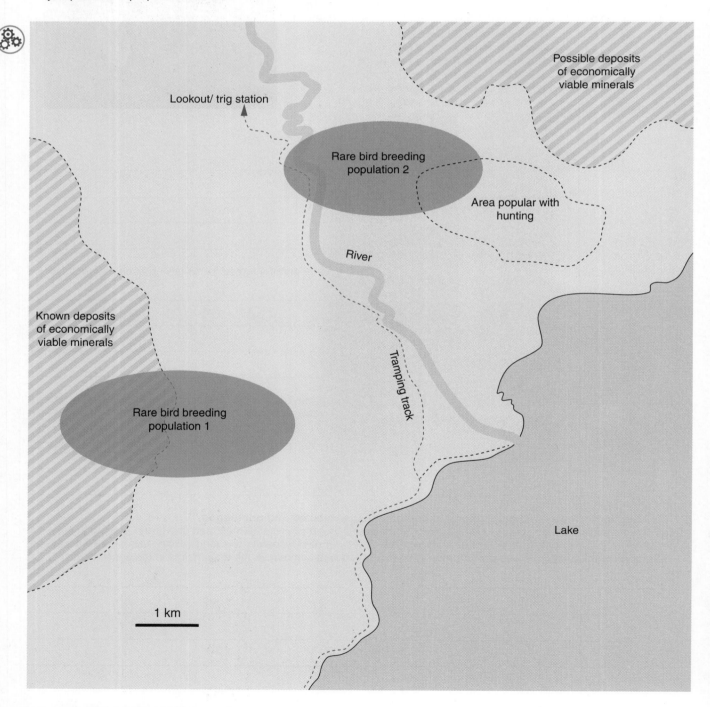

©2022 **BIOZONE** International
ISBN: 978-1-98-856692-4
Photocopying Prohibited

      LS2.C LS4.D SC

171 Review Your Understanding

Key Question: How could bringing back the mammoth restore a lost ecosystem?

1. Earth's northern **ecosystems** have been without mammoths for over 20,000 years. Fill in the chart below with possible consequences of re-introducing the mammoth as a keystone species into the ecosystem. The consequences can be negative or positive. Write two further consequences arising from this to the right and then work in small groups to compare and discuss the various answers.

Action taken	Consequences of action	Further consequences
	(a)	(b)
		(c)

2. You will work in groups to take part in a class debate on the pros and cons of reintroducing mammoth back into the northern European ecosystem as a means to reduce warming from melting permafrost. Divide each larger group into two smaller teams of around 3 students.

   (a) Which side of the debate are you arguing for? For (Pro) or against (con)? _____

   (b) Work together in your smaller group to decide upon at least three key points that you will use to support your side of the debate. Record those points down below.

   _____

   _____

   _____

   (c) Construct a summary statement linking the introduction of mammoth as a means to reduce **climate change**. Draw on your learning from activities you have completed within this chapter to help craft your answer.

   _____

   _____

   _____

   _____

   _____

   _____

   _____

©2022 **BIOZONE** International
ISBN: **978-1-98-856692-4**
Photocopying Prohibited

172 Summing Up

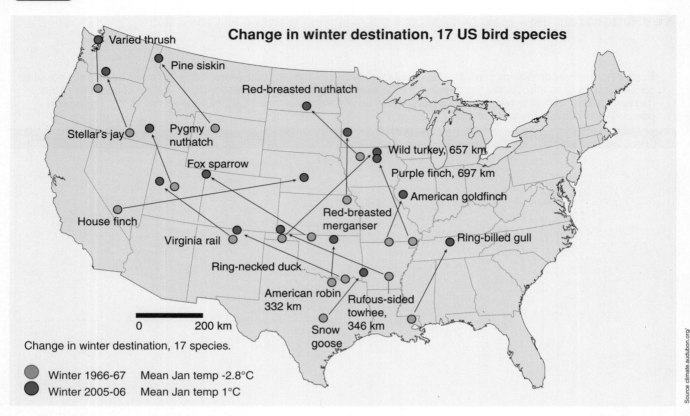

Change in winter destination, 17 US bird species

Varied thrush
Pine siskin
Red-breasted nuthatch
Stellar's jay
Pygmy nuthatch
Fox sparrow
Wild turkey, 657 km
Purple finch, 697 km
American goldfinch
House finch
Red-breasted merganser
Virginia rail
Ring-billed gull
Ring-necked duck
American robin 332 km
Rufous-sided towhee, 346 km
Snow goose

0   200 km

Change in winter destination, 17 species.

Winter 1966-67   Mean Jan temp -2.8°C
Winter 2005-06   Mean Jan temp 1°C

Source climate.audubon.org/

Study the map above showing the change in winter destinations of 17 bird species in the US between 1966-67 (average January temperature -2.8°C) and 2005-06 (average January temperature 1°C).

1. (a) What does the map show? _____

_____

(b) What is the most likely reason for this?_____

_____

(c) What is the likely outcome for these bird populations of a predicted further 1.5°C increase in temperature?

_____

_____

2. What is the likely outcome for the habitable range of Arctic resident bird species under the predicted warmer climate?

_____

3. Under a predicted climate change of +1.5°C or more by 2100 (relative to the 1980–1999 average), what would you predict for the ranges of:

(a) Low altitude adapted species? _____

(b) High altitude adapted species?_____

4. It is predicted that **climate change** will be one of the major drivers of patterns of species abundance and distribution in the future. Changes to temperature ranges and precipitation will alter the composition and distribution of vegetation. How might changes in vegetation patterns affect the distribution and abundance of animal species?

_____

_____

5. Species do not respond to climate change equally. On a separate sheet of paper, discuss some of the likely consequences of climate change to the composition and stability of **ecosystems**. What ecosystems do you think are most at risk and why? Attach your response to this page.

    LS2.C LS4.D   SC

©2022 **BIOZONE** International
ISBN: 978-1-98-856692-4
Photocopying Prohibited

CHAPTER 9

Social Behavior

Anchoring Phenomenon

Internet or Anternet?: What is the purpose of social group behaviour?   173 183

How and why do animals group socially?

☐ 1 Consider the advantages and disadvantages of different types of social groupings, including solitary, schooling, and dominance hierarchy.   174

☐ 2 Interpret models and discuss how flocking and schooling rules assist maintenance of efficient group movement. Describe some benefits of schooling, bird breeding colonies, and herding behaviour, relating them to a survival advantage for the individuals and groups.   175

☐ 3 Discuss how migration can be both an advantage and disadvantage to birds, linking to how this social behaviour might increase overall survival of individuals. Use a variety of evidence to construct an argument about why birds fly in V formation during migration.   176

☐ 4 Distinguish between eusocial and presocial social groups. Discuss how a eusocial group organisation could contribute to increased survival of an individual.   177

☐ 5 Use evidence to determine if the extent of relatedness of a white fronted bee-eater to another increases the amount of assistance given in chick rearing.   178

How do cooperative behaviors increase the chances of survival?

☐ 6 Define altruism in the context of animal behavior. Explain how co-operative behaviour can increase the chance of survival in both individuals and the group.   179

☐ 7 Explain how cooperative defence is beneficial to the survival of individuals, including mob defence and predator vigilance.   180

☐ 8 Utilize case study information on cooperative attack to suggest reasons for why this type of social behaviour might be advantageous to survival. Construct a line graph to show trends on the hunting success of chimpanzees vs group size.   181

☐ 9 Explain how cooperative food gathering is advantageous to survival.   182

☐ 10 Define a eusocial caste social group, as represented by red harvester ants. Discuss the differentiated roles of food gathering in an ant colony. Compare the advantages and disadvantages of flock social grouping in jackdaws.   183

☐ 11 Using cooperative hunting in chimpanzees as an example, discuss how success in hunting relates to hunting success. Analyze evidence on sentinel meerkat social behavior to assess the claim of altruistic behaviour.   184

173 Internet or Anternet?

Key Question: What is the purpose of social group behavior?

▶ Many species of animals live in collective **social groups**. These groups operate with sometimes complex rules that allow the individuals to work together efficiently to move, gather food, and shelter. Animals in these groups must somehow manage to communicate and co-ordinate their behavior with each other.

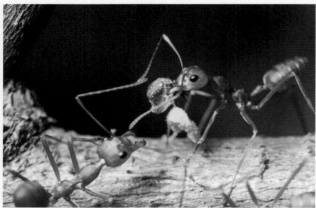

Red harvester ants live in large colonies in Southwestern United States. Individuals work as a social group to forage (find and collect food), dividing roles between leaving the nest to find food and remaining in the nest to collect it. This foraging behavior is governed by strict rules, similar to an algorithm, so that scientists have labelled it the "Anternet".

Jackdaws are members of the crow family and are widespread throughout Europe, Asia, and North Africa. Individuals pair-bond for life, but often form large flocks of up to 40,000 birds that roost (sleep) together at night. They fly in a group, moving around almost as one unit, without crashing into each other.

1. Why do you think animals might work in groups, rather than as individuals, to find food, roost overnight, or travel from one place to another?

   _____

   _____

2. Do you have classroom rules or household rules? How do rules help any group of organisms to live harmoniously, and make the best use of resources?

   _____

   _____

   _____

   _____

3. What prevents individual birds from crashing into one another when they are flying in huge flocks?

   _____

   _____

   _____

   _____

   _____

4 Why do you think groups of animals migrate from one place to another, sometimes over huge distances?

   _____

   _____

   _____

   _____

174 Social Groupings

Key Question: What are the advantages and disadvantages of different types of animal social groupings?

▶ No animal lives completely alone. At some stage in their lives, all animals must interact with others of their species, e.g. to reproduce or to compete for food or other resources.

▶ Generally, animals are either solitary, form loosely associated groups, or form complex groups with clear social structures. Each behavior has its advantages and disadvantages.

Solitary animal

Non-social groups

Social groups

Solitary animals spend the majority of their lives alone, often in defended territories. They may only seek out others of their species for breeding. Offspring are often driven away shortly after they become independent.

Solitary life is often an advantage when resources are scarce or scattered over a large area. Solitary animals include many of the cat family species e.g. tiger (above), as well as bears, and various invertebrates.

Many animals form loose associations but do not interact socially. Each animal is acting directly for its own benefit, with little or no direct cooperation between them. Schools of fish, flocks of birds and many herding mammals exhibit this non-social grouping.

Non-social groups provide protection from predators by reducing the possibility of being preyed upon individually. There may also be benefits during feeding and moving.

Primates form complex social structures which are usually based around a family group. Some animals that form **social groups** also form dominance hierarchies.

Dominance hierarchies help distribute resources and maintain social structure. In some species, members of the group are further divided into specialized roles: some produce offspring or help raise young, others may be workers or help defend the colony, e.g. ants and bees.

Advantages of large social groupings	Disadvantages of large social groups
▶ Protection from physical factors and predators.	▶ Increased competition between group members for resources as group size increases.
▶ Assembly for mate selection.	▶ Increased chance of the spread of disease and parasites.
▶ Locating and obtaining food.	▶ Interference with reproduction, e.g. infanticide by non-parents or cheating in parental care so that non-parents may unknowingly raise another's offspring.
▶ Defending resources against other groups.	
▶ Division of labor amongst specialists.	
▶ Richer learning environment.	
▶ Population regulation, e.g. breeding restricted to a dominant pair.	

1. Give one advantage and one disadvantage of solitary living: _____

_____

2. Explain why group behavior, such as schooling, is more about individual advantage than group advantage:

_____

_____

_____

_____

3. Give one advantage and one disadvantage of a dominance hierarchy: _____

_____

_____

CE LS2.D

175 Schooling, Flocking, and Herding

Key Question: How do schooling, flocking, and herding enhance survival?

Dynamics in a flock, school, or herd

▶ **Schooling** by fish, **flocking** by birds, and **herding** by grazing mammals are essentially all the same behavior. Each individual is behaving in a way that helps its own survival, regardless of the others within that group. Within the group, the application of a few simple rules results in apparently complex behavior.

▶ In a school, flock, or herd, three simple rules tend to apply:

- Move towards the group or others in the group.
- Avoid collision with others in the group and with external objects.
- Align your movement with the movement of the others in the group.

▶ If every individual moves according to these rules, the school, flock, or herd will stay as a dynamic, cohesive unit, changing according to the movement of others and the cues from the environment.

Flock of auklets (a small seabird)

A   B

Schooling dynamics: In schooling fish or flocking birds, every individual behaves according to a set of rules. In diagram A, each fish moves away from the predator while remaining close to each other. The school splits to avoid the predator (A), before moving close together again behind the predator (B).

In a **flash expansion**, each individual moves directly away from the predator. Collisions have never been observed, suggesting each fish is able to sense the direction of movement of the fish next to it.

Why do fish school?

▶ Being part of a group enhances survival by providing protection from predators and by reducing energy expenditure during movement.

▶ Schooling in predatory fish may enhance the ability of any individual to catch its prey. If the prey avoids one predator, it may get caught by the next predator. Prey fish may school for defensive reasons.

Advantages of schooling

▶ **Avoidance of predators**
- Confusion caused by the movement of the school.
- Protection by reduced probability of individual capture.
- Predator satiation (more than the predator could eat).
- Better predator detection (the many eyes effect).

▶ **Better hydrodynamics** within the school, so less energy is used up while swimming.

School of jacks

1. How do the rules for flocking and schooling help to maintain a cohesive group?

_____

_____

_____

_____

_____

©2022 **BIOZONE** International
ISBN: 978-1-98-856692-4
Photocopying Prohibited

Flocking in birds

Flocking in birds follows similar rules to schooling. Each individual maintains a constant distance from others and keeps flying in the average direction of the group. Flocks can be very large, with thousands of birds flying together as a loosely organized unit, such as starlings flocking in the evenings, or queleas (pictured above) flocking over feeding or watering sites. Flocking can be a seasonal activity, usually in winter, with individuals, pairs, or small family groups living independently at other times of the year.

Bird nesting

Some birds, such as gannets (above), occupy nest sites that are spaced close together, often for communal protection and detection of predators. Mobbing behavior of many smaller birds can often deter a much larger predator. Nest sites are often situated in areas close to a sizeable food source. Competition for good sites can be fierce. Male gannets can fight to the death over a spot, as they will not attract a mate without securing one first. There is often conflict to ensure sufficient spacing between birds.

Herding in mammals

Herding is common in hoofed mammals. A herd provides protection because while one animal has its head down feeding, another will have its head up looking for predators. In this way, each individual benefits from a continual supply of lookouts (the many eyes principle). During an attack, individuals move closer to the center of the herd, away from the predator, as those on the outside are more frequently captured. The herd moves as one group, driven by individual needs.

2. Identify some benefits of schooling to a:

   (a) Predatory fish species: _____

   (b) Prey fish species: _____

   _____

3. Explain how herding in grazing mammals provides a survival advantage: _____

   _____

   _____

   _____

   _____

4. Emperor penguins are the only species of penguin that remain in breeding colonies in the Antarctic during winter. Other penguin species are usually very territorial. Females leave the males incubating their eggs on top of their feet while they head out to sea to feed for up to four months. Over the cold winter, males do not eat at all, instead they huddle in huge groups taking turns to move from the outside to the middle of the group. Discuss the advantage of this grouping behavior.

   _____

   _____

   _____

   _____

©2022 **BIOZONE** International
ISBN: 978-1-98-856692-4

176 Migration

Key Question: Why do animals migrate?

▶ **Migration** is the long distance movement of individuals from one place to another. Migration usually occurs on a seasonal basis and for a specific purpose, e.g. feeding, breeding, or over-wintering.

▶ In order for migratory behavior to evolve, the advantages of migration must outweigh the disadvantages. Migration uses up a lot of energy, and animals must spend a lot of time building energy stores before they migrate. The destination provides enough food or shelter to enhance survival of individuals and their offspring.

▶ Although some animals migrate individually, many migrate in large groups (right). Grouping together reduces the energy used and improves survival of all of the migrating individuals in the group.

Benefits of group migration

▶ Group migration helps navigation in what is called the "many wrongs principle," in which the combining of many inaccurate navigational compasses produces a more accurate single compass. If an animal navigates by itself with a slightly inaccurate internal compass, or inaccurately interprets environmental cues, it may arrive in the wrong location. In a group, each member can adjust its heading according to the movement of the others; an average direction is produced and each member is more likely to arrive in the correct place (right).

▶ Birds flying together increase aerodynamic efficiency to each other, saving energy. In schooling fish, individuals in the center of the school use less effort for movement. **Flocking** and **schooling** also provide feeding benefits and decreased risk of predation along the migration route for the individuals in a group.

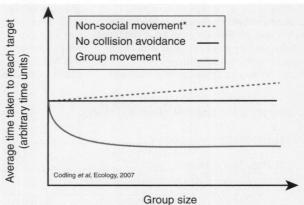

Codling *et al*, Ecology, 2007

Group size

* Individuals do not interact except to avoid collision.

Increasing group size decreases the time taken to reach a navigational target when the group is moving as a social unit. Non-social groups take longer with increasing size because of the need to avoid others in the group.

1. What is migration? _____

2. Describe an advantage and a disadvantage of migration: _____

3. (a) How does grouping together increase navigational efficiency? _____

   (b) How does this enhance individual survival? _____

   LS2.D CE

©2022 **BIOZONE** International
ISBN: 978-1-98-856692-4
Photocopying Prohibited

Why do migrating birds fly in V formation?

▸ Many migrating flocks fly in a V formation (photo, right).

▸ Many species of birds that need to cover large distances to travel to the same location fly by this method. They include ducks, geese, swans, and cormorants, but also larger species of birds such as some pelicans, ibises, and cranes. The rare whooping crane was reduced to just 23 individuals in 1941. Due to conservation success there are now over 800 birds in the wild, and they can again be seen flying in V formation as they migrate.

Migrating geese

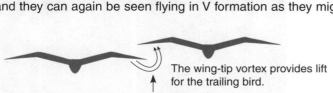
The wing-tip vortex provides lift for the trailing bird.

There are two complementary explanations for why birds fly in formation.

▸ The first reason is to help navigation, coordination, and communication between individuals. The V formation places the birds close to one another, allowing them to hear and see others.

▸ The second reason is that the V formation provides the best aerodynamics for all in the flock except the leader and those trailing at the tips of the V. Each bird gains lift from the movement of the air caused by the bird ahead of it and this saves energy. Moreover, birds regularly take turns at the front.

How do we know?

▸ **Evidence #1:** Research on great white pelicans has shown that the V formation helps the birds conserve energy. As the wing moves down, air rushes from underneath the wing to above it, causing an upward moving vortex behind the wing. This provides lift to the bird flying behind and to one side, requiring it to use less effort to maintain lift. Energy savings come from increased gliding.

▸ **Evidence #2:** Research on the critically endangered Northern bald ibis showed that the flying behavior is cooperative and the amount of time a bird spends leading a formation is correlated with the time it spent flying behind another bird. Overall, individuals spent an average of 32% of their time behind another bird, and a similar amount of time leading a formation (right).

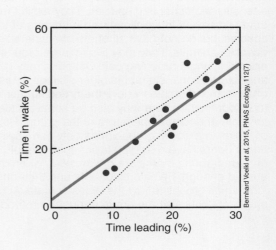
Bernhard Voelkl et al, 2015, PNAS Ecology, 112(7)

4. (a) How does flying in V formation help a bird during migration? _____

_____

_____

_____

_____

(b) How would this help survival in the longer term? _____

_____

_____

_____

_____

5. Discuss how well the provided evidence helps to answer the question "why do migrating birds fly in V formation"?

_____

_____

_____

177 Social Organization

Key Question: How are social groups organized?

In **social groups**, members of the group interact regularly. Social species organize themselves in a way that divides resources and roles between group members.

Eusocial animals

▶ Eusocial animals include ants, honey bees, termites, and naked mole rats. A single female produces the offspring and non-reproductive individuals care for the young. They have the highest form of social organization. Individuals are divided into groups called castes that carry out specific roles. In most cases, there is a queen that produces the young and members of the group are normally directly related to the queen. Non-reproductive members of the group may be involved in care of the young or defense of the nest site.

Honeybees

Termites

Presocial animals

▶ Presocial animals include wolves, many primates, herd animals, and some bird species. They exhibit more than just sexual interactions with members of the same species, but do not have all of the characteristics of eusocial animals. They may live in large groups based around a single breeding pair. Offspring may be looked after by relations, e.g. aunts/older siblings. These groups often form hierarchies where the breeding pair are the most dominant. There may also be separate hierarchies for male and female members of the group. The number of males in a social group varies between species. Wild horse herds have a single stallion that controls a group of mares. Young males are driven away when they are old enough. Female elephants and their offspring form small groups led by the eldest female. Adult male elephants only visit the group during the reproductive season.

1. (a) Identify two examples of eusocial animals: _____

_____

(b) Describe the organization of a eusocial animal group: _____

_____

_____

_____

2. What is the difference between eusocial and presocial groups? _____

_____

_____

_____

3. In eusocial animals, worker and soldier castes never breed but are normally all genetically related. How might their contribution to the group help pass their own genes to the next generation (ensure the survival of their own genes)?

_____

_____

_____

LS2.D

©2022 **BIOZONE** International
ISBN: 978-1-98-856692-4
Photocopying Prohibited

178 How Social Behavior Improves Survival

Key Question: How can social behavioral adaptations in animals improve survival?

By working together, directly or indirectly, members of a group increase each others chances of survival. The level of help given depends on the level of relatedness, e.g. improving foraging success or decreasing the chances of predation. Animals such as meerkats, ground squirrels, and prairie dogs decrease the chances of predation by using sentries, who produce alarm calls when a predator approaches.

Gunnison's prairie dogs

Gunnison's prairie dogs (right) live in large communities called towns in the grasslands of western North America. The towns are divided into territories, which may include up to 20 individuals. During their foraging, above-ground individuals may produce alarm calls if a predator approaches, alerting other prairie dogs to take cover. However, whether or not an alarm call is given, depends on the relatedness of the individuals receiving the call to the individual giving it. Gunnison's prairie dogs put themselves at risk when giving an alarm call by attracting the attention of the predator. Apparently altruistic (self-sacrificing) behavior involving close relatives is called **kin selection**.

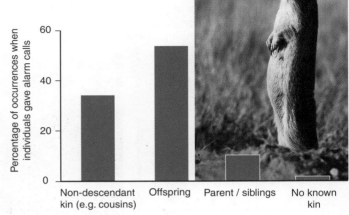

White fronted bee-eaters

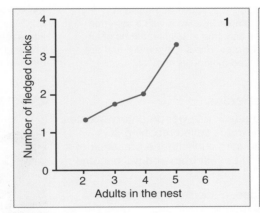

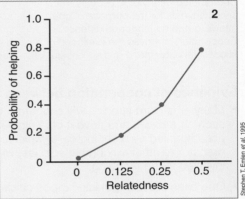

White fronted bee-eaters (left) live in family groups that include a breeding pair and non-breeding pairs. All adults help provide for the chicks. Graph 1 shows the relationship between the number of adults in the nest and the number of chicks fledged. Graph 2 shows how relatedness affects the amount of help the pairs give the chicks.

1. Use an example to explain how living in a group improves survival: _____

_____

_____

_____

2. (a) The level of help between group members often depends on relatedness. Using the examples above, explain how relatedness to the helper affects the level of help given:

_____

_____

_____

(b) With respect to this, what is unexpected about the prairie dog data? _____

_____

_____

©2022 **BIOZONE** International
ISBN: 978-1-98-856692-4
Photocopying Prohibited

CE   LS2.D

179 Cooperative Behaviors

▶ **Cooperative behavior** describes two or more individuals working together to achieve a common goal, such as defense, food acquisition, or rearing young. It increases the probability of survival for all individuals involved. Examples include hunting as a team, e.g. wolf packs, chimpanzee hunts; responding to the actions of others with a similar goal, e.g. migrating mammals; or acting to benefit others, e.g. mobbing in small birds. Cooperation occurs most often between members of the same species.

▶ **Altruism** is an extreme form of cooperative behavior in which one individual disadvantages itself for the benefit of another. Altruism is often seen in highly social animal groups. Most often, the individual who is disadvantaged receives benefit in some non-material form, e.g. increased probability of passing genes onto the next generation.

Coordinated behavior is used by many social animals for the purpose of both attack (group hunting) and defense. Cooperation improves the likelihood of a successful outcome, e.g. a successful kill.

Animals may move en masse in a coordinated way and with a common goal, as in the mass migrations of large herbivores. Risks to the individual are reduced by group behavior.

Kin selection is altruistic behavior towards relatives. Individual meerkats from earlier litters remain to care for new pups instead of breeding themselves. They help more often when more closely related.

Evidence of cooperation between species

▶ Many small bird species will cooperate to attack a larger predatory species, such as a hawk, and drive it off. This behavior is called mobbing. It is accompanied by mobbing calls, which can communicate the presence of a predator to other vulnerable species, which benefit from and will become involved in the mobbing.

▶ One example is the black-capped chickadee, a species that often forms mixed flocks with other species. When its mobbing calls in response to a screech owl were played back, at least ten other species of small bird displayed various degrees of mobbing behaviour. The interspecific communication helps to coordinate the community.

1. (a) What is altruism?_____
    _____

    (b) Why would altruism be more common when individuals are related? _____
    _____
    _____

2. How do cooperative interactions enhance the survival of both individuals and the group they are part of? _____
    _____
    _____

3. What evidence is there that unrelated species can act cooperatively? Why would they do this?_____
    _____
    _____
    _____

LS2.D   CE

©2022 **BIOZONE** International
ISBN: 978-1-98-856692-4
Photocopying Prohibited

180 Cooperative Defense

Key Question: How is cooperative defense used to enhance the survival of individuals in a group?

▶ Group defense is a key strategy for survival in social or herding mammals. Forming groups during an attack by a predator decreases the chances of being singled out, while increasing the chances of a successful defense.

Group defense in musk oxen

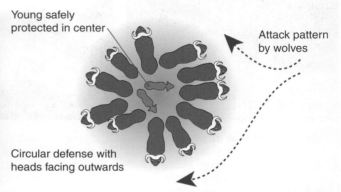

Young safely protected in center

Attack pattern by wolves

Circular defense with heads facing outwards

In the Siberian steppes, which are extensive grasslands, musk oxen must find novel ways to protect themselves from predators. There is often no natural cover, so they must make their own barrier in the form of a defensive circle. When wolves (their most common predator) attack, they shield the young inside the circle. Lone animals have little chance of surviving an attack, as wolves hunt in packs.

Red colobus monkey defense

Red colobus monkeys are a common target during chimpanzee hunts. They counter these attacks by fleeing (especially females with young), hiding, or mounting a group defense. The group defense is usually the job of the males and the more defenders there are, the greater the likelihood of the defense being successful.

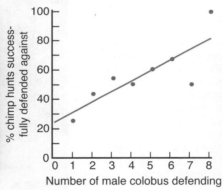

y-axis: % chimp hunts successfully defended against
x-axis: Number of male colobus defending

Japanese honeybees

Japanese honeybees are often attacked by the Asian giant hornet. When a hornet scout enters the honeybee hive, the honeybees mob it with more than 100 bees, forming a bee-ball. The center of the ball can reach 50°C, literally baking the scout to death.

A bee-ball formed around a hornet

1. Describe two benefits of cooperative defense: _____

_____

_____

2. How many colobus males are needed to effectively guarantee a successful defense against chimpanzees?

_____

3. Sheep need to spend most of their day feeding on grass. They form mobs both naturally in the wild as well as on farms.

   (a) Explain why sheep form mobs: _____

   _____

   _____

   (b) Explain how this might enhance an individual sheep's ability to feed: _____

   _____

   _____

CE   LS2.D

181 Cooperative Attack

Key Question: How can working together in attack help increase the chance of success?

Lions hunting buffalo in Africa

Slavemaker ant with 'slave' ants

Asian giant hornet

Lionesses hunt as a coordinated group. Several lionesses hide downwind of the prey, while others circle upwind and stampede the prey towards the lionesses in wait. Group cooperation reduces the risk of injury and increases the chance of a kill. Only 15% of hunts by a solitary lioness are successful. Those hunting in a group are successful 40% of the time.

Some ant species, known as slavemaker ants, raid other ant nests (called slave-raiding), killing workers and capturing grubs. The grubs are carried back to the home nest where they grow and tend the slavemaker ants' own young. Sometimes however, the slaves rebel and can destroy the slavemaker nest.

Wasps and some species of bees, e.g. Africanized honeybees, can be extremely aggressive, although this aggressive behavior is an extension of the defense of the hive. Asian giant hornets often attack honeybee nests to collect grubs to feed their own young. An attack by 50 giant hornets can destroy an entire honeybee colony within a few hours.

The Gombe Chimpanzee War

Group attacks between members of the same species, and even the same **social groups,** do occur. They usually involve disputes over resources or territory, but may be due simply to rifts in social groups. One of the most well recorded and startling examples of group fighting is the Gombe Chimpanzee War. Observed by Jane Goodall, the violence began in 1974, after a split in a group of chimpanzees in the Gombe Stream National Park, Tanzania. The group divided into two, the Kasakela in the northern part of the former territory and the Kahama in the south. Over the course of four years the Kasakela systematically destroyed the Kahama, killing all six males and one female and kidnapping three more females. The Kasakela then took over the Kahama territory. However, ironically, the territorial gains made by the Kasakela were quickly lost as their new territory bordered a larger more powerful group of chimpanzees, the Kalande. After a few violent skirmishes along this border, the Kasakela were pushed back into their former territory.

1. (a) Suggest two reasons for cooperative attacks: _____

_____

_____

(b) Suggest why cooperative attacks are more likely to be successful than individual attacks:

_____

_____

_____

_____

2. Use the data below to draw a graph of the hunting success of chimpanzees vs number of hunters in a group:

No. hunters	1	2	3	4	5	6
Hunt success (%)	13	29	49	72	75	42

NEED HELP? See Activities 17 & 18

  LS2.D CE

©2022 **BIOZONE** International
ISBN: 978-1-98-856692-4
Photocopying Prohibited

182 Cooperative Food Gathering

Key Question: How can cooperative behavior increase the chances of obtaining enough food?

▶ Cooperating to gather food can be much more efficient than finding it alone. It increases the chances of finding food or capturing prey.

▶ Cooperative hunting will evolve in a species if the following circumstances apply:
- There is a sustained benefit to the hunting participants.
- The benefit for a single hunter is less than that of the benefit of hunting in a group.
- Cooperation within the group is guaranteed.

Worker castes in army ants

Orcas hunting

Wild beehive

Cooperative food gathering in ants often involves division of labor. Leaf-cutter ants harvest parts of leaves and use them to cultivate a fungus, which they eat. Workers that tend the fungus gardens have smaller heads than the foragers, who cut and transport the leaves. Similarly, army ants have several distinct worker castes. The smaller castes collect small prey, and larger, porter ants collect larger prey. The largest workers defend the nest.

Dolphins herd fish into shallow water and trap them against the shore where they can be easily caught. When a pod of orcas (killer whales) spots a seal on an iceberg, they swim towards it at high speed before ducking under the ice. This causes a large wave to wash over the iceberg, knocking the seal into the sea where it can be captured. If the seal fails to fall into the water, one of the whales will land itself onto the iceberg, tipping the seal into the water.

Sometimes, different species work together to gather food. The greater honeyguide, a bird that is found in the sub-Sahara, is notable for its behavior of guiding humans (either deliberately or not) to wild beehives. When the humans retrieve the honey, they leave behind some of the wax comb which the honeyguide eats. It is not known why this behavior evolved because the honeyguide can still enter a beehive without human help.

Army ants foraging

▶ There are two species of army ant that have quite different raiding patterns (right): *Eciton hamatum*, whose columns go in many directions, and *Eciton burchelli*, which is a swarm-raider, forming a broad front. Both species cache food at various points along the way.

▶ Through group cooperation, the tiny ants are able to subdue prey much larger than themselves, even managing to kill and devour animals such as lizards and small mammals. This would not be possible if they hunted as individuals.

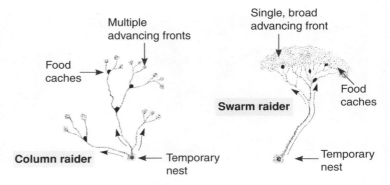

Multiple advancing fronts

Food caches

Column raider

Temporary nest

Single, broad advancing front

Food caches

Swarm raider

Temporary nest

1. Using examples, describe how cooperative food gathering provides an advantage to survival or reproduction: _____

_____

_____

_____

_____

2. What conditions favor cooperative food gathering? _____

_____

_____

_____

 CE LS2.D

183 Review Your Understanding

Key Question: What is the purpose of social group behavior?

▸ Survival of individual animals can depend on how well they are able to gain their food supply while still being able to detect predators and escape them or, if they are predators, how well they can capture prey. Survival can also be influenced by how efficiently animals utilize the energy gained from food, once eaten. **Social groups** can form, and enhance survival of the species, if it is of benefit.

1. The frequency of food collection by red harvester ants is controlled by rules, yet this task is only performed by sterile (cannot reproduce) female ants. The only purpose of male ants is reproduction. The ants live in a eusocial caste social group. Define this term:

_____

_____

_____

_____

_____

2. Food gathering rules are based on a cost-benefit calculation and, if the energy cost is determined to be too high, the ants will stop foraging for a while. Why would this social behavior not occur in solitary insects?

_____

_____

_____

_____

3. Flying in large flocks comes at an energetic cost for jackdaws, as they constantly need to be aware of other jackdaws' positions so they do not collide. They also need to fly for longer to find sufficient food to feed greater numbers of birds. What is the advantage of this flocking social behavior of jackdaws, over remaining as pairs?

_____

_____

_____

_____

_____

_____

©2022 **BIOZONE** International
ISBN: 978-1-98-856692-4

184 Summing Up

Cooperative hunting in chimpanzees

Chimpanzees benefit from cooperative hunting. Although they may hunt alone, they also form hunting groups of up to six members or more. Chimpanzee hunts differ from the cooperative hunting of most other animals in that each chimpanzee in the hunt has a specific role in the hunt, such as a blocker or ambusher. Studies of chimpanzee hunting show that different groups employ different hunting strategies.

Hunt information in table 1 was gathered from chimpanzees in the Tai National Park in Ivory Coast, Africa.

Number of hunters	Number of hunts	Hunting success (%)	Meat per hunt (kg)	Net benefit per hunter (kJ)
1	30	13	1.23	4015
2	34	29	0.82	1250
3	39	49	3.12	3804
4	25	72	5.47	5166
5	12	75	4.65	3471
6	12	42	3.17	1851
>6	10	90	9.27	5020

Christophe Boesch 1994

Hunt information in table 2 was gathered from chimpanzees in the Gombe Stream National Park in Tanzania, Africa.

Number of hunters	Number of hunts	Hunting success (%)	Meat per hunt (kg)	Net benefit per hunter (kJ)
1	30	50	1.23	4245
2	13	61	1.85	3201
3	9	78	1.61	1837
4	7	100	2.86	2494
5	1	100	3.00	2189
6	2	50	2.00	861

Christophe Boesch 1994

NEED HELP?
See Activities
17 & 18

1. Use the information in the table to discuss the differences between the two groups of chimpanzees in the extent of cooperation and how it relates to hunting success. You should plot graphs to help illustrate reasons for differences:

_____

_____

_____

_____

_____

CE LS2.D

_____

_____

_____

_____

_____

Sharing and bonding in chimpanzees

In Tai chimpanzees, hunting is a chance to form social bonds. Study the information below, showing the number of chimpanzees taking part in a hunt and eating afterwards, and the mean (average) number of bystanders during the hunt and eating afterwards.

2. Explain what the information is showing and discuss the reasons why this might occur:

Number of hunters	Mean number of hunters eating	Mean number of bystanders	Mean number of bystanders eating
1	0.7	3.5	3.0
2	1.6	3.6	2.6
3	2.5	3.6	3.0
4	2.5	2.7	2.1
5	3.5	2.7	2.3
6	4.7	2.4	2.2

Christophe Boesch 1994

_____

_____

_____

_____

_____

Sentinel behavior in meerkats

Meerkats are highly social carnivores that live in mobs consisting of a dominant (alpha) breeding pair and up to 40 subordinate helpers of both sexes who do not normally breed but are usually related to the alpha pair. They are known for their sentinel behavior, watching for predators and giving alarm calls when they appear.

The graphs (right) show the likelihood of female or male meerkats standing sentinel when pups are either in the burrow or outside in the sentinel's group. The scale represents a statistical measure from a large number of observations. Error bars are ± SE.

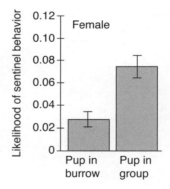
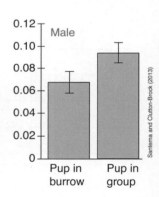

Santema and Clutton-Brock (2013)

3. Discuss the evidence that meerkat sentinel behavior is altruistic in its nature: _____

_____

_____

_____

_____

_____

_____

Heredity: Inheritance and Variation of Traits

Concepts and connections
Use arrows to make connections between related concepts in this section of the book

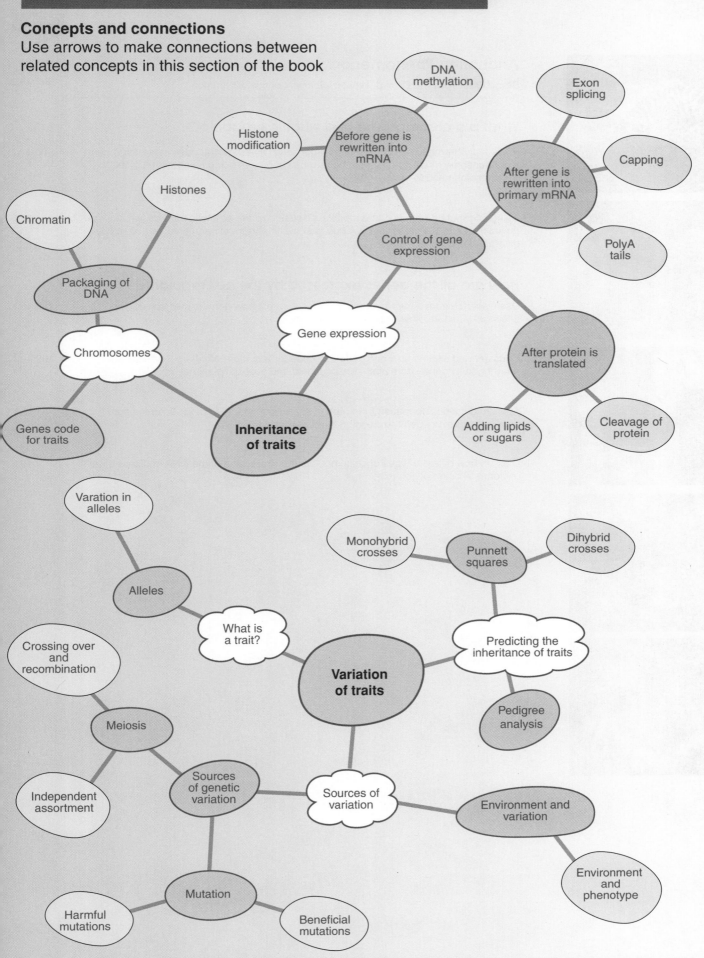

DNA methylation

Exon splicing

Histone modification

Before gene is rewritten into mRNA

After gene is rewritten into primary mRNA

Capping

Histones

Chromatin

After protein is translated

Packaging of DNA

Control of gene expression

PolyA tails

Chromosomes

Gene expression

Genes code for traits

Inheritance of traits

Adding lipids or sugars

Cleavage of protein

Varation in alleles

Monohybrid crosses

Punnett squares

Dihybrid crosses

Crossing over and recombination

Alleles

What is a trait?

Predicting the inheritance of traits

Meiosis

Variation of traits

Pedigree analysis

Independent assortment

Sources of genetic variation

Sources of variation

Environment and variation

Mutation

Environment and phenotype

Harmful mutations

Beneficial mutations

Inheritance of Traits

Toby Hudson CC 3.0

Activity number

Anchoring Phenomenon

Stand Out From the Crowd: What is albinism, and how common is it?

185 192

What is a chromosome and what is it made of?

☐ 1 Know that chromosomes consist of a single, long DNA molecule. Use information from a chromosome model to describe the purpose of histones and components of the chromosome used to make protein.

186 193

☐ 2 Using information from text and models, discuss how the scientific method was able to reveal the heritable nature of DNA due to some important experiments carried out by different groups of scientists.

187

How are all the genes expressed by the cell regulated?

☐ 3 Use evidence from a chromosome model to suggest how gene expression can be affected by factors other than DNA sequence.

188 193

☐ 4 Consider and explain how one gene can code for a number of different proteins, referring to evidence from post-transcriptional and post-translational modeling.

189 193

5 Analyze detailed information from a gene expression model to describe some non-protein products of gene expression and their likely function.

190

☐ 6 Discuss how different types of cells regulate the expression of the DNA in different ways, exploring a range of contexts.

191

185 Stand Out From the Crowd

Key Question: What is albinism, and how common is it?

Albinism is widespread throughout the animal kingdom

Albinism is an inherited genetic disease resulting in the absence of pigmentation or coloration. The condition is widespread throughout the animal kingdom, but more common in birds, reptiles, and amphibians than it is in mammals. Affected individuals have a characteristic appearance with white or extremely pale skin and hair, if present. In many animals, this is accompanied by red or pink eye color. Individuals with albinism are subject to different selection pressures from non-albinos.

1. From your own observations and using your prior knowledge, how common do you think albinism is within populations?

   _____

   _____

   _____

   _____

2. (a) Albinism is an inherited condition. Later, you will learn about dominant and recessive genes. From your knowledge of the condition, do you think it is a dominant or recessive trait?

   _____

   (b) Explain your answer: _____

   _____

3. What environmental conditions might negatively affect an individual with albinism? _____

   _____

   _____

4. What can an albino human do to protect themselves from environmental pressures, that a wild animal can't?

   _____

   _____

186 Chromosomes

Key Question: What are genes and what are they made of?

- In eukaryotes, **DNA** is associated with certain proteins to form **chromatin**. The proteins in the chromatin are responsible for packaging the chromatin into discrete linear structures called **chromosomes**. The extent of packaging changes during the life cycle of the cell, with the chromosomes becoming visible during mitosis.

- Each chromosome includes protein-coding regions called **genes**. Segments of DNA called promoters and enhancers are found at the start of eukaryotic genes. These are involved in beginning the process of transcribing the gene into mRNA.

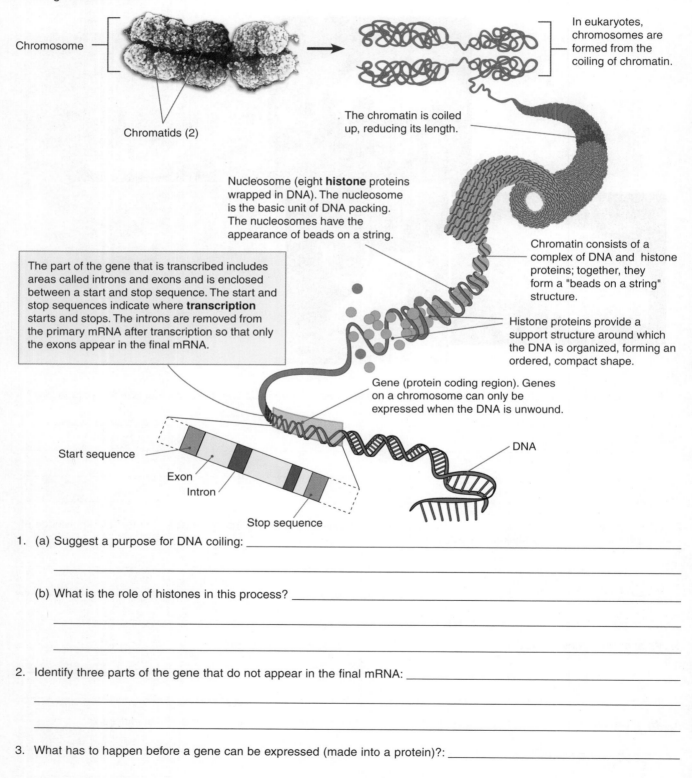

Chromosome

Chromatids (2)

In eukaryotes, chromosomes are formed from the coiling of chromatin.

The chromatin is coiled up, reducing its length.

Nucleosome (eight **histone** proteins wrapped in DNA). The nucleosome is the basic unit of DNA packing. The nucleosomes have the appearance of beads on a string.

Chromatin consists of a complex of DNA and histone proteins; together, they form a "beads on a string" structure.

The part of the gene that is transcribed includes areas called introns and exons and is enclosed between a start and stop sequence. The start and stop sequences indicate where **transcription** starts and stops. The introns are removed from the primary mRNA after transcription so that only the exons appear in the final mRNA.

Histone proteins provide a support structure around which the DNA is organized, forming an ordered, compact shape.

Gene (protein coding region). Genes on a chromosome can only be expressed when the DNA is unwound.

DNA

Start sequence

Exon

Intron

Stop sequence

1. (a) Suggest a purpose for DNA coiling: _____

   _____

   (b) What is the role of histones in this process? _____

   _____

   _____

2. Identify three parts of the gene that do not appear in the final mRNA: _____

   _____

   _____

3. What has to happen before a gene can be expressed (made into a protein)?: _____

   _____

  LS3.A

©2022 **BIOZONE** International
ISBN: 978-1-98-856692-4
Photocopying Prohibited

187 DNA Carries the Code

Key Question: How do we know that DNA is the heritable material responsible for the characteristics we see in organisms?

▸ Many years before Watson and Crick discovered the structure of DNA, biologists had discovered, through experimentation, that **DNA** carried the information that was responsible for the heritable traits we see in organisms.

▸ Prior to the 1940s, it was thought that proteins carried the code. The variety of protein structures and functions suggested they could account for the many traits we see in organisms.

▸ Two early experiments, one by Griffith and another by Avery, MacLeod, and McCarty, provided important information about how traits could be passed on and what cellular material was responsible. The experiments involved strains of the bacterium *Streptococcus pneumoniae*. The S strain is pathogenic (causes disease). The R strain is harmless.

Griffith (1928)

▸ Griffith found that, when he mixed heat-killed pathogenic bacteria with living harmless cells, some of the living cells became pathogenic. Moreover, the newly acquired trait of pathogenicity was inherited by all descendants of the transformed bacteria. He concluded that the living R cells had been transformed into pathogenic cells by a heritable substance from the dead S cells.

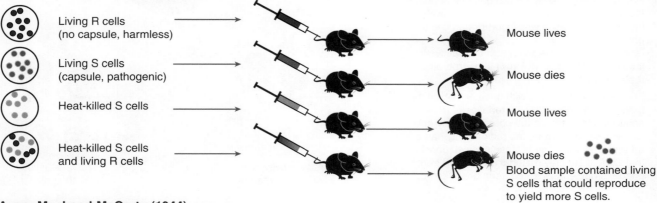

Avery-MacLeod-McCarty (1944)

▸ What was the unknown transformation factor in Griffith's experiment? Avery designed an experiment to determine if it was RNA, DNA, or protein. He broke open the heat-killed pathogenic cells and treated samples with agents that inactivated either protein, DNA, or RNA. He then tested the samples for their ability to transform harmless bacteria.

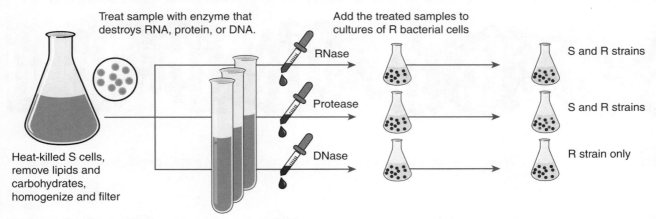

1. Griffith did not predict transformation in his experiment. What results was he expecting? Explain: _____

_____

_____

2. (a) What did Avery's experimental results show? _____

_____

(b) How did Avery's experiment build on Griffith's findings? _____

_____

_____

Hershey and Chase (1952)

▸ Despite the findings of Avery and his colleagues, the scientific community were slow to accept the role of **DNA** as the carrier of the code. The approaches they used were not fashionable and some scientists criticized the results, saying the procedures they used led to protein contamination. At the time, protein was still favored as the carrier of the code because nucleic acids were not believed to have any biological activity and their structure was not defined. Despite the importance of the work, Avery and his colleagues were overlooked for a Nobel Prize.

▸ The work of Hershey and Chase followed the work of Avery and his colleagues and was instrumental in the acceptance of DNA as the hereditary material. Hershey and Chase worked on viruses called **phages**, which infect bacteria. Phages are composed of only DNA and protein. When they infect, they inject their DNA into the bacterial cell, leaving their protein coat stuck to the outside.

▸ Hershey and Chase used two batches of phage. Batch 1 phage were grown with radioactively labeled sulfur, which was incorporated into the phage protein coat.
Batch 2 phage were grown with radioactively labeled phosphorus, which was incorporated into the phage DNA.

▸ The phage were mixed with bacteria, which they then infected. When the bacterial cells were separated from the phage coats, Hershey and Chase looked at where the radioactivity had ended up (right). Hershey and Chase showed conclusively that DNA is the only material transferred from phage to bacteria when bacteria are infected.

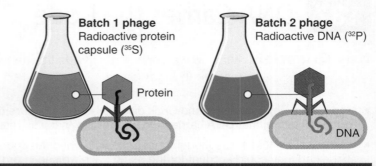

Batch 1 phage
Radioactive protein capsule (^{35}S)

Batch 2 phage
Radioactive DNA (^{32}P)

Homogenization (blending) separates phage outside the bacteria from the cells and their contents. After centrifugation (spinning down), the cells and their DNA form a pellet. Viral protein coats are left in the supernatant (liquid).

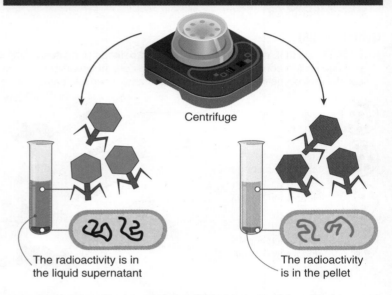

Centrifuge

The radioactivity is in the liquid supernatant

The radioactivity is in the pellet

▸ The combined efforts of these three important teams of scientists illustrates that science is an ever evolving process in which teams of researchers build on the work of others, often over decades, or longer.

▸ In 1953, just one year after Hershey and Chase demonstrated that DNA carried hereditary information, James Watson and Frances Crick published their work on the actual structure of DNA: the now familiar double helix.

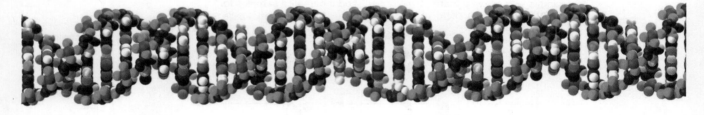

3. (a) How did the Hershey-Chase experiment provide evidence that nucleic acids, not protein are the hereditary material?

_____

_____

(b) How would the results of the experiment have differed if proteins carried the genetic information? _____

_____

(c) Why do you think the Hershey-Chase experiment was so successful in convincing the scientific community at the time that DNA was the material that carried the genetic code? Why weren't Avery's experiments equally successful?

_____

_____

_____

_____

©2022 **BIOZONE** International
ISBN: 978-1-98-856692-4
Photocopying Prohibited

188 DNA Packaging and Control of Transcription

Key Question: Does the way in which DNA is packed affect gene expression?

▶ Regulation of **gene expression** in eukaryotes is a complex process, beginning before the **DNA** is even transcribed to RNA. The packaging of DNA regulates gene expression either by making the **chromatin** pack together tightly or more loosely. This affects whether or not RNA polymerase can attach to the DNA and transcribe it into messenger RNA and eventually translate it into proteins.

▶ Packaging of DNA is affected by **histone modification** and DNA methylation. These modifications alter how the DNA is packaged and determine whether or not a **gene** can be transcribed. The modifications are called **epigenetic** ("above genetics") because they do not involve changes to the DNA sequence itself. Epigenetic tags help to regulate gene expression as the cell differentiates, and record a cell's history as it experiences different, changing environments.

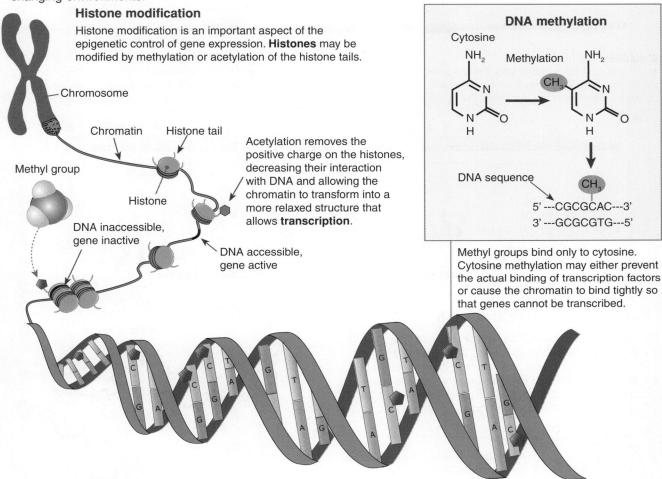

Histone modification

Histone modification is an important aspect of the epigenetic control of gene expression. **Histones** may be modified by methylation or acetylation of the histone tails.

Chromosome

Chromatin   Histone tail

Methyl group

Histone

Acetylation removes the positive charge on the histones, decreasing their interaction with DNA and allowing the chromatin to transform into a more relaxed structure that allows **transcription**.

DNA inaccessible, gene inactive

DNA accessible, gene active

DNA methylation

Cytosine

NH₂   Methylation   NH₂

DNA sequence

5' ---CGCGCAC---3'
3' ---GCGCGTG---5'

Methyl groups bind only to cytosine. Cytosine methylation may either prevent the actual binding of transcription factors or cause the chromatin to bind tightly so that genes cannot be transcribed.

1. (a) Describe the effect of histone modification and DNA methylation on DNA packaging: _____

_____

_____

(b) How do these processes affect transcription of the DNA? _____

_____

_____

2. When a zygote forms at fertilization, most of the epigenetic tags are erased so that cells return to a genetic "blank slate", ready for development to begin. However, some epigenetic tags are retained and inherited. Why do you think it might be advantageous to inherit some epigenetic tags from a parent?

_____

_____

CE   LS3.A

189 Changes after Transcription and Translation

Key Question: How are primary mRNA molecules modified in the nucleus before being translated into proteins, and how are proteins modified after translation?

▶ Human **DNA** contains 25,000 genes, but produces up to 1 million different proteins. Each **gene** must therefore produce more than one protein. This is achieved by both post transcriptional and post translational modification.

Post transcriptional modification

▶ Primary mRNA contains exons and introns. Introns are usually removed after **transcription** and exons are spliced together. However, there are many alternative ways to splice the exons and the alternatives create variations in the translated proteins. In mammals, the most common method of alternative splicing involves exon skipping, in which not all exons are spliced into the final mRNA (below). Other alternative splicing options create further variants.

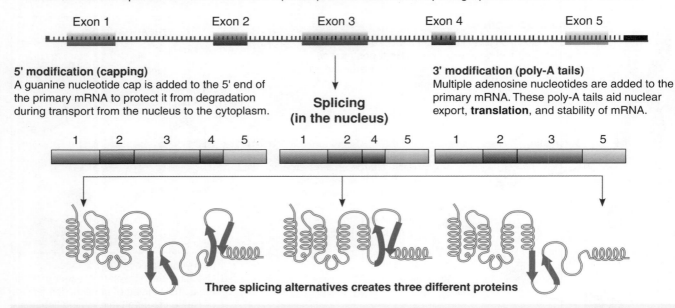

5' modification (capping)
A guanine nucleotide cap is added to the 5' end of the primary mRNA to protect it from degradation during transport from the nucleus to the cytoplasm.

Splicing (in the nucleus)

3' modification (poly-A tails)
Multiple adenosine nucleotides are added to the primary mRNA. These poly-A tails aid nuclear export, **translation**, and stability of mRNA.

Three splicing alternatives creates three different proteins

Post translational modification

▶ After proteins have been made, they can be modified by adding carbohydrate, lipid, or phosphate groups. These modifications are involved in completing the protein's functional role. Some proteins are first made as a single polypeptide and then cleaved and reformed into a functional form, e.g. insulin.

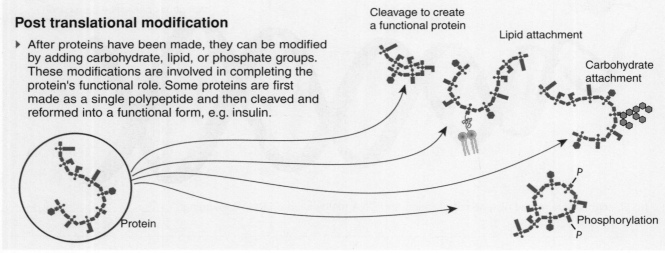

Cleavage to create a functional protein

Lipid attachment

Carbohydrate attachment

Protein

Phosphorylation

1. How can so many proteins be produced from so few genes? _____

_____

_____

2. What is the advantage of being able to modify the mRNA to produce different proteins? _____

_____

3. If a human produces 1 million proteins, but human DNA only codes for 25,000 genes, on average how many proteins are produced per gene?

_____

LS3.A | CE

©2022 **BIOZONE** International
ISBN: 978-1-98-856692-4
Photocopying Prohibited

190 Not All DNA Codes for Protein

Key Question: How much of an organism's DNA actually codes for protein, and what is the function of the non protein-coding DNA?

▶ The sequencing of the human genome, completed in 2003, revealed that humans had only 20,000-25,000 protein-coding genes, far fewer than the predictions of 100,000 or more. With improvement in sequencing technology, scientists now know that multitasking genes make more than one protein and about 98% of the genome does not code for proteins at all! These non protein-coding **DNA** sequences include introns and regulatory sequences, and DNA that encodes RNA. Among eukaryotes, greater complexity is associated with a higher proportion of non protein-coding DNA. This makes sense if the non protein-coding DNA is involved in regulating genomic function.

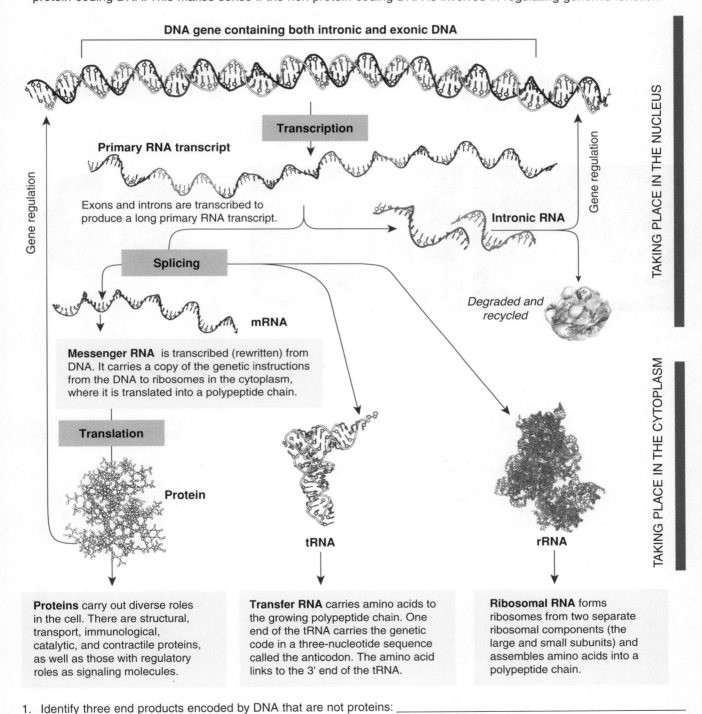

DNA gene containing both intronic and exonic DNA

Transcription

Primary RNA transcript

Exons and introns are transcribed to produce a long primary RNA transcript.

Intronic RNA

Splicing

Gene regulation

Gene regulation

TAKING PLACE IN THE NUCLEUS

mRNA

Degraded and recycled

Messenger RNA is transcribed (rewritten) from DNA. It carries a copy of the genetic instructions from the DNA to ribosomes in the cytoplasm, where it is translated into a polypeptide chain.

Translation

Protein

tRNA

rRNA

TAKING PLACE IN THE CYTOPLASM

Proteins carry out diverse roles in the cell. There are structural, transport, immunological, catalytic, and contractile proteins, as well as those with regulatory roles as signaling molecules.

Transfer RNA carries amino acids to the growing polypeptide chain. One end of the tRNA carries the genetic code in a three-nucleotide sequence called the anticodon. The amino acid links to the 3' end of the tRNA.

Ribosomal RNA forms ribosomes from two separate ribosomal components (the large and small subunits) and assembles amino acids into a polypeptide chain.

1. Identify three end products encoded by DNA that are not proteins: _____

2. Briefly describe the roles of these three products: _____

_____

_____

©2022 **BIOZONE** International
ISBN: 978-1-98-856692-4
Photocopying Prohibited

LS3.A

191 The Outcomes of Differing Gene Expression

Key Question: How do variations in the way genes are expressed cause significant differences between cells or organisms, even if their DNA is identical?

Same genes, different result

All the cells in a multicellular organism have identical **DNA**. As an organism develops from the zygote, differences in the way the DNA is expressed in developing cells cause them to differentiate into different types (right). Cells of the same type can express different proteins or amounts of protein depending on the environment they are exposed to during development, or due to random **gene** inactivations (below).

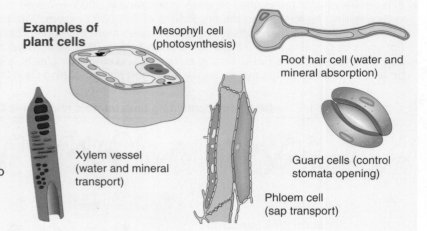

Examples of plant cells

Mesophyll cell (photosynthesis)

Root hair cell (water and mineral absorption)

Xylem vessel (water and mineral transport)

Guard cells (control stomata opening)

Phloem cell (sap transport)

Female mammals have two X **chromosomes** in every cell but one of them is inactivated. In cats, the gene for coat color is found on the X chromosome. Cats with two different alleles for coat color will only have one active allele per pigment cell, so each cell will produce one or the other color, giving a patchwork (calico) coat.

All worker bees and the queen bee (circled above) in a hive have the same genome, yet the queen looks and behaves very differently from the workers. Only bee larvae fed a substance called royal jelly will develop into queens. Royal jelly contains factors that silence the activation of a specific gene (Dnmt3), which then silences other genes.

Twins have the same genome, but over their life become different in both appearance and behavior. Studies have found that 35% of twins have significant differences in **gene expression**. More importantly, the older the twins, the more difference there is in the gene expression.

1. How do cells with identical DNA differentiate into different types? _____

_____

2. Why are patchwork (calico) cats always female? _____

_____

_____

_____

3. Studies on bee development focused on the Dnmt3 gene. One study switched off the Dnmt3 gene in 100 bee larvae. All the larvae developed into queens. Leaving the gene switched on in larvae causes them to develop into workers. Compare these results to feeding larvae royal jelly. Which results mimic feeding larvae royal jelly? Explain:

_____

_____

_____

_____

_____

LS3.A CE

©2022 **BIOZONE** International
ISBN: 978-1-98-856692-4
Photocopying Prohibited

192 Review Your Understanding

Key Question: What is albinism, and how common is it?

▸ The anchoring phenomenon for this chapter was albinism. Now, you can apply the knowledge you have gained in this chapter to answer the following questions.

▸ The most common form of albinism results from a mutation to the TYR **gene**. The TYR gene produces an enzyme called tyrosinase, which is needed to make a pigment called melanin. Melanin gives the skin, hair, and eyes their color. People with albinism have either a partial or complete lack of pigment. Medical conditions associated with albinism include skin cancer and vision problems.

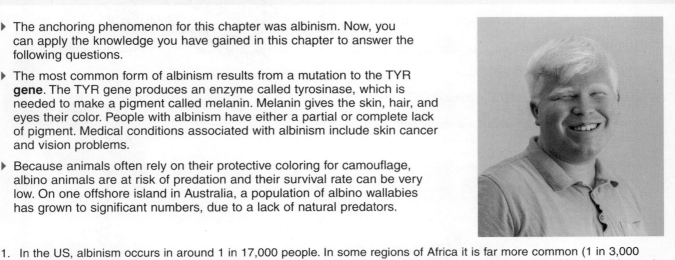

▸ Because animals often rely on their protective coloring for camouflage, albino animals are at risk of predation and their survival rate can be very low. On one offshore island in Australia, a population of albino wallabies has grown to significant numbers, due to a lack of natural predators.

1. In the US, albinism occurs in around 1 in 17,000 people. In some regions of Africa it is far more common (1 in 3,000 people). One hypothesis suggests that albinism protects against a debilitating bacterial disease called leprosy (Hansen's disease), which has a high prevalence in some African countries. Some researchers have suggested that the bacterium is killed by UV light, which penetrates further into skin lacking melanin than skin with normal pigmentation. In 2011, the World Health Organization reported 12,673 new cases of leprosy in Africa, and 173 cases in the USA. New infection rates for leprosy are declining slowly worldwide, following the WHO's resolution to reduce rates to < 1 per 10,000 by the year 2000 (achieved).

    (a) Do you think the data above support the hypothesis that albinism protects against leprosy? Why or why not?

    _____

    _____

    (b) What other factors could account for the lower number of cases in the US? _____

    _____

    _____

    (c) If albinism protects against leprosy, predict what might happen to the rates of albinism in Africa as leprosy rates fall:

    _____

2. Albino animals in the wild are under threat from predation, but other factors related to albinism can pose dangers to them as well. What do you think these might be?

    _____

    _____

    _____

    _____

    _____

3. Why should parents, who suspect that a very blonde haired baby is albino, have it tested for the albino gene mutation?

    _____

    _____

    _____

193 Summing Up

Complete the following questions, with reference to the diagram on the right:

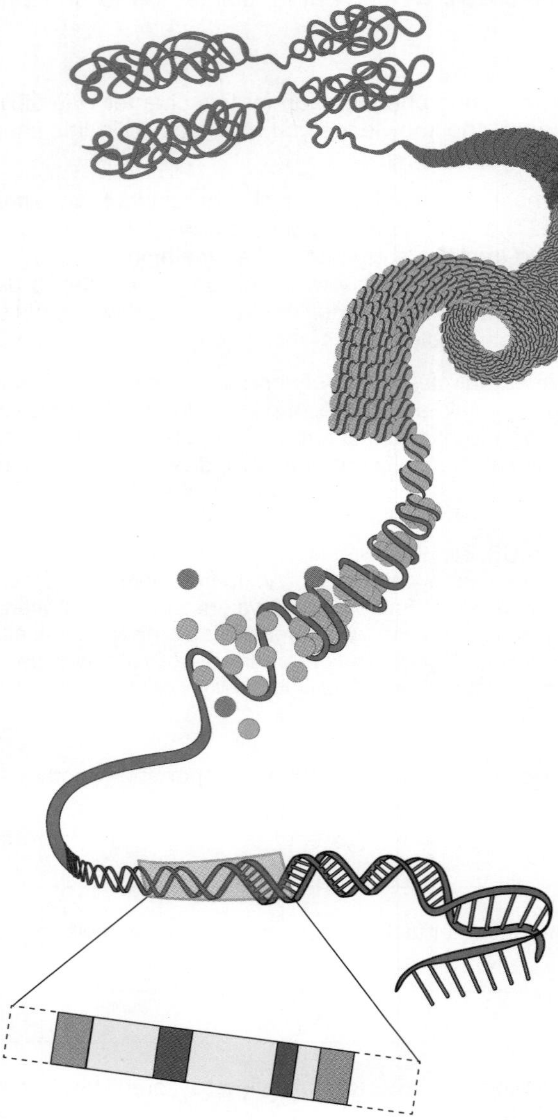

1. (a) Circle the chromosome and label the chomatids.

    (b) What is a chromatid?_____

    _____

2   (a) Label the histone proteins and the DNA.

    (b) What is the name given to the material formed by these two components?

    _____

    (c) What is the role of the histone proteins? _____

    _____

    _____

3. (a) Label the gene, the start sequence, stop sequence, intron, and exons.

    (b) Which regions of the gene are expressed?

    _____

4. (a) In what state must the DNA be in order to be transcribed?

    _____

    (b) This is achieved by (select the correct answer):

    (i) DNA methylation

    (ii) Histone acetylation

    (iii) Histone methylation

5. (a) How does the cell prevent transcription of a gene?

    _____

    _____

    (b) How is this achieved? _____

    _____

    _____

6. Describe how messenger RNA can be modified after being transcribed from DNA? _____

    _____

    _____

    _____

7. How can proteins be modified to have different functions, after translation? _____

    _____

    _____

8. What is the name given to the differences in gene expression that take place over the course of a person's lifetime?

    _____

©2022 **BIOZONE** International
ISBN: 978-1-98-856692-4
Photocopying Prohibited

CHAPTER 11

Variation of Traits

Activity number

Toby Hudson CC 3.0

Marc King

Marc King

Marc King

Anchoring Phenomenon

Anyone for chocolate? Can we get a chocolate labrador puppy from black parents?

194 218

How does variation result from genetic processes?

☐ 1 Define the terms trait and true-breeding. Use information from a pedigree chart model of Mendel's pea experiment to predict inherited traits in successive generations. Define the terms homozygous, heterozygous, recessive, and dominant, in the context of genetic inheritance.

195 196

☐ 2 Use evidence from inheritance models of sexual and asexual reproduction to argue that mutations can be an important source of variation.

197

☐ 3 Define the terms mutation, genotype, and phenotype. Explain how different factors can influence phenotype.

198

☐ 4 Distinguish between continuous and discontinuous variation and investigate several phenotypic traits in your class.

199

☐ 5 Explain how meiosis, and how the processes of independent assortment, crossing over, and recombination introduce variation in a population.

200 201 202

☐ 6 Explain how mutation introduces variation, and can be beneficial to an organism. Identify different types of mutations in a DNA segment. Compare and contrast beneficial and harmful mutations in organisms.

203 204

☐ 7 Understand how antibiotic resistance evolves and can cause harmful impacts to humans. Investigate the spread of antibiotic resistance by using a spreadsheet to model bacterial growth data.

205

☐ 8 Explain how beneficial human mutations can spread in a population. Explain the impact on humans of a mutation causing Huntington's disease.

206 207

How does variation result from environmental influences?

☐ 9 Discuss how a phenotype is the expression of both genetic and environmental influences.

208 209

☐ 10 Explain how environmental influences experienced by one generation may affect subsequent generations.

210

How can Punnett squares and pedigree charts be used as tools to predict inheritance?

☐ 11 Use Punnett squares to predict the phenotype and genotype of offspring in a monohybrid cross. Demonstrate, using Punnett squares, how a recessive phenotype can be used as a test cross to determine the genotype in an individual displaying a dominant phenotype.

211 212 213

☐ 12 Use Punnett squares to predict the genotype and phenotype of offspring in a dihybrid cross. Use the chi-squared test to test the outcome of dihybrid crosses compared to a predicted ratio.

214 215 216

☐ 13 Use pedigree charts to predict and analyze the genotype of parents and offspring, and trace genetic disorders to its origin.

217

☐ 14 Use evidence from a pedigree chart to determine if lactose intolerance is a dominant or recessive trait. Understand that heritable genetic variation arises through meiosis or mutation.

219

194 Anyone for Chocolate?

Key Question: Can we get a chocolate labrador puppy from black parents?.

▸ Labrador dogs are popular family pets and come in three main colors: black, brown (usually called chocolate), and yellow. There are various shades of each of these three base colors.

▸ You might expect that if two black labradors mate, you would always get black puppies, but that's not necessarily the case. you can cross two black animals and get different colored puppies in the litter.

▸ Puppy color is determined by a number of different genes. If you know what the adults' genotypes are, you will know what color of puppies they might be able to produce.

▸ The gene for yellow color is in a different location from the black/brown gene.

1. What do you think is the likelihood of crossing two black adults and producing a different colored puppy? _____

_____

_____

_____

2. Do you think it might be possible to cross a black labrador with a chocolate labrador and produce yellow puppies?

_____

_____

3. Labrador retrievers can sometimes suffer from a hip disorder. How do you think we could we tell if this condition was caused by genetic factors or dietary and environmental factors?

_____

_____

_____

4. Some labrador breeders import dogs from other regions or even countries. Why do you think they do this?

_____

_____

_____

_____

_____

_____

_____

©2022 **BIOZONE** International
ISBN: 978-1-98-856692-4
Photocopying Prohibited

195 What is a Trait?

Key Question: What are traits, and how are they inherited and passed from one generation to the next?

Traits are inherited

▶ **Traits** are particular variants of phenotypic (observed) characteristics, e.g. eye color. Traits may be controlled by one gene or many genes and can show continuous variation, e.g. height in humans, or discontinuous variation, e.g. flower color in pea plants.

▶ Gregor Mendel, an Austrian monk (1822-1884), used pea plants to study inheritance. Using several phenotypic characteristics he was able to show that traits were inherited in predictable ways.

Mendel's experiments

Mendel studied seven phenotypic characteristics of the pea plant:

- Flower color (violet or white)
- Pod color (green or yellow)
- Height (tall or short)
- Position of the flowers on the stem (axial or terminal)
- Pod shape (inflated or constricted)
- Seed shape (round or wrinkled)
- Seed color (yellow or green)

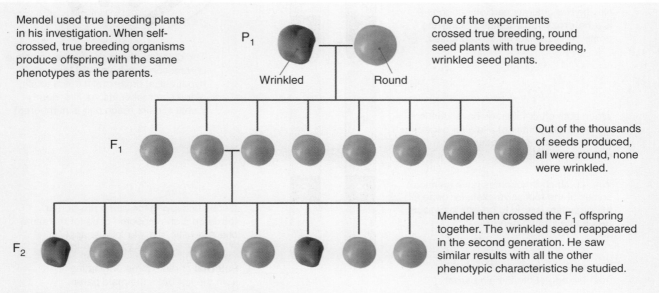

Mendel used true breeding plants in his investigation. When self-crossed, true breeding organisms produce offspring with the same phenotypes as the parents.

P_1 — Wrinkled — Round

One of the experiments crossed true breeding, round seed plants with true breeding, wrinkled seed plants.

F_1

Out of the thousands of seeds produced, all were round, none were wrinkled.

F_2

Mendel then crossed the F_1 offspring together. The wrinkled seed reappeared in the second generation. He saw similar results with all the other phenotypic characteristics he studied.

Three conclusions could be made from these results:

- Traits are determined by a unit that passes unchanged from parent to offspring. These units are now called genes.
- Each individual inherits one gene for each trait from each parent, so each individual has two genes for the trait.
- Traits may not physically appear in an individual, but the genes for them can still be passed to offspring.

1. Define a trait: _____

_____

2. Define true breeding: _____

_____

_____

3. (a) What was the ratio of smooth seeds to wrinkled seeds in the F_2 (second) generation? _____

(b) Suggest why the wrinkled seed trait did not appear in the F_1 (first) generation: _____

_____

_____

_____

©2022 **BIOZONE** International
ISBN: 978-1-98-856692-4
Photocopying Prohibited

CE LS3.B

196 Different Alleles For Different Traits

Key Question: What are alleles, and what determines whether a trait will be passed to an organism's offspring?

Homologous chromosomes

In sexually reproducing organisms, chromosomes are generally found in pairs. Each parent contributes one chromosome to the pair. The pairs are called **homologues** or **homologous pairs**. Each homologue carries an identical assortment of genes, but the version of the gene, known as the **allele,** from each parent may differ. This diagram shows the position of three different genes on the same chromosome that control three different traits (A, B and C).

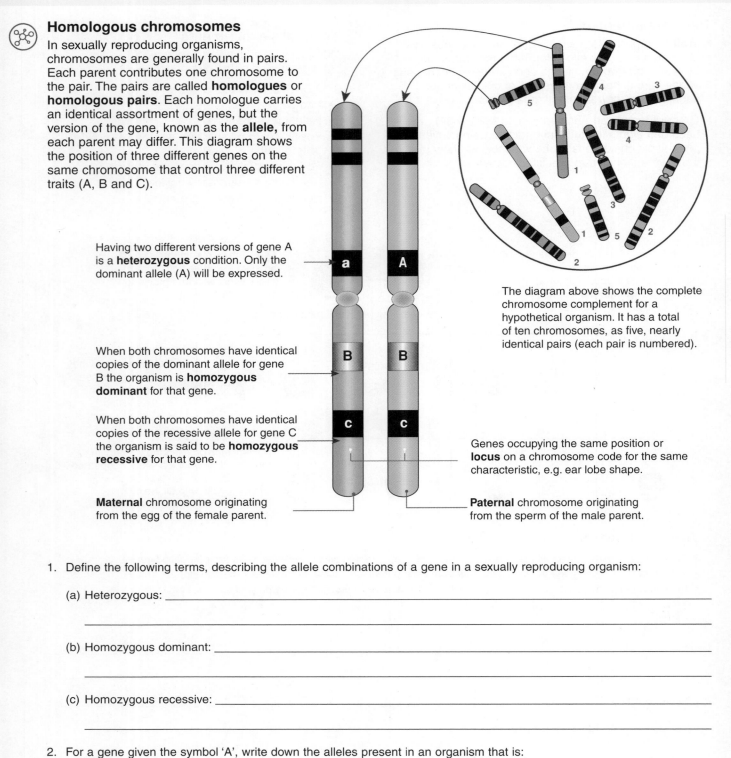

Having two different versions of gene A is a **heterozygous** condition. Only the dominant allele (A) will be expressed.

When both chromosomes have identical copies of the dominant allele for gene B the organism is **homozygous dominant** for that gene.

When both chromosomes have identical copies of the recessive allele for gene C the organism is said to be **homozygous recessive** for that gene.

Maternal chromosome originating from the egg of the female parent.

The diagram above shows the complete chromosome complement for a hypothetical organism. It has a total of ten chromosomes, as five, nearly identical pairs (each pair is numbered).

Genes occupying the same position or **locus** on a chromosome code for the same characteristic, e.g. ear lobe shape.

Paternal chromosome originating from the sperm of the male parent.

1. Define the following terms, describing the allele combinations of a gene in a sexually reproducing organism:

(a) Heterozygous: _____

_____

(b) Homozygous dominant: _____

_____

(c) Homozygous recessive: _____

_____

2. For a gene given the symbol 'A', write down the alleles present in an organism that is:

(a) Heterozygous:_____ (b) Homozygous dominant: _____ (c) Homozygous recessive: _____

3. What is a homologous pair of chromosomes? _____

_____

_____

_____

LS3.B

©2022 **BIOZONE** International
ISBN: 978-1-98-856692-4

197 Why is Variation Important?

Key Question: Why is variation in a population or species important, and what strategies do both sexually and asexually reproducing species have to increase variation?

▸ **Variation** refers to the diversity of **phenotypes** or genotypes within a population or species. Variation helps organisms survive in a changing environment.

▸ Sexual reproduction produces variability, which provides the ability to adapt to a changing physical environment. However, environments can change very slowly and it may take millions of years for a mountain range to rise from the seabed. This is more than enough time for even asexually reproducing species to acquire the variability needed to adapt. However, changes in the biotic environment, such as the appearance of new strains of disease, require a fast response.

▸ Variation is important for defending against disease. Species that evolve to survive a disease, flourish. Those that do not, die out. It is thought that sexual reproduction is an adaptation to increase variability in offspring and so provide a greater chance that any one of the offspring will survive a given disease.

▸ Even species that reproduce asexually for much of the time can show a large amount of variation within a population.

Aphids can reproduce sexually and asexually. Females hatch in spring and give birth to clones. Many generations are produced asexually. Just before fall, the aphids reproduce sexually. The males and females mate and the females produce eggs which hatch the following spring. This increases variability in the next generation.

Diagrams model how three beneficial mutations could be combined through sexual or asexual reproduction.

Variation by sexual reproduction

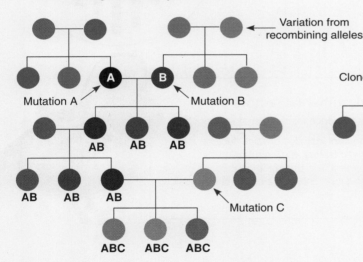

Variation by asexual reproduction

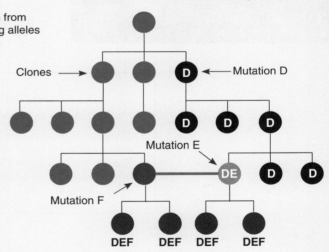

During meiosis, alleles are recombined in new combinations. Some combinations of alleles may be better suited to a particular environment than others. This variability is produced without the need for mutation. Beneficial mutations in separate lineages can be quickly combined through sexual reproduction.

Some asexually reproducing organisms are able to exchange genes occasionally. Bacteria exchange genes with other bacteria during a process called conjugation (thicker red line). This allows mutations that arise in one lineage to be passed to another.

1. Why is variation important in populations or species? _____

_____

_____

_____

©2022 **BIOZONE** International
ISBN: 978-1-98-856692-4
Photocopying Prohibited

CE LS3.B

198 Sources of Variation

Key Question: What are some of the ways in which variation arises?

Mutations
changes to the DNA
Changes to the DNA modifies existing **genes**. Mutations can create new **alleles**.

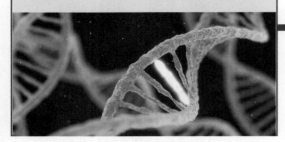

Sexual reproduction
fertilization, and mate selection
Sexual reproduction rearranges and reshuffles the genetic material into new combinations.

Phenotype
The phenotype describes the physical characteristics we see in an organism. It is the result of the expression of the genotype in a particular environment.

Genes interact with each other and with the environment to influence the phenotype.

Genotype
An individual's genetic makeup. It determines that individual's genetic potential.

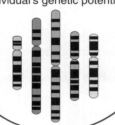

Environmental factors
Environmental factors influence expression of the genotype.

• The external environment includes physical factors, e.g. temperature, and biotic factors, e.g. competition.
• The internal environment, e.g. presence or absence of hormones during development, may also affect expression of the genotype.

1. Define the following terms:

(a) Mutation: _____

(b) Genotype: _____

(c) Phenotype: _____

2. What factors determine the phenotype? _____

_____

_____

3. How could two individuals with the same genotype have a different phenotype? _____

_____

_____

 LS3.B

©2022 **BIOZONE** International
ISBN: 978-1-98-856692-4

199 Examples of Genetic Variation

Key Question: What is continuous and discontinuous variation, and what is the difference between quantitative and qualitative traits?

▸ Individuals show particular variants of phenotypic characters called traits, e.g. eye color.

▸ Traits that show continuous variation are called quantitative traits.

▸ Traits that show discontinuous variation are called qualitative traits.

Quantitative traits

Quantitative traits are determined by a large number of genes. For example, skin color has a continuous number of variants from very pale to very dark. Individuals fall somewhere on a normal distribution curve of the phenotypic range. Other examples include height in humans for any given age group, length of leaves in plants, grain yield in corn, growth in pigs, and milk production in cattle. Most quantitative traits are also influenced by environmental factors.

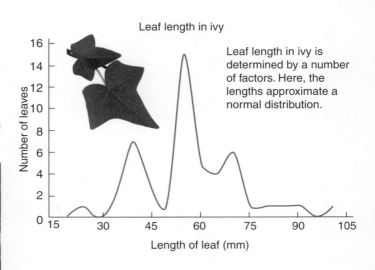

Leaf length in ivy is determined by a number of factors. Here, the lengths approximate a normal distribution.

Grain yield in corn

Growth in piglets

Qualitative traits

Qualitative traits are determined by one or two genes with a very limited number of variants present in the population. For example, blood type in humans has four discontinuous traits A, B, AB or O. Individuals fall into separate categories. Comb shape in poultry (right) is a qualitative trait and birds have one of four phenotypes depending on which combination of four alleles they inherit. The dash (missing allele) indicates that the allele may be recessive or dominant. Albinism is the result of the inheritance of recessive alleles for melanin production. Those with the albino phenotype lack melanin pigment in the eyes, skin, and hair.

Single comb rrpp

Walnut comb R_P_

Pea comb rrP_

Rose comb R_pp

Photos courtesy Marc King

1. What is the difference between continuous and discontinuous variation? _____

_____

_____

_____

2. Identify each of the following phenotypic traits as continuous (quantitative) or discontinuous (qualitative):

(a) Wool production in sheep: _____

(d) Albinism in mammals: _____

(b) Hand span in humans: _____

(e) Body weight in mice: _____

(c) Blood groups in humans: _____

(f) Flower color in snapdragons: _____

SPQ LS3.B

Investigation 11.1   Phenotypic variation in your class

See appendix for equipment list.

1. Choose a phenotypic variable in your class you would like to investigate. This could be a quantitative variable (one that can be measured, such as height) or a qualitative (categorical) variable to which you can assign a ranking, e.g. eye color.

2. Record a value for the variable for each person in your class. Tabulate the data in the space below and plot a histogram of the number in each category.

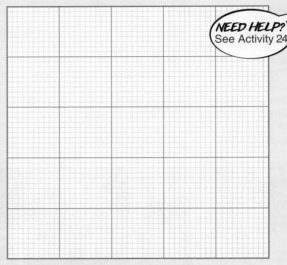

NEED HELP? See Activity 24

NEED HELP? See Activity 8

1. The data in the table below show foot length for 20 adults.

    (a) In the space, construct a tally chart for the data.

    (b) Plot the data as a histogram on the grid below.

Adult foot length (mm)			
265	272	257	315
300	320	250	250
215	330	240	270
252	270	265	350
315	300	290	310

Tally chart

©2022 **BIOZONE** International
ISBN: **978-1-98-856692-4**
Photocopying Prohibited

200 Meiosis

Key Question: What is meiosis, and how does it produce haploid cells for the purposes of sexual reproduction?

Meiosis

▶ **Meiosis** is a special type of cell division necessary for the production of gametes (sex cells) for the purpose of sexual reproduction.

▶ DNA replication precedes meiosis. If mutations (genetic errors) occur here, they will be passed on and inherited by the offspring.

▶ Meiosis involves a single chromosomal duplication followed by two successive nuclear divisions, and it halves the diploid chromosome number.

▶ An overview of meiosis is shown on the right. Meiosis occurs in the sex organs of plants and animals.

Meiosis produces variation

▶ During meiosis, a process called **crossing over** may occur when homologous chromosomes may exchange **genes**. This further adds to the variation in the gametes.

▶ Meiosis is an important way of introducing genetic **variation**. The assortment of chromosomes into the gametes (the proportion from the father or mother) is random and can produce a huge number of possible chromosome combinations.

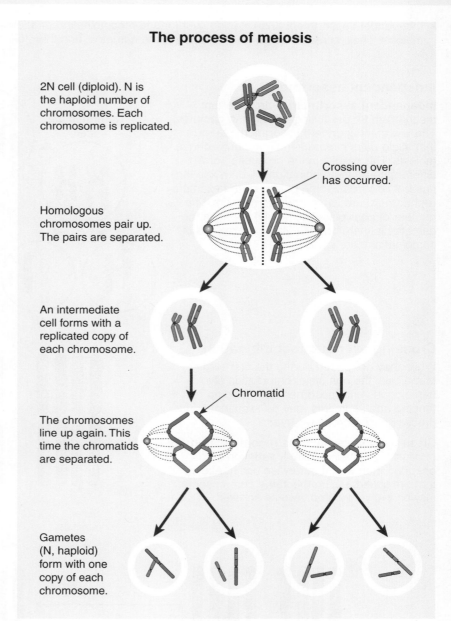

The process of meiosis

2N cell (diploid). N is the haploid number of chromosomes. Each chromosome is replicated.

Crossing over has occurred.

Homologous chromosomes pair up. The pairs are separated.

An intermediate cell forms with a replicated copy of each chromosome.

Chromatid

The chromosomes line up again. This time the chromatids are separated.

Gametes (N, haploid) form with one copy of each chromosome.

---

1. (a) What is the purpose of meiosis? _____

    (b) Where does meiosis take place? _____

    _____

    _____

    _____

2. Describe how variation can arise during meiosis: _____

    _____

    _____

    _____

    _____

201 Meiosis and Variation

Key Question: What are the important ways of introducing variation into the gametes formed during meiosis?

▸ Independent assortment and crossing over lead to recombination of **alleles** and are mechanisms that occur during meiosis. They increase the genetic variation in the gametes, and therefore the offspring.

Independent assortment

Independent assortment is an important mechanism for producing variation in gametes. The law of independent assortment states that allele pairs separate independently during meiosis. This results in the production of 2^x different possible combinations, where x is the number of chromosome pairs. For the example (right), there are two chromosome pairs. The number of possible allele combinations in the gametes is therefore $2^2 = 4$. Only two possible combinations are shown.

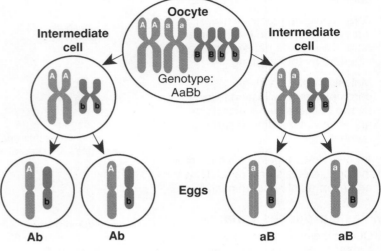

Crossing over and recombination

While they are paired during the first stage of meiosis, the non-sister chromatids of homologous chromosomes may become tangled and segments may be exchanged in a process called **crossing over**.

Crossing over results in the **recombination** of alleles, producing greater variation in the offspring than would otherwise occur. Alleles that are linked, i.e. on the same chromosome, may be exchanged and become unlinked.

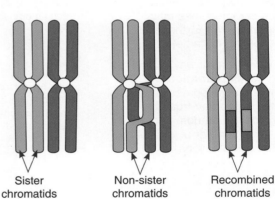

Sister chromatids

Non-sister chromatids

Recombined chromatids

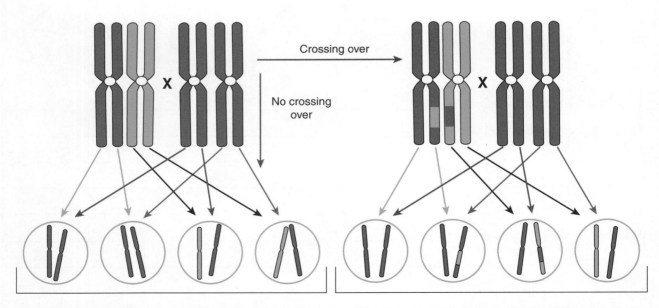

No crossing over, therefore no recombination in this cross results in all the offspring having the same genotypes as the parents.

Crossing over results in recombination. Although half of the offspring are the same as the parents, half have a new genetic combination.

©2022 **BIOZONE** International
ISBN: 978-1-98-856692-4
Photocopying Prohibited

1. (a) Using the diagram on independent assortment (previous page), draw the other two gamete combinations not shown in the diagram:

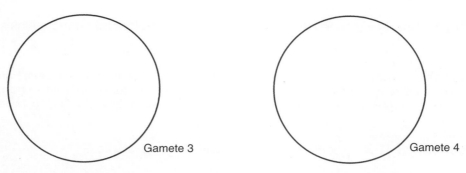

Gamete 3   Gamete 4

(b) For each of the following chromosome numbers, calculate the number of possible gamete combinations:

i.   8 chromosomes: _____

ii. 24 chromosomes: _____

iii 64 chromosomes: _____

2. What are sister and non-sister chromatids? _____

_____

_____

_____

3. (a) What is crossing over? _____

_____

_____

_____

(b) How does crossing over increase the variation in the gametes (and hence the offspring)? _____

_____

_____

_____

4. Crossing over occurs at a single point between the chromosomes below.

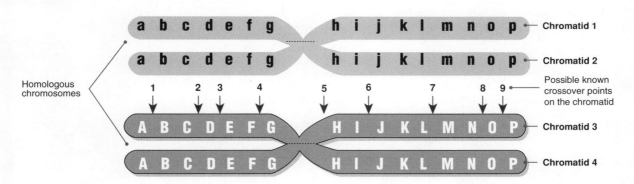

(a) Draw the gene sequences for the four chromatids (above), after crossing over has occurred at crossover point 2:

(b) Which genes have been exchanged between the homologous chromosomes?

_____

202 Modeling Meiosis

Key Question: How is variation introduced into the gametes formed during meiosis?

Modeling meiosis using popsicle sticks can help to understand how meiosis creates variation. Each of your somatic (body) cells contains 46 chromosomes: 23 maternal and 23 paternal. Therefore, you have 23 **homologous pairs**. For simplicity, the number of chromosomes studied in this exercise has been reduced to four, i.e. two homologous pairs.

Investigation 11.2 Modelling meiosis using popsicle sticks

See appendix for equipment list.

To study the effect of crossing over on genetic variation, you will work in pairs to simulate the inheritance of two of your own traits: ability to tongue roll and handedness. This activity will take 25-45 minutes.

1. Record your phenotype and genotype for each trait in the table (right). If you have a dominant trait, you will not know if you are heterozygous or homozygous for that trait, so you can choose either genotype.

Chromosome number	Phenotype	Genotype
10	Tongue roller	TT, Tt
10	Non-tongue roller	tt
2	Right handed	RR, Rr
2	Left handed	rr

Step 1

Trait	Phenotype	Genotype
Handedness		
Tongue rolling		

2. Before you start the simulation, partner up with a classmate. Your gametes will combine with theirs (fertilization) at the end of the activity to produce a "child". Decide who will be female, and who will be male. You will need to work with this person again at step 7.

3. Collect four popsicle sticks. These represent four chromosomes. Color two sticks blue or mark them with a P. for paternal chromosomes. The plain sticks are the maternal chromosomes. Write your initials on each of the four sticks. Label each chromosome with its number. Label four sticky dots with alleles to describe your phenotype and stick each onto the appropriate chromosome. In the example shown (right), the person is heterozygous for tongue rolling so sticky dots with alleles T and t are placed on chromosome 10. The person is also left handed, so alleles r and r are placed on chromosome 2.

Step 2

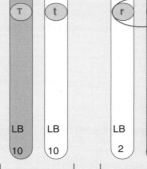
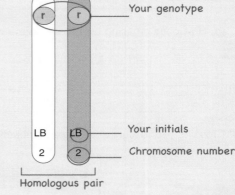

Your genotype
Your initials
Chromosome number

Homologous pair   Homologous pair

4. Randomly drop the chromosomes onto a table. This represents a cell in either the testes or ovaries.
Duplicate your chromosomes by adding four more identical popsicle sticks to the table (right).
What are you simulating with this action?

Step 3

Simulate the first stage of meiosis by lining the duplicated chromosome pair with their homologous pair (below). For each chromosome number, you will have four sticks touching side-by-side (A, below).
At this stage crossing over occurs. Simulate this by swapping sticky dots from adjoining homologues (B, below).

Step 4

(A) 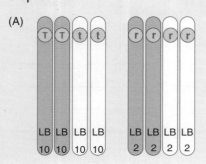   (B)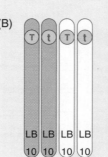

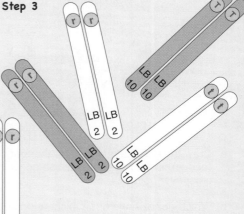

5. Randomly align the homologous chromosome pairs to simulate alignment across the cell's equator (center), as occurs in the next phase of meiosis. Simulate the separation of the chromosome pairs.
For each group of four sticks, two are pulled to each pole of the cell.

Step 5

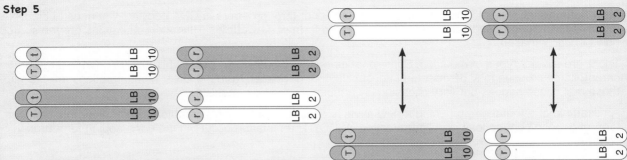

6. Two intermediate cells are formed. If you have completed step 5 correctly, each intermediate cell will be haploid (half the diploid chromosome number shown in step 3) with a mixture of maternal and paternal chromosomes. This is the end of the first division of meiosis. Your cells now need to divide for a second time. Repeat steps 4 and 5 but this time there is no crossing over and you are now separating replicated chromosomes, not homologues. At the end of this process each intermediate cell will have produced two haploid gametes. Each will have a maternal chromosome (white) and a paternal chromosome (blue) (below).

Step 6

Intermediate cells

Haploid gametes

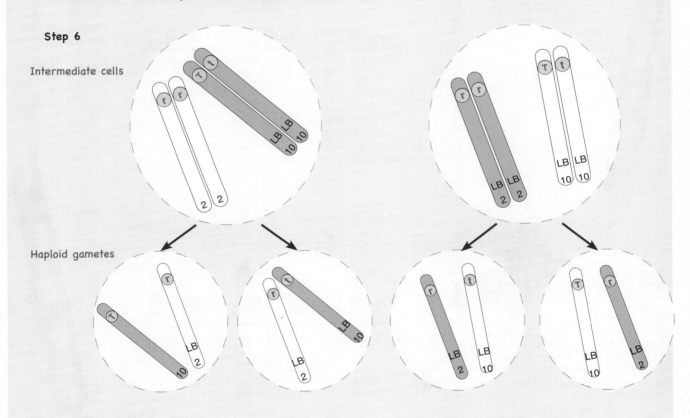

7. Pair up with the partner you chose at the beginning of the exercise to carry out fertilisation. Randomly select one sperm and one egg cell. The unsuccessful gametes can be removed from the table. Combine the chromosomes of the successful gametes. You have created a child! Fill in the following chart to describe your child's genotype and phenotype for tongue rolling and handedness.

Trait	Phenotype	Genotype
Handedness		
Tongue rolling		

©2022 **BIOZONE** International
ISBN: **978-1-98-856692-4**
Photocopying Prohibited

203 Mutations

Key Question: What are mutations, and why are they the ultimate source of new alleles, and therefore genetic variation?

▶ **Mutations** are changes to the DNA sequence and occur through errors in DNA copying. Changes to the DNA modify existing genes and can create variation in the form of new **alleles**. Ultimately, mutations are the source of all new genetic variation.

▶ There are several types of mutation. Some change only one nucleotide base, while others change large parts of chromosomes. Bases may be inserted into, substituted, or deleted from the DNA. Most mutations are harmful. e.g. those that cause cancer, but very occasionally a mutation can be beneficial.

▶ An example of a mutation producing a new allele is described below. This mutation causes the most common form of genetic hearing loss in children (called NSRD). The mutation occurs in the gene coding for a protein called connexin 26.

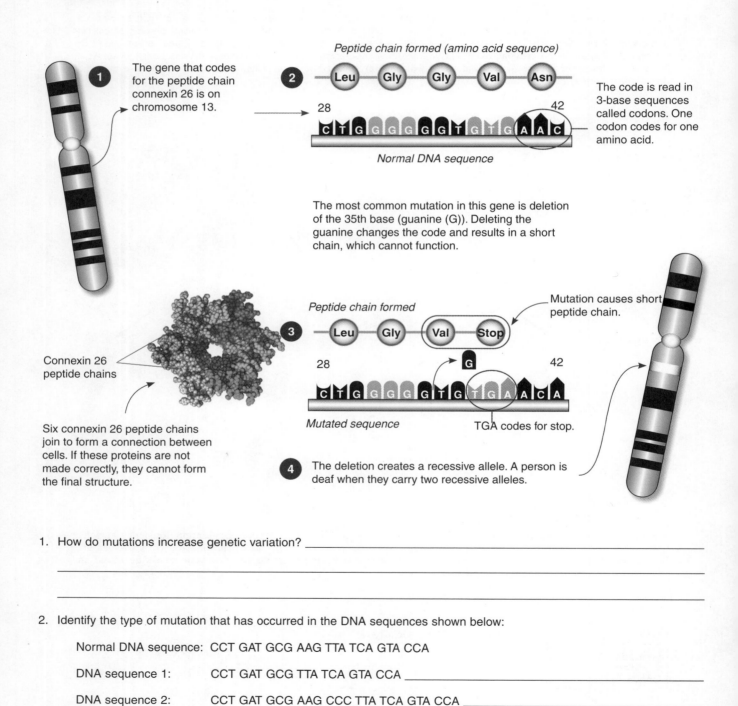

1 The gene that codes for the peptide chain connexin 26 is on chromosome 13.

Peptide chain formed (amino acid sequence)

2 Leu — Gly — Gly — Val — Asn

28   42
C T G G G G G G T G T G A A C

Normal DNA sequence

The code is read in 3-base sequences called codons. One codon codes for one amino acid.

The most common mutation in this gene is deletion of the 35th base (guanine (G)). Deleting the guanine changes the code and results in a short chain, which cannot function.

Connexin 26 peptide chains

Six connexin 26 peptide chains join to form a connection between cells. If these proteins are not made correctly, they cannot form the final structure.

Peptide chain formed

3 Leu — Gly — Val — Stop

Mutation causes short peptide chain.

28   G   42
C T G G G G G T G T G A A C A

Mutated sequence   TGA codes for stop.

4 The deletion creates a recessive allele. A person is deaf when they carry two recessive alleles.

1. How do mutations increase genetic variation? _____

_____

_____

2. Identify the type of mutation that has occurred in the DNA sequences shown below:

Normal DNA sequence: CCT GAT GCG AAG TTA TCA GTA CCA

DNA sequence 1:   CCT GAT GCG TTA TCA GTA CCA _____

DNA sequence 2:   CCT GAT GCG AAG CCC TTA TCA GTA CCA _____

DNA sequence 3:   CCT GAT GCG AAG TTA TGA GTA CCA _____

  LS3.B   CE

©2022 **BIOZONE** International
ISBN: 978-1-98-856692-4
Photocopying Prohibited

204 The Effects of Mutations

▸ Most mutations have a harmful effect on the organism. This is because changes to the DNA sequence of a gene can potentially change the amino acid chain encoded by the gene. Proteins need to fold into a precise shape to function properly. A mutation may change the way the protein folds and prevent it from carrying out its usual biological function.

▸ However, sometimes a mutation can be beneficial. The mutation may result in a more efficient protein, or produce an entirely different protein that can improve the survival of the organism.

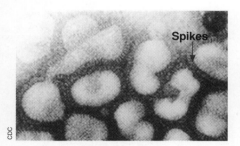

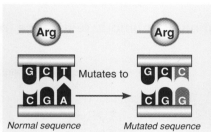

Normal sequence — Mutated sequence

Beneficial mutations

Some mutations aid survival. In viruses, e.g. influenzavirus (above), genes coding for the glycoprotein spikes (arrowed) are constantly mutating, producing new strains that avoid detection by our immune system. This has also happened with the virus responsible for Covid-19.

Silent mutations

Silent mutations do not change the amino acid sequence or the final protein. In the genetic code, several codons may code for the same amino acid. Silent mutations may be neutral if they do not alter an organism's fitness.

Harmful mutations

Most mutations cause harmful effects, usually because they stop or alter the production of a protein, often an enzyme. Albinism (above) is one of the more common mutations in nature, and leaves an animal with no pigmentation.

Beneficial mutations in *E. coli*

An experiment known as the *E.coli* long term evolution experiment has incubated 12 lines of *E. coli* bacteria for more than 20 years.

After 31,000 generations, a mutation in one of the *E. coli* populations enabled it to feed off citrate, a component of the medium they were grown in, (*E. coli* are usually unable to do this). This ability gave these *E. coli* an advantage because they could use another food source.

The mutation was noticed when the optical density (cloudiness) of the flask containing the *E. coli* increased (right), indicating an increase in bacterial numbers.

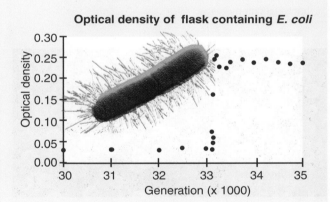

Optical density of flask containing *E. coli*

1. How might a mutation cause a beneficial effect on protein function? _____

   _____

2. How might a mutation have a harmful effect on protein function? _____

   _____

   _____

3. Why does a silent mutation have no apparent affect on an organism? _____

   _____

   _____

4. How would the citrate mutation in *E. coli* give it an advantage over other *E. coli* populations? _____

   _____

   _____

CE LS3.B

205 The Evolution of Antibiotic Resistance

Key Question: How does resistance to antibiotics arise, and how do antibiotic resistant bacteria pass this resistance on to the next generation and to other populations?

▸ Antibiotic resistance arises when a genetic change allows bacteria to tolerate levels of an antibiotic that might normally kill it or stop its growth. This resistance may arise spontaneously through **mutation** or by transfer of genetic material between microbes.

▸ Genomic analyses from 30,000 year old permafrost sediments show that antibiotic resistant genes are not new. They have long been present in the bacterial genome, pre-dating modern antibiotic use. Because of current practices, these genes have proliferated and antibiotic resistance has spread.

The evolution of antibiotic resistance in bacteria

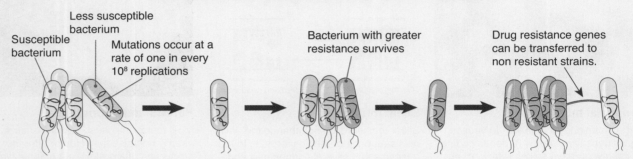

Susceptible bacterium

Less susceptible bacterium

Mutations occur at a rate of one in every 10^8 replications

Bacterium with greater resistance survives

Drug resistance genes can be transferred to non resistant strains.

Any population includes variants with unusual **traits**, in this case reduced sensitivity to an antibiotic. These variants arise as a result of mutations in the bacterial chromosome.

When a person takes an antibiotic, only the most susceptible bacteria will die. The more resistant cells remain and continue dividing.

If the amount of antibiotic taken is too low or not potent enough, the resistant cells survive and divide to produce a population with a higher than normal antibiotic resistance.

The antibiotic initially used against this bacterial strain will now be ineffective. The resistant cells can exchange genetic material with other bacteria, or pass on the genes for resistance to their descendants.

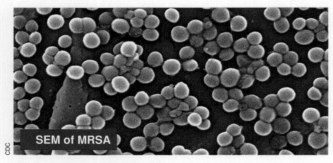

SEM of MRSA

Staphylococcus aureus is a common bacterium causing minor skin infections in humans. Methicillin resistant *S. aureus* (MRSA) is a variant strain that is resistant to penicillin and related antibiotics. MRSA is troublesome in hospital-associated infections because patients with open wounds, invasive devices, e.g. catheters, or poor immunity are at greater risk of infection.

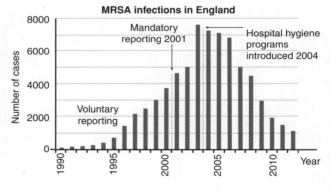

MRSA infections in England

Mandatory reporting 2001

Hospital hygiene programs introduced 2004

Voluntary reporting

Number of cases

Year

In the UK, MRSA cases rose during the 1990s, reflecting the increase in antibiotic resistance. They are now declining as a result of mandatory reporting and stringent hospital hygiene programs. A similar pattern has been observed in the USA.

1. (a) Why can bacterial strains such as MRSA be so harmful? _____

_____

   (b) How do bacterial strains such as MRSA arise? _____

_____

_____

2. How can the resistance become widespread? _____

_____

LS3.B CE

©2022 **BIOZONE** International
ISBN: 978-1-98-856692-4
Photocopying Prohibited

Resistance changes with the seasons

▸ Antibiotic resistance comes at a cost to bacteria. For example, they may be slower to reproduce or have thicker cell walls and slower nutrient uptake. When the selective environment (the antibiotic) is removed, the fitness of resistant strains decreases compared to susceptible strains.

Modeling the spread of antibiotic resistance

Antibiotic resistance occurs through a mutation within a cell's DNA or being acquired from another bacterium. An antibiotic gene will stop an antibiotic from working to prevent bacterial growth. A bacterial population that is exposed to an antibiotic is growing in an environment that selects for antibiotic resistance. As you saw with your models of natural selection, organisms with characteristics that help them survive and reproduce will become more common.

In this activity, you will develop a spreadsheet model to visualize the change in the number of antibiotic resistant bacteria in a population. The antibiotic used in this model kills 75% of the bacterial population each generation. The model assumes all surviving bacteria reproduce once per generation, that the antibiotic is present in the environment every generation, and that all resistant bacteria are fully resistant. Replicate this spreadsheet yourself by following the instructions below (or go **BIOZONE's Resource Hub** for a copy).

 Investigation 11.3 Modeling antibiotic resistance

See appendix for equipment list.

1. Open a new spreadsheet and set up row 1 with the headings shown in the image below.

2. Fill in A2 – F2 as shown in the image below. This information tells you the total number of the starting population is 100,001 bacteria and one of these individuals is resistant to the antibiotic in question.

3. Row 3 simulates the action of the antibiotic. In B3 enter the formula =(B2*0.25). This simulates 75% of the susceptible bacteria dying. In D3 enter the formula =D2*1, this calculates the number of resistant bacteria remaining.

4. Row 4 simulates the bacterial population reproducing (second generation). Fill in the cells as shown below to calculate the number of susceptible bacteria, number of resistant bacteria, and total population.

5. Repeat the steps for several more generations.

	A	B	C	D	E	F
1	Time	Number of susceptible bacteria		Number of resistant bacteria		Total number
2	1	100000		1		=B2+D2
3	Antibiotic	=(B2*0.25)		=D2*1		
4	2	=B3*2		=D3*2		=B4+D4
5	Antibiotic	=(B4*0.25)		=D4*1		
6	3	=B5*2		=D5*2		=B6+D6
7	Antibiotic	=(B6*0.25)		=D6*1		
8	4	=B7*2		=D7*2		=B8+D8
9	Antibiotic	=(B8*0.25)		=D8*1		
10	5	=B9*2		=D9*2		=B10+D10
11	Antibiotic	=(B10*0.25)		=D10*1		
12	6	=B11*2		=D11*2		=B12+D12
13	Antibiotic	=(B12*0.25)		=D12*1		
14	7	=B13*2		=D13*2		=B14+D14
15	Antibiotic	=(B14*0.25)		=D14*1		
16	8	=B15*2		=D15*2		=B16+D16
17	Antibiotic	=(B16*0.25)		=D16*1		
18	9	=B17*2		=D17*2		=B18+D18
19	Antibiotic	=(B18*0.25)		=D18*1		

3. (a) What happens to the susceptible population over time? _____

_____

(b) What happens to the total population of bacteria over time? _____

_____

(c) In reality, there is an energetic cost to a bacterial cell in maintaining resistance to an antibiotic. What might happen to the bacterial population if the antibiotics were removed halfway through the treatment?

_____

_____

_____

206 Beneficial Mutations in Humans

Key Question: How do beneficial mutations increase the fitness of the organisms that possess them, and how common are they?

▸ Beneficial **mutations** are mutations that increase the fitness of the organisms that possess them. Although beneficial mutations are rare compared to those that are harmful, there are a number of well documented beneficial mutations in humans.

▸ Some of these mutations are not very common in the human population. This is because the mutations have been in existence for a relatively short time, so the mutations have not had time to become widespread in the human population.

▸ Scientists often study mutations that cause disease. By understanding the genetic origin of various diseases, it may be possible to develop targeted medical drugs and therapies against them.

The village of Limone, Italy

Apolipoprotein A1-Milano is a well documented mutation to apolipoprotein A1 that helps transport cholesterol through the blood. The mutation causes a change to one amino acid and increases the protein's effectiveness by ten times, dramatically reducing incidence of heart disease. The mutation can be traced back to its origin in Limone, Italy, in 1644. Another mutation, to a gene called PCSK9, has a similar effect, lowering the risk of heart disease by 88%.

Lactose is a sugar found in milk. All infant mammals produce an enzyme called lactase that breaks the lactose into the smaller sugars glucose and galactose. As mammals become older, their production of lactase declines and they lose the ability to digest lactose. As adults, they become lactose intolerant and feel bloated after drinking milk. About 10,000 years ago, a mutation appeared in humans that maintained lactase production into adulthood. This mutation is now carried in people of mainly European, African, and Indian descent.

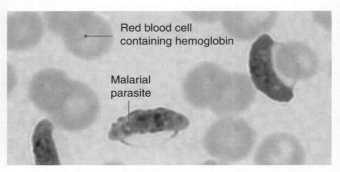

Red blood cell containing hemoglobin

Malarial parasite

Malaria resistance results from a mutation to the hemoglobin gene (Hb^S) that also causes sickle cell disease. This mutation in beneficial in regions where malaria is common. A less well known mutation (Hb^C) to the same gene, discovered in populations in Burkina Faso, Africa, results in a 29% reduction in the likelihood of contracting malaria if the person has one copy of the mutated gene, and a 93% reduction if the person has two copies. In addition, the anemia that person suffers as a result of the mutation is much less pronounced than in the Hb^S mutation.

1. Why is it that many of the recent beneficial mutations recorded in humans have not spread throughout the entire human population?

_____

_____

2. What selection pressure could act on Apolipoprotein A1-Milano to help it spread through a population? _____

_____

_____

3. Why would it be beneficial to be able to digest milk in adulthood? _____

_____

_____

_____

LS3.B   CE

©2022 **BIOZONE** International
ISBN: **978-1-98-856692-4**
Photocopying Prohibited

207 Harmful Mutations in Humans

Key Question: What are some examples of harmful mutations in humans, and what changes to the DNA are responsible for them?

Cystic fibrosis

▶ Cystic fibrosis (CF) is an inherited disorder caused by a mutation of the CFTR gene. It is one of the most common genetic conditions affecting white skinned people of European descent.

▶ The CFTR gene's protein product is a membrane-based protein that regulates chloride transport in cells. A specific mutation produces an abnormal CFTR protein, which cannot take its position in the plasma membrane (below, right), or perform its transport function. This mutation is the most common one causing CF.

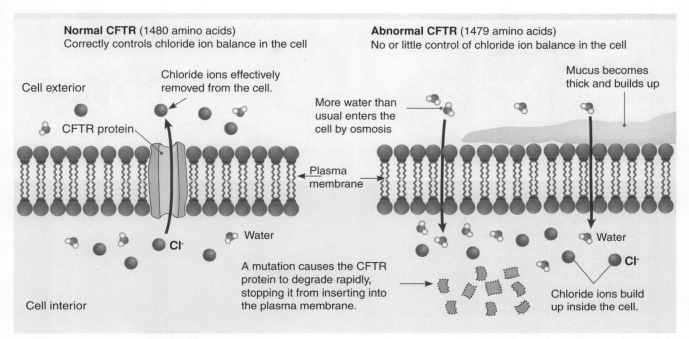

Normal CFTR (1480 amino acids)
Correctly controls chloride ion balance in the cell

Cell exterior

Chloride ions effectively removed from the cell.

CFTR protein

Plasma membrane

Cell interior

Water

Cl⁻

Abnormal CFTR (1479 amino acids)
No or little control of chloride ion balance in the cell

Mucus becomes thick and builds up

More water than usual enters the cell by osmosis

A mutation causes the CFTR protein to degrade rapidly, stopping it from inserting into the plasma membrane.

Water

Cl⁻

Chloride ions build up inside the cell.

Huntington's disease

▶ Huntington's disease is a progressive genetic disorder in which nerve cells in certain parts of the brain waste away, or degenerate. Symptoms include shaky hands and an awkward gait.

▶ Huntington's disease is caused by a mutation of a specific gene on chromosome 4. This gene has a repeating base sequence CAG. Normally, this section repeats between 10 and 28 times but in people with Huntington's disease, this sequence repeated between 36 to 120 times. The greater the number of repeats, the greater the effects appear to be, and the earlier the onset of the disease.

▶ Woody Guthrie (right) was an influential folk singer-songwriter who died in 1967 due to complications related to Huntington's disease.

Al Aumuller, NY World Telegram and the Sun, Public Domain

1. How does a mutation affect the amino acid sequence for the CFTR protein? _____

_____

2. (a) What causes Huntington's disease? _____

_____

   (b) How does the extent of the mutation affect the symptoms and onset of the disease? _____

_____

_____

CE | LS3.B

208 Influences on Phenotype

Key Question: How is an organism's phenotype influenced by the effects of the environment during and after development, even though the genotype remains unaffected?

The phenotype encoded by genes is a product not only of the genes themselves, but of their internal and external environment and the variations in the way those genes are controlled. This is know as epigenetics.

Even identical twins have minor differences in their appearance due to epigenetic and environmental factors such as diet and intrauterine environment. Genes, together with epigenetic and environmental factors, determine the unique phenotype that is produced.

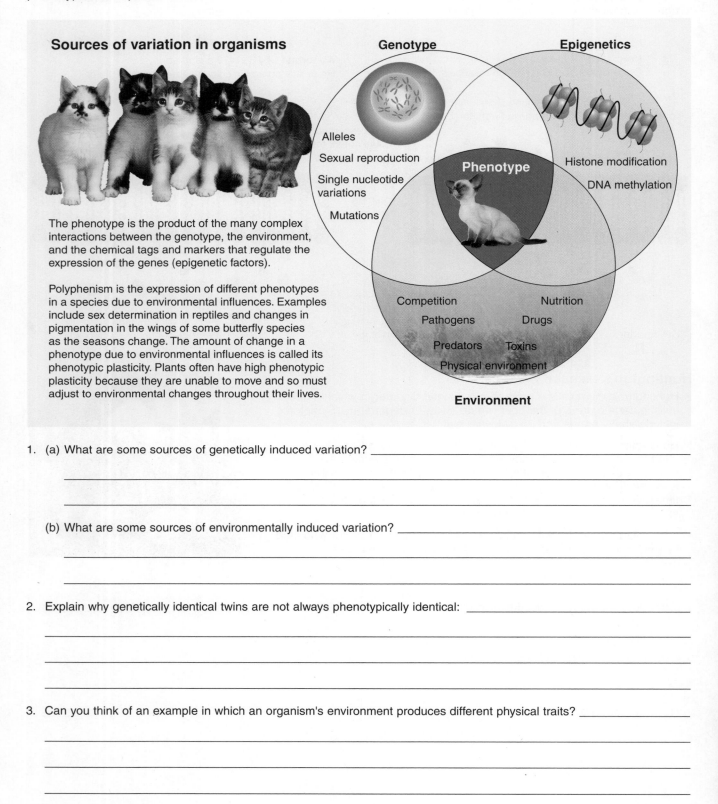

Sources of variation in organisms

The phenotype is the product of the many complex interactions between the genotype, the environment, and the chemical tags and markers that regulate the expression of the genes (epigenetic factors).

Polyphenism is the expression of different phenotypes in a species due to environmental influences. Examples include sex determination in reptiles and changes in pigmentation in the wings of some butterfly species as the seasons change. The amount of change in a phenotype due to environmental influences is called its phenotypic plasticity. Plants often have high phenotypic plasticity because they are unable to move and so must adjust to environmental changes throughout their lives.

Genotype
Alleles
Sexual reproduction
Single nucleotide variations
Mutations

Phenotype

Epigenetics
Histone modification
DNA methylation

Environment
Competition
Pathogens
Predators
Nutrition
Drugs
Toxins
Physical environment

1. (a) What are some sources of genetically induced variation? _____

_____

_____

(b) What are some sources of environmentally induced variation? _____

_____

_____

2. Explain why genetically identical twins are not always phenotypically identical: _____

_____

_____

_____

3. Can you think of an example in which an organism's environment produces different physical traits? _____

_____

_____

_____

©2022 **BIOZONE** International
ISBN: 978-1-98-856692-4
Photocopying Prohibited

209 Environment and Variation

Key Question: How can the environment affect an organism's phenotype?

▶ Environmental factors can modify the phenotype encoded by genes without changing the genotype. This can occur both during development and later in life. Environmental factors that affect the phenotype of plants and animals include nutrients or diet, temperature, altitude or latitude, and the presence of other organisms.

The effect of temperature

▶ The sex of some animals is determined by the incubation temperature during their embryonic development. Examples include turtles, crocodiles, and the American alligator. In some species, high incubation temperatures produce males and low temperatures produce females. In other species, the opposite is true. Temperature regulated sex determination may provide an advantage by preventing inbreeding, since all siblings will tend to be of the same sex.

▶ Color-pointing is a result of a temperature sensitive mutation to one of the melanin-producing enzymes. The dark pigment is only produced in the cooler areas of the body (face, ears, feet, and tail), while the rest of the body is a pale color, or white. Color-pointing is seen in some breeds of cats and rabbits, e.g. Siamese cats and Himalayan rabbits.

The effect of other organisms

▶ The presence of other individuals of the same species may control sex determination for some animals. Some fish species, including Sandager's wrasse (right), show this characteristic. The fish live in groups consisting of a single male with attendant females and juveniles. In the presence of a male, all juvenile fish of this species grow into females. When the male dies, the dominant female will undergo physiological changes to become a male. The male and female look very different.

Female   Male

▶ Some organisms respond to the presence of other, potentially harmful, organisms by changing their body shape. Invertebrates, such as some *Daphnia* species, grow a helmet when invertebrate predators are present. The helmet makes *Daphnia* more difficult to attack and handle. Such changes are usually in response to chemicals produced by the predator (or competitor) and are common in plants as well as animals.

Non-helmeted *Daphnia*   Helmeted *Daphnia*

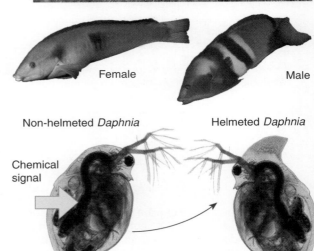

Chemical signal

1. (a) Give two examples of how temperature affects a phenotypic characteristic in an organism: _____

_____

(b) Why are the darker patches of fur in color-pointed cats and rabbits found only on the face, paws, and tail? _____

_____

_____

2. How is helmet development in *Daphnia* an adaptive response to environment? _____

_____

_____

CE   LS3.B

314

The effect of altitude

Severe stunting

Growth to genetic potential

Cline

Increasing altitude can stunt the phenotype of plants with the same genotype. In some conifers, e.g. Engelmann spruce, plants at low altitude grow to their full genetic potential, but growth becomes progressively more stunted as elevation increases. Growth is gnarled and bushy at the highest, most severe sites. Gradual change in phenotype over an environmental gradient is called a cline.

The effect of chemical environment

The chemical environment can influence the phenotype in plants and animals. The color of hydrangea flowers varies with soil pH. Blue flowers (due to the presence of aluminium compounds in the flowers) occur in more acidic soils (pH 5.0-5.5) in which aluminium is more readily available. In less acidic soils (pH 6.0-6.5), the flowers are pink.

3. (a) What is a cline? _____

_____

(b) What physical factors associated with altitude could affect plant phenotype? _____

_____

4. Describe an example of how the chemical environment of a plant can influence phenotype: _____

_____

_____

_____

_____

_____

5. Vegetable growers can produce enormous vegetables for competition. How could you improve the chance that a vegetable would reach its maximum genetic potential?

_____

_____

_____

_____

6. Two different species of plant (A and B) were found growing together on a windswept portion of a coast, Both have a prostrate (low growing) phenotype. One of each plant type was transferred to a greenhouse where "ideal" conditions were provided to allow maximum growth. In this controlled environment, species B continued to grow in its original prostrate form, but species A changed its growing pattern and became erect in form. Identify the cause of the prostrate phenotype in each of the coastal grown plant species and explain your answer:

Plant species A: _____

_____

_____

_____

Plant species B: _____

_____

_____

_____

©2022 **BIOZONE** International
ISBN: **978-1-98-856692-4**
Photocopying Prohibited

210 Genes and Environment Interact

Key Question: How can the environment or experiences of an individual affect the development of following generations?

Studies of heredity have found that the environment or lifestyle of an ancestor can have an effect on future generations. Certain environments or diets can affect the methylation and packaging of the DNA (rather than the DNA itself), determining which genes are switched on or off, and affecting the development of the individual. These effects can be passed on to offspring, and even to future generations. It is thought that these inherited effects may provide a rapid way to adapt to particular environmental situations.

The destruction of New York's Twin Towers on September 11, 2001, traumatized thousands of people. In those thousands were 1700 pregnant women. Some of them suffered severe, post-traumatic stress disorder (PTSD), others did not. Studies on the mothers who developed PTSD found very low levels of the stress-related hormone cortisol in their saliva. Low levels of cortisol can be caused by very high stress, as the body uses it faster than it can be produced. The children of these mothers also had much lower levels of cortisol than those whose mothers had not suffered PTSD, particularly those who had been in the third trimester of pregnancy, indicating a developmental response to severe stress.

In the winter of 1944-45, towards the end of the Second World War, widespread starvation affected people in the western Netherlands, including pregnant women. In the following year, the war ended and diets quickly returned to normal, but children who had been fetuses during the famine showed long-term effects. Those whose mothers were in the first trimester of pregnancy in the famine period had higher than average rates of obesity later in life, as well as higher incidence rates of cardiovascular disease. In the first trimester of fetal development, DNA methylation had occurred on genes responsible for energy metabolism. Malnutrition in mothers caused epigenetic changes to their offspring.

Our ancestors' environment can have a long lasting effect

▶ How much do genes contribute to phenotype? What about environment? This question is often called nature versus nurture.

▶ A 2004 study of the grooming of rat pups by their mothers provides some insight. In this study, the quality of care by a pup's mother affected how the pup behaved when it reached adulthood.

▶ Rat pups that were groomed more often by their mother were better at coping with stress than pups that received less grooming. The effect was caused by changes in the expression of a hormone receptor with a role in the response to stress.

▶ DNA analysis found differences in the way the DNA was chemically tagged to regulate gene expression. Rats that received a lot of grooming had higher expression of the gene for the hormone receptor. The opposite was true for rats who received little grooming.

1. (a) Describe how grooming by mother rats on their pups affect the pups in the long term: _____

_____

(b) How was this achieved?_____

_____

_____

Studying the effect of environment in generations of rats

The effect of the environment and diet of mothers on later generations exposed to a breast cancer trigger (a cancer-causing chemical) was investigated in rats fed a diet high in estrogen. The length of time taken for breast cancer to develop in later generations after the trigger for breast cancer was given was recorded and compared. The data are presented below.

F_1= daughters, F_2= granddaughters, F_3 = great granddaughters.

	Cumulative percentage rats with breast cancer (rat mothers on high estrogen diet (HED))					
	F_1%		F_2%		F_3%	
Weeks since trigger	Mothers on high estrogen diet	Control	Mothers on high estrogen diet	Control	Mothers on high estrogen diet	Control
6	5	0	10	0	0	0
8	10	0	10	0	15	10
10	30	15	15	20	30	20
12	38	19	30	30	40	20
14	50	22	30	40	50	20
16	50	22	30	40	50	30
18	60	35	40	40	75	40
20	60	42	50	50	80	45
22	80	55	50	50	80	60

Data source: S. De Assis: Nature Communications 3 (Article 1053) (2012)

2. Plot the tabulated data above on to the grids below. Include a key in the box provided:

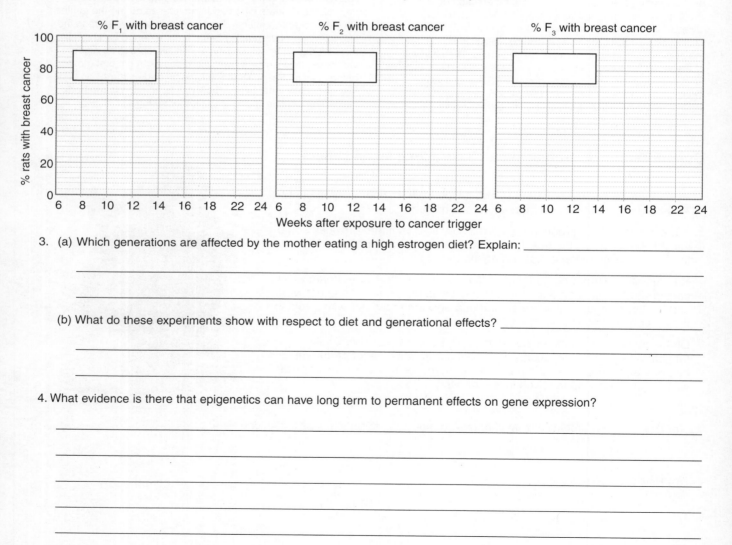

% F_1 with breast cancer

% F_2 with breast cancer

% F_3 with breast cancer

% rats with breast cancer

Weeks after exposure to cancer trigger

3. (a) Which generations are affected by the mother eating a high estrogen diet? Explain: _____

_____

_____

(b) What do these experiments show with respect to diet and generational effects? _____

_____

_____

4. What evidence is there that epigenetics can have long term to permanent effects on gene expression?

_____

_____

_____

_____

_____

211 Predicting Traits: The Monohybrid Cross

Key Question: How does the parental genotype affect the outcome of a cross?

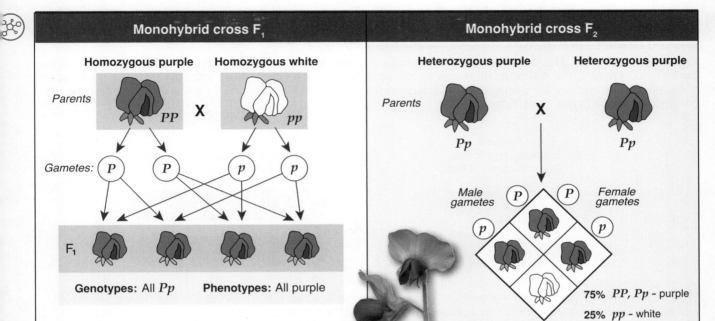

Monohybrid cross F₁

Homozygous purple   Homozygous white

Parents   *PP*   X   *pp*

Gametes: P   P   p   p

F₁

Genotypes: All *Pp*   **Phenotypes: All purple**

A true-breeding organism is homozygous for the gene involved. The F₁ (first filial generation) offspring of a cross between two **true breeding** parent plants are all purple (*Pp*).

Monohybrid cross F₂

Heterozygous purple   Heterozygous purple

Parents   *Pp*   X   *Pp*

Male gametes   P   P   Female gametes

p   p

75% *PP, Pp* - purple

25% *pp* - white

A cross between the F₁ offspring (*Pp* x *Pp*) would yield a 3:1 ratio in the F₂ of purple (*PP, Pp, Pp*) to white (*pp*).

---

1. Study the diagrams above and explain why white flower color does not appear in the F₁ generation but reappears in the F₂ generation:

_____

_____

_____

_____

2. Complete the crosses below:

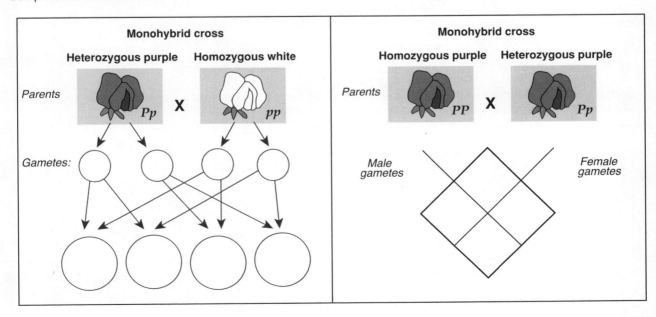

Monohybrid cross

Heterozygous purple   Homozygous white

Parents   *Pp*   X   *pp*

Gametes:

Monohybrid cross

Homozygous purple   Heterozygous purple

Parents   *PP*   X   *Pp*

Male gametes   Female gametes

SPQ   LS3.B

212 Predicting Traits: The Test Cross

Key Question: How can we use a cross to determine an individual's genotype?

It is not always possible to determine an organism's genotype by its appearance because gene expression is complicated by patterns of dominance and by gene interactions. The test cross was developed by Mendel as a way to establish the genotype of an organism with the dominant phenotype for a particular trait.

The principle is simple. The individual with the unknown genotype is bred with a homozygous recessive individual for the trait(s) of interest. The homozygous recessive can produce only one type of allele (recessive), so the phenotypes of the offspring will reveal the genotype of the unknown parent (below). The test cross can be used to determine the genotype of single genes or multiple genes.

Parent 1
Unknown genotype
(but with dominant traits)

Parent 2
Homozygous recessive genotype
(no dominant traits)

X

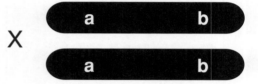

The common fruit fly (*Drosophila melanogaster*) is often used to illustrate basic principles of inheritance because it has several genetic markers whose phenotypes are easily identified. One such phenotype is body color. Wild type (normal) *Drosophila* have yellow-brown bodies. The allele for yellow-brown body color (E) is dominant. The allele for an ebony colored body (e) is recessive. The test crosses below show possible outcomes for an individual with homozygous and heterozygous alleles for ebony body color.

A. A homozygous recessive female (ee) with an ebony body is crossed with a homozygous dominant male (EE).

	Female gametes	
	e	e
E	Ee	Ee
E	Ee	Ee

Male gametes

Cross A:
(a) Genotype frequency: _100% Ee_
(b) Phenotype frequency: _100% yellow-brown_

A. A homozygous recessive female (ee) with an ebony body is crossed with a heterozygous male (Ee).

	Female gametes	
	e	e
E	Ee	Ee
e	ee	ee

Male gametes

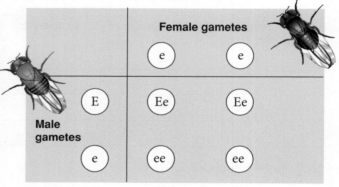

Cross B:
(a) Genotype frequency: _50% Ee, 50% ee_
(b) Phenotype frequency: _50% yellow-brown, 50% ebony_

1. In *Drosophila*, the allele for brown eyes (b) is recessive, while the red eye allele (B) is dominant. Set up and carry out a test cross to determine the genotype of a male who has red eyes:

2. 50% of the resulting progeny have red eyes, and 50% have brown eyes. What is the genotype of the male *Drosophila*?

   LS3.B

©2022 **BIOZONE** International
ISBN: 978-1-98-856692-4
Photocopying Prohibited

213 Practicing Monohybrid Crosses

Key Question: How can we use a monohybrid cross to study the inheritance pattern of one gene, and what are the predictable ratios in the offspring from this cross?

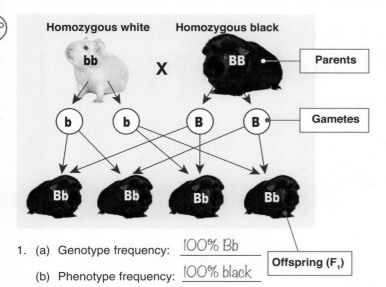

Homozygous white   Homozygous black

bb   X   BB   → Parents

b   b   B   B   → Gametes

Bb   Bb   Bb   Bb

Offspring (F₁)

1. (a) Genotype frequency: 100% Bb

   (b) Phenotype frequency: 100% black

Monohybrid crosses can be used to determine the genotype and phenotype outcomes for coat color in guinea pigs. Complete the monohybrid crosses below by determining the gametes and phenotypic and genotypic frequencies of the offspring. Question one has been done for you.

2. (a) Which coat color is dominant?

   _____

   (b) Which is the dominant allele? _____

   (c) Which coat color is recessive?

   _____

   (d) Which is the recessive allele? _____

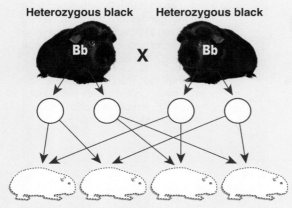

Heterozygous black   Heterozygous black

Bb   X   Bb

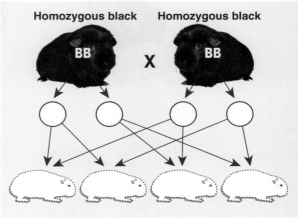

Homozygous black   Homozygous black

BB   X   BB

3. (a) Genotype frequency: _____

   (b) Phenotype frequency: _____

4. (a) Genotype frequency: _____

   (b) Phenotype frequency: _____

5. Two parent guinea pigs with the genotypes Bb and BB are crossed:

   (a) What is the probability that any one offspring is BB? _____

   (b) What is the probability that any one offspring is black? _____

6. Two parent guinea pigs with genotypes bb and Bb are crossed:

   (a) What is the probability that any one offspring is Bb? _____

   (b) What is the probability that any one offspring is black? _____

7. A white guinea pig and a black guinea pig are crossed. All of the guinea pigs that are born are white.

   (a) What is the genotype of the black guinea pig? _____

   (b) Explain the result: _____

   _____

   _____

   _____

©2022 **BIOZONE** International
ISBN: 978-1-98-856692-4
Photocopying Prohibited

LS3.B

214 Predicting Traits: the Dihybrid Cross

Key Question: How can we use dihybrid crosses to study the inheritance pattern of two unlinked genes, and what are their predictable ratios?

There are four types of gamete produced in a cross involving two genes, where the genes are carried on separate chromosomes and are sorted independently of each other during meiosis.

The two genes in the example below are on separate chromosomes, i.e. they are unlinked, and control two unrelated characteristics, hair color and coat length. Black (B) and short (L) are dominant to white and long.

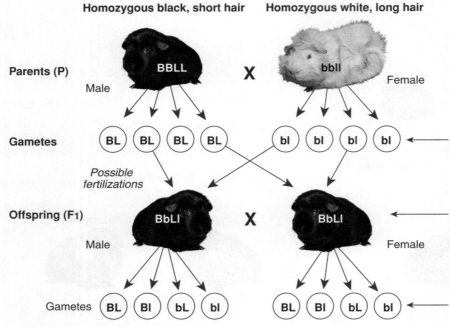

Homozygous black, short hair   **Homozygous white, long hair**

Parents (P)   BBLL   X   bbll
Male   Female

Gametes   BL BL BL BL   bl bl bl bl

Possible fertilizations

Offspring (F₁)   BbLl   X   BbLl
Male   Female

Gametes   BL Bl bL bl   BL Bl bL bl

Parents: The notation P is only used for a cross between true breeding (homozygous) parents.

Gametes: Only one type of gamete is produced from each parent (although they will produce four gametes from each oocyte or spermatocyte). This is because each parent is homozygous for both traits.

F₁ offspring: There is only one kind of gamete from each parent, therefore only one kind of offspring produced in the first generation. The notation F₁ is only used to denote the heterozygous offspring of a cross between two true breeding parents.

F₂ offspring: The F₁ were mated with each other (selfed). Each individual from the F₁ is able to produce four different kinds of gamete.

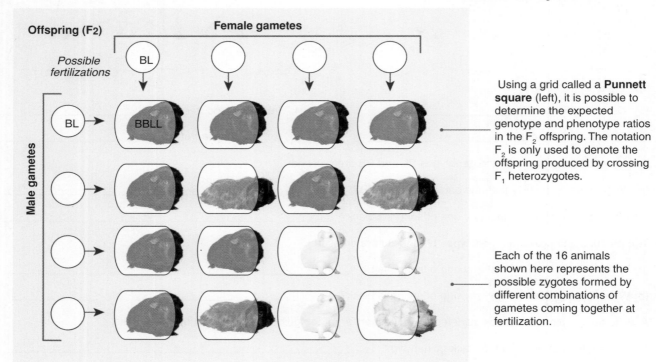

Offspring (F2)   **Female gametes**

Possible fertilizations   BL

Male gametes   BL → BBLL

Using a grid called a **Punnett square** (left), it is possible to determine the expected genotype and phenotype ratios in the F₂ offspring. The notation F₂ is only used to denote the offspring produced by crossing F₁ heterozygotes.

Each of the 16 animals shown here represents the possible zygotes formed by different combinations of gametes coming together at fertilization.

1. Fill in the gametes and complete the Punnett square above.

2. Use the Punnett square to identify the number of each phenotype in the offspring: _____

_____

    LS3.B SPQ

©2022 **BIOZONE** International
ISBN: 978-1-98-856692-4
Photocopying Prohibited

215 Practicing Dihybrid Crosses

Key Question: How can we use a Punnett square to predict the outcome of dihybrid crosses?

1. In guinea pigs, rough coat **R** is dominant over smooth coat **r** and black coat **B** is dominant over white **b**. The genes are not linked. A homozygous rough, black animal was crossed with a homozygous smooth, white animal.

(a) State the genotype of the F_1: _____

(b) State the phenotype of the F_1: _____

(c) Use the Punnett square (top right) to show the outcome of a cross between the F_1 (the F_2):

(d) Using ratios, state the phenotypes of the F_2 generation:

_____

(e) Use the Punnett square (middle right) to show the outcome of a back cross of the F_1 to the rough, black parent:

(f) Using ratios, state the phenotype of the offspring of this back cross:

_____

(g) A rough, black guinea pig was crossed with a rough, white guinea pig and produced the following offspring: 3 rough black, 2 rough white, and 1 smooth white. What are genotypes of the parents?

_____

_____

_____

_____

2. In humans, two genes affecting the appearance of the hands are the gene for thumb hyperextension (curving) and the gene for mid-digit hair. The allele for curved thumb, **H**, is dominant to the allele for straight thumb, **h**. The allele for mid digit hair, **M**, is dominant to that for an absence of hair, **m**.

(a) Give all the genotypes of individuals who are able to curve their thumbs, but have no mid-digit hair:

_____

(b) Complete the Punnett square (bottom right), to show the possible genotypes from a cross between two individuals heterozygous for both alleles:

(c) State the phenotype ratios of the F_1 progeny:

_____

_____

_____

(d) What is the probability one of the offspring would have mid-digit hair?

_____

_____

LS3.B

3. In rabbits, spotted coat **S** is dominant to solid color **s**, while for coat color, black **B** is dominant to brown **b**. A brown spotted rabbit is mated with a solid black one and all the offspring are black spotted (the genes are not linked).

(a) State the genotypes:

Parent 1: _____

Parent 2: _____

Offspring: _____

(b) Use the Punnett square (top right) to show the outcome of a cross between the F_1 (i.e. the F_2):

(c) Using ratios, state the phenotypes of the F_2 generation:

_____

_____

_____

4. The Himalayan color-pointed, long-haired cat is a breed developed by crossing a pedigree (true-breeding), uniform-colored, long-haired Persian with a pedigree color-pointed (darker face, ears, paws, and tail), short-haired Siamese.

Persian   Siamese   Himalayan

The genes controlling hair coloring and length are on separate chromosomes: uniform color **U**, color pointed **u**, short hair **S**, long hair **s**.

(a) State the genotype of the F_1 (Siamese X Persian): _____

(b) State the phenotype of the F_1: _____

(c) Use the Punnett square (right) to show the outcome of a cross between the F_1 (the F_2):

(d) What ratio of the F_2 will be Himalayan? _____

(e) State whether the Himalayan would be true breeding:

_____

(f) What ratio of the F_2 will be color-point, short-haired cats?

_____

5. In cats, the following alleles are present for coat characteristics: black (**B**), brown (**b**), short (**L**), long (**l**). The genes are not linked. Use the information to complete the dihybrid crosses below:

A black short haired (**BBLl**) male is crossed with a black long haired (**Bbll**) female. Determine the genotypic and phenotypic ratios of the offspring:

Genotype ratio: _____

_____

Phenotype ratio: _____

_____

©2022 **BIOZONE** International
ISBN: **978-1-98-856692-4**
Photocopying Prohibited

216 Testing the Outcome of Genetic Crosses

Key Question: How do we use the chi-squared test for goodness of fit (χ^2) for testing the outcome of dihybrid crosses against a predicted Mendelian ratio?

Using χ^2 in Mendelian genetics

In genetic crosses, certain ratios of offspring can be predicted based on the known genotypes of the parents. The chi-squared test is a statistical test to determine how well observed numbers match (or fit) expected numbers. Raw counts should be used, and a large sample size is required for the test to be valid.

In a chi-squared test, the null hypothesis predicts the ratio of offspring of different phenotypes is the same as the expected Mendelian ratio for the cross, assuming independent assortment of alleles (no linkage, i.e. the genes involved are on different chromosomes).

Significant departures from the predicted Mendelian ratio indicate linkage (the genes are on the same chromosome) of the alleles in question.

In a *Drosophila* genetics experiment, two individuals were crossed (the details of the cross are not relevant here). The predicted Mendelian ratios for the offspring of this cross were 1:1:1:1 for each of the four following phenotypes: gray body-long wing, gray body-vestigial wing, ebony body-long wing, ebony body-vestigial wing.

The observed results of the cross were not exactly as predicted. The following numbers for each phenotype were observed in the offspring of the cross:

Gray body, vestigial wing	Gray body, long wing	Ebony body, long wing	Ebony body, vestigial wing
88	**98**	**102**	**112**

Table 1: Critical values of χ^2 at different levels of probability. By convention, the critical probability for rejecting the null hypothesis (H_0) is 5%. If the test statistic is less than the tabulated critical value for P = 0.05 we cannot reject H_0 and the result is not significant. If the statistic is greater than the tabulated value for P = 0.05 we reject (H_0) in favor of the alternative hypothesis.

Degrees of freedom	Level of probability (P)					
	0.50	**0.20**	**0.10**	**0.05**	**0.02**	**0.01**
1	0.455	1.64	2.71	3.84	5.41	6.64
2	1.386	3.22	4.61	5.99	7.82	9.21
3	2.366	4.64	6.25	7.82	9.84	11.35
4	3.357	5.99	7.78	9.49	11.67	13.28
5	4.351	7.29	9.24	11.07	13.39	15.09

Do not reject H_0 ←   Reject H_0 →

Steps in performing a χ^2 test

1 Enter the observed value (O).

Enter the values of the offspring into the table in the appropriate category (column 1).

2 Calculate the expected value (E).

In this case the expected ratio is 1:1:1:1. Therefore the number of offspring in each category should be the same (i.e. total offspring/ no. categories). 400 / 4 = 100 (column 2).

3 Calculate O-E and $(O-E)^2$

The difference between the observed and expected values is calculated as a measure of the deviation from a predicted result. Since some deviations are negative, they are all squared to give positive values (column 3 and 4).

4 Calculate χ^2

For each category calculate $(O - E)^2 / E$. Then sum these values to produce the χ^2 value (column 5).

$$\chi^2 = \sum \frac{(O - E)^2}{E}$$

5 Calculate degrees of freedom

The probability that any particular χ^2 value could be exceeded by chance depends on the number of degrees of freedom. This is simply one less than the total number of categories (this is the number that could vary independently without affecting the last value) In this case 4 - 1 = 3.

6 Use χ^2 table

On the χ^2 table with 3 degrees of freedom, the calculated χ^2 value correspond to a probability between 0.2 and 0.5. By chance alone, a χ^2 value of **2.96** will happen 20% to 50% of the time. The probability of 0.0 to 0.5 is higher than 0.05 (i.e 5% of the time) and therefore the null hypothesis cannot be rejected. We have no reason to believe the observed values differ significantly from the expected values.

	1	2	3	4	5
Category	O	E	O-E	$(O\_E)^2$	$(O\_E)^2$/E
GB, LW	98	100	-2	4	0.04
GB, VW	88	100	-12	144	1.44
EB, LW	102	100	2	4	0.04
EB, VW	112	100	12	144	1.44
				$\chi^2 \longrightarrow$	2.96

©2022 **BIOZONE** International
ISBN: 978-1-98-856692-4
Photocopying Prohibited

SPQ   LS3.B

1. Students carried out a pea plant experiment, where two heterozygous individuals were crossed. The predicted Mendelian ratios for the offspring were **9:3:3:1** for each of the four following phenotypes: round-yellow seed, round-green seed, wrinkled-yellow seed, wrinkled-green seed.

   The observed results of the cross were not exactly as predicted. The numbers of offspring with each phenotype are provided below:

Observed results of the pea plant cross			
Round-yellow seed	441	Wrinkled-yellow seed	143
Round-green seed	159	Wrinkled-green seed	57

   (a) State your null hypothesis for this investigation (H_0)_____

   _____

   _____

   (b) State the alternative hypothesis (H_A): _____

   _____

   Use the chi-squared test to determine if the differences between the observed and expected phenotypic ratios are significant. Use the table of critical values for χ^2 at different *P* values on the previous page.

   (c) Enter the observed and expected values (number of individuals) and complete the table to calculate the χ^2 value.

Category	O	E	O – E	$(O - E)^2$	$\dfrac{(O - E)^2}{E}$
Round-yellow seed					
Round-green seed					
Wrinkled-yellow seed					
Wrinkled-green seed					Σ

   (d) Calculate the χ^2 value using the equation $\chi^2 = \sum \dfrac{(O - E)^2}{E}$: _____

   (e) Calculate the degrees of freedom: _____

   (f) Using the χ^2 table, state the P value corresponding to your calculated χ^2 value: _____

   (g) State your decision (circle one):   reject H_0 / do not reject H_0

2. In another experiment a group of students bred two corn plants together. The first corn plant was known to grown from a kernel that was colorless (c) and did not have a waxy endosperm (w). The second corn plant was grown from a seed that was colored (C) but with a waxy endosperm (W). When the corn ear was mature, the students removed it and counted the different phenotypes in the corn kernels.

Observed results of corn kernels			
Colored - waxy	201	Colorless - waxy	86
Colored - not waxy	85	Colorless - not waxy	210

   From the observed results the students argued two points:
   (1) The plant with the dominant phenotype must have been heterozygous for both traits.
   (2) The genes for kernel color and endosperm waxiness must be linked (on the same chromosome).

   (a) Defend the students' first argument: _____

   _____

   _____

   (b) On a separate sheet, use a chi-squared test to prove or disprove the students' second argument:

©2022 **BIOZONE** International
ISBN: **978-1-98-856692-4**
Photocopying Prohibited

217 Pedigree Analysis

Key Question: How do pedigree charts illustrate inheritance patterns over a number of generations and allow us to trace a genetic disorder back to its origin?

Sample pedigree chart

Pedigree charts are a way of showing inheritance patterns over a number of generations. They are often used to study the inheritance of genetic disorders. The key should be consulted to decode the symbols. Individuals are identified by their generation number and their order number in that generation. For example, **II-6** is the sixth person in the second row. The arrow indicates the person through whom the pedigree was discovered, i.e. who reported the condition.

If the chart on the right were illustrating a human family tree, it would represent three generations: grandparents (I-1 and I-2) with three sons and one daughter. Two of the sons (II-3 and II-4) are identical twins, but did not marry or have any children. The other son (II-1) married and had a daughter and another child (sex unknown). The daughter (II-5) married and had two sons and two daughters (plus a child that died in infancy).

For the particular trait being studied, the grandfather was expressing the phenotype (showing the trait) and the grandmother was a carrier. One of their sons and one of their daughters also show the trait, together with one of their granddaughters.

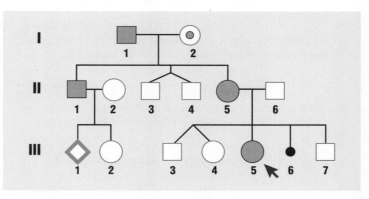

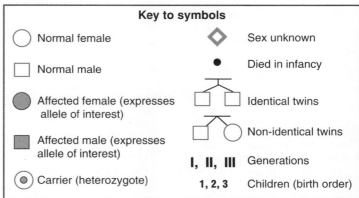

Key to symbols

- ○ Normal female
- □ Normal male
- ● Affected female (expresses allele of interest)
- ■ Affected male (expresses allele of interest)
- ◉ Carrier (heterozygote)
- ◇ Sex unknown
- • Died in infancy
- ⊐⊏ Identical twins
- ⊓ Non-identical twins
- **I, II, III** Generations
- **1, 2, 3** Children (birth order)

1. **Pedigree chart of your family**

   Using the symbols in the key above and the example illustrated as a guide, construct a pedigree chart of your own family (or one that you know of) starting with the parents of your mother and/or father on the first line. Your parents will appear on the second line (II) and you will appear on the third line (III). There may be a fourth generation line (IV) if one of your brothers or sisters has had a child. Use a ruler to draw up the chart carefully.

©2022 **BIOZONE** International
ISBN: 978-1-98-856692-4
Photocopying Prohibited

2. The pedigree chart (right) shows the inheritance of allele A in a flower that can be blue or white. Blue flower = affected:

(a) Which color is produced by the dominant allele?

_____

(b) Write on the chart the genotype for each of the generation I individuals.

(c) III4 is crossed with a white flower. What is the probability that any one offspring also has a white flower?

_____

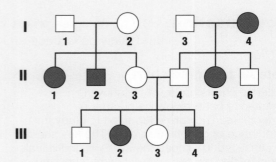

3. The pedigree chart (right) shows the inheritance of the allele B in a mammal that can have a coat color of black or white. Black coat = affected.

(a) Which color is produced by the dominant allele?

_____

(b) Explain how you know this: _____

_____

_____

_____

_____

4. **Autosomal recessive traits**
Albinos lack pigment in the hair, skin and eyes. This trait is inherited as an autosomal recessive allele, i.e. it is not carried on the sex chromosome.

(a) Write the genotype for each of the individuals on the chart using the following letter codes:
PP normal skin color; **P-** normal (but unknown if homozygous), **Pp** carrier, **pp** albino.

(b) Why must the parents (II-3) and (II-4) be **carriers** of a **recessive** allele?

_____

_____

_____

_____

Albinism in humans

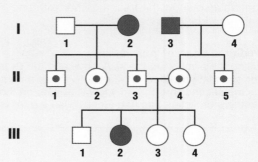

5. **Autosomal dominant traits**
An unusual trait found in some humans is woolly hair (not to be confused with curly hair). Each affected individual will have at least one affected parent.

(a) Write the genotype for each of the individuals on the chart (right) using the following letter codes:
WW woolly hair, **Ww** woolly hair (heterozygous), **W-** woolly hair, but unknown if homozygous, **ww** normal hair.

(b) Describe a feature of this inheritance pattern that suggests the trait is the result of a **dominant** allele:

_____

_____

_____

Woolly hair in humans

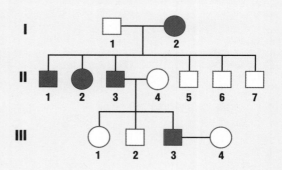

218 Review Your Understanding

Key Question: Can we get a chocolate labrador puppy from black parents?.

At the beginning of the chapter you were asked whether a pair of black labradors could be crossed to give a chocolate (brown) colored puppy.

▸ The gene for black coat color is represented by 'B' and is dominant over the recessive form, 'b', which gives a chocolate brown coat color.

▸ Yellow coat color is determined by a different, unlinked gene known as 'E'. This time, the recessive form of the gene 'e' can interfere with the black/brown coat color and mask it, resulting in a yellow puppy. If the dominant form of the gene 'E' is present, black/brown coloration is not masked and dogs will always be black or chocolate colored.

1. Complete the following crosses for labradors, including genotype and phenotype ratios: Black dogs (B) are dominant to brown dogs (b):

   (a) Bb x Bb:

   (b) Bb x bb

2. Two chocolate colored labradors with genotypes bbEe are mated. Fill in the Punnett square below and circle any yellow puppies that might be expected in the litter:

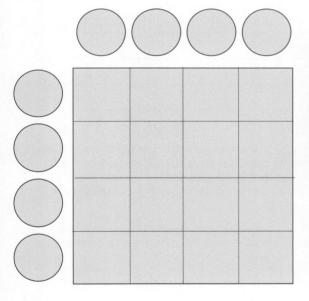

3. Explain whether mating two yellow labradors would result in any black or chocolate colored puppies:

_____

_____

_____

_____

219 Summing Up

The pedigree of lactose intolerance

Lactose intolerance is the inability to digest the milk sugar lactose. It occurs because some people do not produce lactase, the enzyme needed to break down lactose. The pedigree chart below was one of the original studies to determine the inheritance pattern of lactose intolerance.

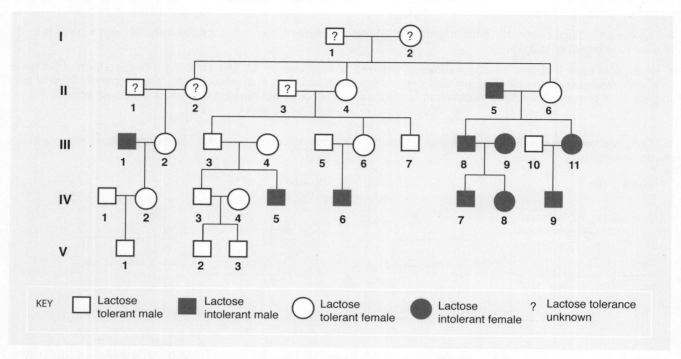

KEY

☐ Lactose tolerant male   ■ Lactose intolerant male   ○ Lactose tolerant female   ● Lactose intolerant female   ? Lactose tolerance unknown

1. Use the pedigree chart above to determine if lactose intolerance is a dominant trait or a recessive trait:

_____

2. Explain your answer: _____

_____

_____

_____

Linked genes

Shortly after the rediscovery of Mendel's work early in the 20th century, it became apparent that his ratios of 9:3:3:1 for heterozygous dihybrid crosses did not always hold true. Experiments on sweet peas by William Bateson and Reginald Punnett, and on *Drosophila* by Thomas Hunt Morgan, showed that there appeared to be some kind of coupling between genes. This coupling, which we now know to be linkage, did not follow any genetic relationship known at the time.

3. The data below is for a cross of sweet peas carried out by Bateson and Punnett. Purple flowers (P) are dominant to red (p), and long pollen grains (L) are dominant to round (l). If these genes were unlinked, the outcome of a cross between two heterozygous sweet peas should be a 9:3:3:1 ratio.

Study the data and use it to test the null hypothesis that the genes for flower color and grain shape are unlinked:

Table 1: Sweet pea cross results

	Observed
Purple long (P_L_)	284
Purple round (P_ll)	21
Red long (ppL_)	21
Red round (ppll)	55
Total	381

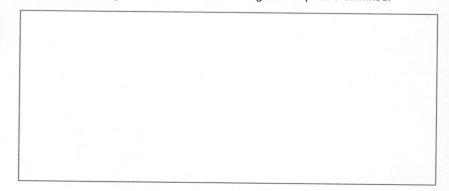

©2022 **BIOZONE** International
ISBN: 978-1-98-856692-4
Photocopying Prohibited

Members of your immediate family share common features but do not look exactly the same, unless there are identical twins in the family.

Copying errors during DNA replication and errors during the division of chromosomes can result in inheritable genetic variations. Down syndrome (above) occurs when chromosomes fail to separate during meiosis.

4. Inheritable genetic variation can result from 1) new genetic combinations arising through meiosis or 2) viable errors occurring during replication. Defend this claim using evidence and examples provided in this chapter. You can use diagrams to support your answer if you wish. Attach any extra work to this page.

_____

_____

_____

_____

_____

_____

_____

_____

_____

_____

_____

_____

_____

_____

_____

_____

_____

_____

_____

_____

_____

_____

_____

Biological Evolution: Unity and Diversity

Concepts and connections

Use arrows to make connections between related concepts in this section of the book

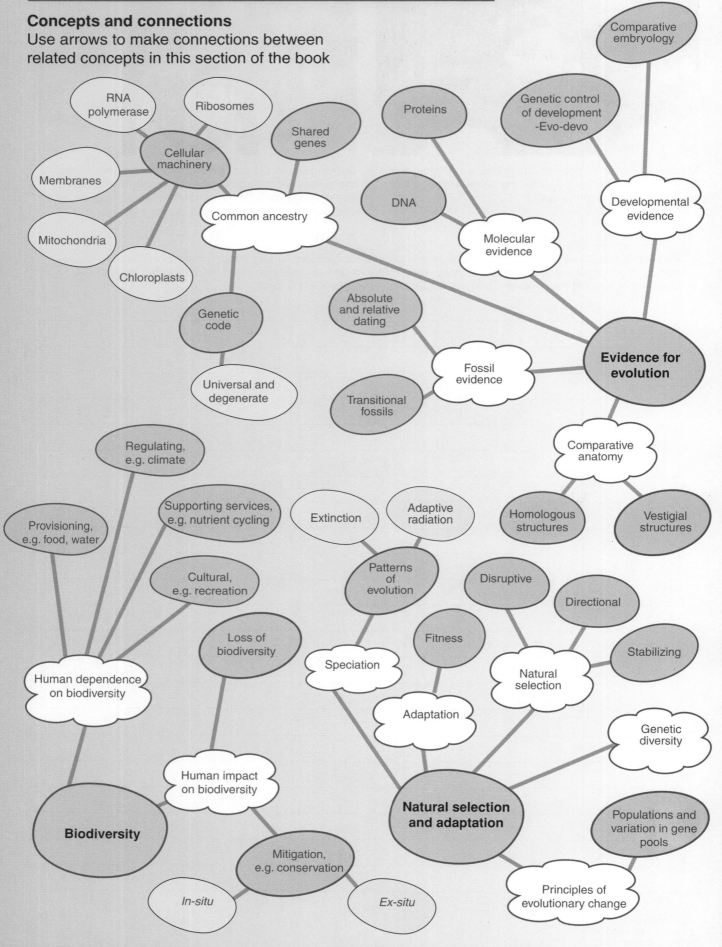

CHAPTER 12
Evidence for Evolution

Activity number

Anchoring Phenomenon

Dinosaur or Bird? How does scientific evidence help us to continually build ideas?   220 231

What evidence can we use to investigate evolutionary processes?

☐ 1 Understand how evolution is defined. Understand that scientists use a wide variety of evidence to continually build their ideas and theories of evolution.   221

☐ 2 Use a model representing how all living organisms are related to each other to identify features in cells that provide evidence for evolution. Use a model to explain the ordering of cellular features as evidence for evolution.   222

How can we use fossils as evidence for evolution?

☐ 3 Examine a fossil-embedded rock profile to discuss how the ordering of layers can be used as evidence for evolution. Consider some difficulties that the fossil record may present when used as evidence for evolution.   223

☐ 4 Interpret data from a rock profile model to order layers, thereby identifying the relative age of rocks. Explain how similar aged rock layers can be identified by the presence of fossil species. Link the rock type with the presence or absence of fossils.   224

☐ 5 Explain what is meant by the term transitional fossil. Discuss how transitional fossils can be used as evidence for evolution.   225

☐ 6 Examine the case study of whale evolution and transitional fossils to discuss how the fossil record provides evidence for evolution. Link the features of transitional fossils to the increased ability of the whale group to swim, over time.   226

How does anatomy, DNA, protein, and developmental similarities provide evidence for evolution?

☐ 7 Identify features of a pentadactyl limb in tetrapods. Discuss how homology in anatomical structures of organisms can provide evidence for evolution and shared ancestry.   227

☐ 8 Use models of DNA sequences and a phylogenetic tree diagram to identify relatedness of different species. Discuss how new genetic evidence can change science ideas about the relatedness and the evolution of different species.   228

☐ 9 In the context of evolution, define the term "highly conserved protein". Discuss how using protein structure homology can be a useful tool to construct phylogenetic trees. Compare the amino acid sequence in hemoglobin to a model, to discuss the relative relatedness of different organisms.   229

☐ 10 Explain how embryonic development can be used as evidence for evolution, using information from data and diagrams. Discuss how the formation of differentiated specialized limbs can originate from the same pentadactyl structure.   230

☐ 11 In the context of the Tiktaalik fossil discovery, use information from fossil features to discuss the evidence used to place the species on a timeline of vertebrate evolution.   232

220 Dinosaur or Bird?

Key Question: How does scientific evidence allow us to continually build ideas of what dinosaurs looked like?

The "old" velociraptor

The *Velociraptor* genus of dinosaurs is well known, thanks to movies such as Jurassic Park. These sleek, hairless hunters lived in packs and had a distinctly reptilian, featherless scaly skin. What evidence allowed us to form an impression of these dinosaurs? Fossil remains of a damaged skull and a toe claw were found in the Mongolian dessert in 1923. This was the first piece of the puzzle.

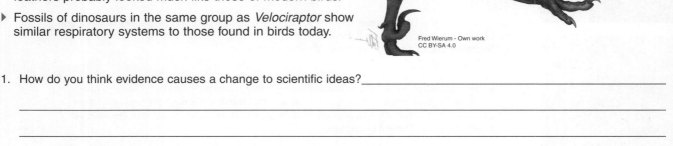

The "new" velociraptor

Recent discoveries came about in 2007, when scientists took another look at a previously uncovered arm bone fossil to locate "quill knobs", typically found in birds for feather attachment to bone.

▸ Increasingly, more detailed fossils of dromaeosaurids (the family of dinosaurs that *Velociraptor* belongs to) that show feather impressions are being uncovered.

▸ The presence of quill knobs suggest that *Velociraptor*'s feathers probably looked much like those of modern birds.

▸ Fossils of dinosaurs in the same group as *Velociraptor* show similar respiratory systems to those found in birds today.

Fred Wierum - Own work
CC BY-SA 4.0

1. How do you think evidence causes a change to scientific ideas?_____

_____

_____

_____

2. Some scientists believe that birds should be classified as dinosaurs. What evidence might they use for their claim?

_____

_____

_____

_____

221 Evidence for Evolution

Key Question: Where does evidence for evolution come from?

What is evolution?

Evolution describes the heritable changes in a population's gene pool over time. Evidence for the fact that populations evolve comes from many fields of science. Evolution is defined as the heritable genetic changes seen in a population over time. There are two important points to take from this definition. The first is that evolution refers to populations, not individuals. The second is that the changes must be passed on to the next generation, i.e. be inherited. The evidence for evolution comes from many branches of science (below) and includes evidence from living populations, as well as from the past.

Comparative anatomy

Comparative anatomy examines the similarities and differences in the anatomy of different species. Similarities in anatomy, e.g. the bones forming the arms in humans and the wings in birds and bats, indicate descent from a common ancestor.

Geology

Geological strata (the layers of rock, soil, and other deposits such as volcanic ash) can be used to determine the relative order of past events, and therefore the relative dates of fossils. Fossils in lower strata are older than fossils in higher (newer) strata, unless strata have been disturbed.

DNA comparisons

DNA can be used to determine how closely organisms are related to each other. More closely related species have greater similarities in their **DNA sequences**.

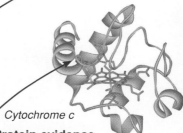

Cytochrome c

Protein evidence

Similarities (and differences) between proteins provides evidence for determining shared ancestry. Fewer differences in amino acid sequences reflects closer genetic relatedness.

EVOLUTION

Fossil record

Fossils, like this shark's tooth (left), are the remains of long-dead organisms. They provide a record of the appearance and extinction of organisms.

Developmental evidence

The study of developmental processes and the genes that control them gives insight into evolutionary processes. This field of study is called evolutionary developmental biology (**evo-devo**).

Biogeography

The geographical distribution of living and extinct organisms provides evidence of common ancestry and can be explained by speciation, extinction, and continental drift. The biogeography of islands, e.g the Galápagos Islands, provides evidence of how species evolve when separated from their ancestral population on the mainland.

Chronometric dating

Radiometric dating techniques, such as carbon dating, allow scientists to determine an absolute date for a fossil by dating it or the rocks around it. Absolute dating has been used to assign ages to strata, and construct the geological time scale.

 P LS4.A

222 The Common Ancestry of Life

DNA encodes the genetic instructions of all life. The form of these genetic instructions, the **genetic code**, is effectively universal, i.e. the same combination of three DNA bases code for the same amino acid in almost all organisms. The very few exceptions in which there are minor coding alternatives occur only in some bacteria and mitochondrial DNA.

How do we know about the relatedness of organisms?

▸ Traditionally, the phylogeny (evolutionary history) of organisms was established by comparing morphology (physical structure). In recent decades, molecular techniques involving the analysis of DNA, RNA, and proteins have provided more information about how all life on Earth is related.

▸ These newer methods have enabled scientists to clarify the origin of the eukaryotes and to recognize two prokaryote domains. The universality of the genetic code and the similarities in the molecular machinery of all cells provide powerful evidence for a common ancestor to all life on Earth.

▸ In recent decades, molecular techniques involving the analysis of DNA, RNA, and proteins have provided more information about how all life on Earth is related.

All known living organisms use ATP. It is the universal energy-carrying molecule in cells.

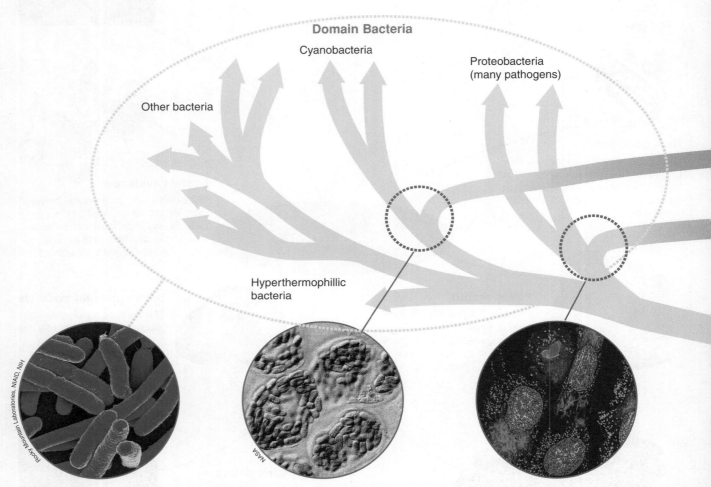

Domain Bacteria

Cyanobacteria

Proteobacteria (many pathogens)

Other bacteria

Hyperthermophillic bacteria

Bacteria lack a distinct nucleus and cell organelles. Features of the cell wall are unique to bacteria and are not found among archaea or eukaryotes. Typically found in less extreme environments than archaea.

Chloroplasts have a bacterial origin

Cyanobacteria (above) are considered to be the ancestors of chloroplasts. The evidence for this comes from similarities in the ribosomes and membrane organization, as well as from genetic studies. Chloroplasts were acquired independently of mitochondria, from a different bacterial lineage, but by a similar process.

Mitochondria have a bacterial origin

Evidence from mitochondrial gene sequences, ribosomes, and protein synthesis indicate that mitochondria (stained purple above) came from prokaryotes. Mitochondria were probably symbiotic inclusions in an early eukaryotic ancestor.

©2022 **BIOZONE** International
ISBN: 978-1-98-856692-4
Photocopying Prohibited

1. Identify three features of the metabolic machinery of cells that support a common ancestry of life:

   (a) _____

   (b) _____

   (c) _____

2. Suggest why scientists believe that mitochondria were acquired before chloroplasts: _____

   _____

   _____

   _____

Eukarya (the eukaryotes) are characterized by complex cells with organelles and a membrane-bound nucleus. This domain is made up of the four kingdoms, recognized under a traditional classification scheme.

Archaea resemble bacteria but the make-up of the membrane and cell wall, as well as aspects of metabolism are very different. Many live in extreme environments similar to those on primeval Earth (see center image below).

Domain Eukarya

Animals   Fungi   Plants

Algae

Domain Archaea

Bacteria that gave rise to chloroplasts

Ciliates

Bacteria that gave rise to mitochondria

RCN

Eukaryotes have linear chromosomes

Eukaryotic cells all have large linear chromosomes (above) within the cell nucleus. The evolution of linear chromosomes was related to the appearance of mitosis and meiosis (eukaryotic types of cell division).

Eukaryotes have an archaean origin

Archaea superficially resemble bacteria but similarities in the molecular machinery, e.g. RNA polymerase and ribosome proteins, show that they are more closely related to eukaryotes.

Last Universal
Common Ancestor
(LUCA)

Living systems share the same molecular machinery

In all living systems, the genetic machinery consists of self-replicating DNA molecules. Some DNA is transcribed into RNA, some of which is translated into proteins. The machinery for translation (left) involves proteins and RNA. Ribosomal RNA analysis supports a universal common ancestor.

©2022 **BIOZONE** International
ISBN: 978-1-98-856692-4
Photocopying Prohibited

223 The Fossil Record

Key Question: How can the fossil record be used as evidence for evolution?

The importance of the fossil record

▶ **Fossils** are the remains of long-dead plants and animals that have become preserved in the Earth's crust.

▶ Fossils provide a record of the appearance and extinction of organisms, from species to whole taxonomic groups.

▶ The **fossil record** can be used to establish the relative order of past events.

▶ The fossil record can be calibrated against a time scale, using dating techniques, to build up a picture of the evolutionary changes that have taken place.

Fossilized fern frond

Gaps in the fossil record

The fossil record contains gaps and without a complete record, it can sometimes be difficult to determine an evolutionary sequence. Scientists also use other information, to produce an order of events that best fits all the evidence.

Gaps in the fossil record can occur because:

▶ Fossils are destroyed.

▶ Some organisms do not fossilize well.

▶ Fossils have not yet been found.

Profile with sedimentary rocks containing fossils

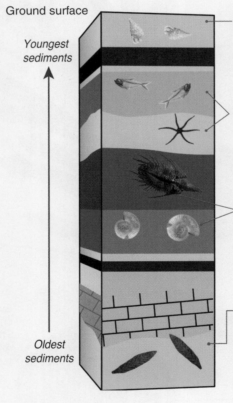

Ground surface

Youngest sediments

Oldest sediments

Rock strata are layered through time

Rock strata are arranged in the order that they were deposited, unless they have been disturbed by geological events. The most recent layers are near the surface and the oldest are at the bottom. Fossils can be used to establish the sequential order of past events in a rock profile.

New fossil types mark changes in environment

In the strata at the end of one geological period, it is common to find many new fossils that become dominant in the next. Each geological period had a different environment from the others. Their boundaries coincided with drastic environmental changes and the appearance of new niches. These produced new selection pressures, resulting in new adaptive features in the surviving species as they responded to the changes.

Recent fossils are found in more recent sediments

The more recent the layer of rock, the more resemblance there is between the fossils found in it and living organisms.

Fossil types differ in each stratum

Fossils found in a given layer of sedimentary rock are generally significantly different from fossils in other layers.

Extinct species

The number of extinct species is far greater than the number of species living today.

More primitive fossils are found in older sediments

Fossils in older layers tend to have quite generalized forms. In contrast, organisms alive today have specialized forms.

---

1. Discuss the importance of fossils as a record of evolutionary change over time: _____

_____

_____

_____

2. Why can gaps in the fossil record make it difficult to determine an evolutionary history? _____

_____

_____

_____

©2022 **BIOZONE** International
ISBN: 978-1-98-856692-4
Photocopying Prohibited

224 Interpreting the Fossil Record

Key Question: How are we able to analyze the fossils within rock strata to order past events in a rock profile, from oldest to most recent?

The diagram below shows a hypothetical rock profile from two locations separated by a distance of 67 km. There are differences between the rock layers at the two locations. Apart from layers D and L, which are volcanic ash deposits, all other layers are composed of sedimentary rock. Use the information on the diagram to answer the questions below.

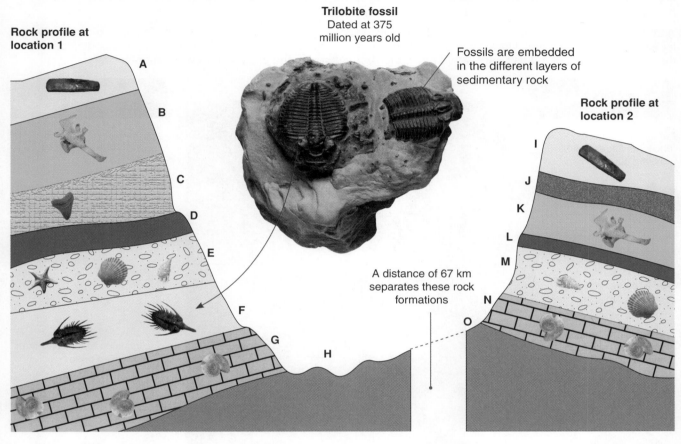

Trilobite fossil
Dated at 375 million years old

Fossils are embedded in the different layers of sedimentary rock

Rock profile at location 1

Rock profile at location 2

A distance of 67 km separates these rock formations

1. Assuming there has been no geological activity to disturb the order of the rock layers, state in which rock layer (A-O) you would find:

    (a) The youngest rocks at Location 1: _____   (c) The youngest rocks at Location 2: _____

    (b) The oldest rocks at Location 1: _____   (d) The oldest rocks at Location 2: _____

2. (a) State which layer at location 1 is of the same age as layer M at location 2: _____

    (b) Explain the reason for your answer in 2 (a): _____

    _____

    _____

    _____

3. (a) State which layers present at location 1 are missing at location 2: _____

    (b) State which layers present at location 2 are missing at location 1: _____

4. The rocks in layer H and O are sedimentary rocks. Why are there no visible fossils in these layers? _____

    _____

    _____

    _____

©2022 **BIOZONE** International
ISBN: 978-1-98-856692-4
Photocopying Prohibited

P LS4.A

225 Transitional Fossils

Key Question: How do transitional fossils provide important links in the fossil record?

Transitional fossils are **fossils** which have a mixture of features, showing intermediate states, that are found in two different, but related, groups. Transitional fossils provide important links in the fossil record and provide evidence to support how one group may have given rise to the other by evolutionary processes.

Important examples of transitional fossils include horses, whales, and *Archaeopteryx* (below), a transitional form between birds and non-avian dinosaurs.

Archaeopteryx was crow-sized (50 cm length) and lived about 150 million years ago. It had a number of birdlike (avian) features, including feathers. However, it also had many non-avian features, which it shared with theropod dinosaurs of the time. Although not a direct ancestor of birds, the *Archaeopteryx* and birds shared a common ancestor.

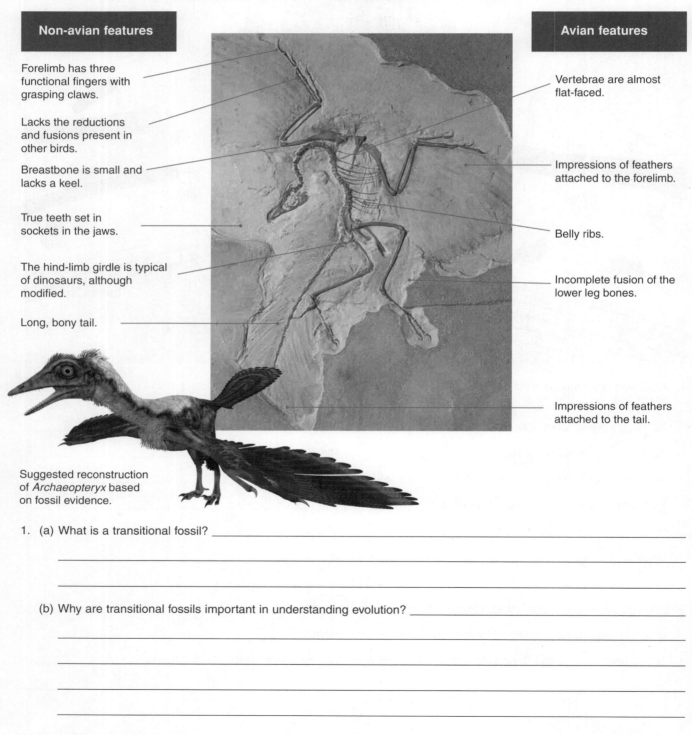

Non-avian features

Forelimb has three functional fingers with grasping claws.

Lacks the reductions and fusions present in other birds.

Breastbone is small and lacks a keel.

True teeth set in sockets in the jaws.

The hind-limb girdle is typical of dinosaurs, although modified.

Long, bony tail.

Avian features

Vertebrae are almost flat-faced.

Impressions of feathers attached to the forelimb.

Belly ribs.

Incomplete fusion of the lower leg bones.

Impressions of feathers attached to the tail.

Suggested reconstruction of *Archaeopteryx* based on fossil evidence.

1. (a) What is a transitional fossil? _____

    _____

    _____

    (b) Why are transitional fossils important in understanding evolution? _____

    _____

    _____

    _____

    _____

    _____

LS4.A P

©2022 **BIOZONE** International
ISBN: 978-1-98-856692-4

226 Case Study: Whale Evolution

Key Question: What fossil evidence do we have to show the evolution of whales?

Whale evolution

The evolution of modern whales from an ancestral land mammal is well documented in the **fossil record**. The fossil record of whales includes many transitional forms, recording the shift from a terrestrial to an aquatic life, which has enabled scientists to develop an excellent model of whale evolution. The evolution of the whales (below) shows a gradual accumulation of adaptive features that have equipped them for life in the open ocean.

Modern whales are categorized into two groups.

▸ Toothed whales have full sets of teeth throughout their lives, e.g. sperm whales and orca.

▸ Baleen whales. These are toothless whales and they use a comb-like structure called baleen to filter food, e.g. humpback whale.

Humpback whale

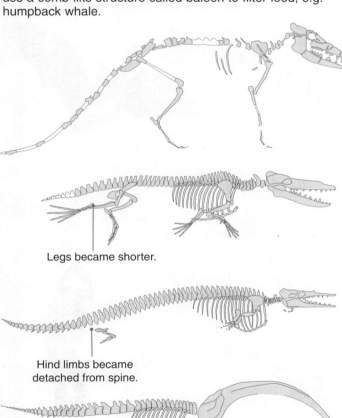

Legs became shorter.

Hind limbs became detached from spine.

Hind limbs are internal and vestigial (have lost their original function).

Redrawn from de Muizon Nature 2001 413 pp259-260

50 mya *Pakicetus*

Pakicetus was a transitional species between carnivorous land mammals and the earliest true whales. It was mainly terrestrial (land dwelling), but foraged for food in water. It had four, long limbs. Its eyes were near the top of the head and its nostrils were at the end of the snout. It had external ears, but they showed features of both terrestrial mammals and fully aquatic mammals.

45 mya *Rodhocetus*

Rodhocetus was mainly aquatic (water living). It had adaptations for swimming, including shorter legs and a shorter tail. Its eyes had moved to the side of the skull, and the nostrils were located further up the skull. The ear showed specializations for hearing in water.

40 mya *Dorudon*

Dorudon was fully aquatic. Its adaptations for swimming included a long, streamlined body, a broad, powerful, muscular tail, development of flippers, and webbing. It had very small hind limbs, not attached to the spine, which would no longer bear weight on land.

Balaena (recent whale ancestor)

The tiny hind limbs became fully internal. Studies of modern whales show that limb development begins, but stops at the limb bud stage. The nostrils became modified as blowholes. This recent ancestor to modern whales diverged into two groups, toothed and baleen, about 36 million years ago. Baleen whales have teeth in their early fetal stage, but lose them before birth.

1. Why does the whale fossil record provide a good example of the evolutionary process? _____

_____

_____

2. Briefly describe the adaptations of whales for swimming that evolved over time: _____

_____

_____

 P LS4.A

227 Anatomical Evidence for Evolution

Key Question: How do homologous structures indicate the evolutionary relationship between groups of organisms?

Homologous structures

Homologous structures are structures found in different organisms that are the result of their inheritance from a common ancestor. Their presence indicates the evolutionary relationship between organisms. Homologous structures have a common origin, but they may have different functions.

For example, the forelimbs of birds and seals are homologous structures. They have the same basic skeletal structure, but have different functions. A bird's wings have been adapted for flight, and a seal's flippers are modified as paddles for swimming.

The pentadactyl limb

A **pentadactyl limb** is a limb with five fingers or toes, e.g. hands and feet, with the bones arranged in a specific pattern (below, left).

Early land vertebrates were amphibians with pentadactyl limbs. All vertebrates that descended from these early amphibians have limbs that have evolved from this same basic pentadactyl pattern. The pentadactyl limb is a good illustration of adaptive radiation (the diversification of an ancestral form into many different forms). The generalized limb plan has been adapted to meet the requirements of organisms in many different niches.

Generalized pentadactyl limb

The forelimbs and hind limbs have the same arrangement of bones. In many cases bones in different parts of the limb have been modified for a specialized locomotory function.

Forelimb	Hind limb
Humerus (upper arm)	Femur (thigh)
	Fibula
	Tibia
Radius	
Ulna	
Carpals (wrist)	Tarsals (ankle)
Metatarsals (palm)	Metatarsals (sole)
Phalanges (fingers)	Phalanges (toes)

Specializations of pentadactyl limbs

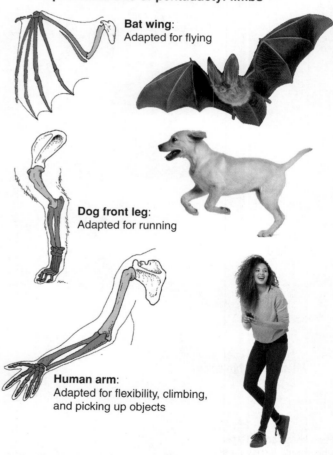

Bat wing: Adapted for flying

Dog front leg: Adapted for running

Human arm: Adapted for flexibility, climbing, and picking up objects

1. What is a pentadactyl limb? _____

_____

_____

2. Explain how homology in the pentadactyl limb provides evidence for adaptive radiation: _____

_____

_____

_____

_____

 LS4.A | P

228 DNA Evidence for Evolution

Key Question: How can DNA sequencing and comparison, and the use of computer databases be used to locate evidence for evolution?

The advancement of techniques in molecular biology is providing increasingly large amounts of information about the genetic makeup of organisms. **DNA sequencing** and comparison, and the use of computer databases are now frequently used when analyzing the evolutionary histories and relationships of different species.

Bioinformatics involves the collection, analysis, and storage of biochemical information, e.g. DNA sequences, using computer science and mathematics.

Bioinformatics allows biological information to be stored in databases where it can be easily retrieved, analyzed, and compared. Comparison of DNA or protein sequences between species enables researchers to investigate and better understand their evolutionary relationships.

An overview of the bioinformatics process

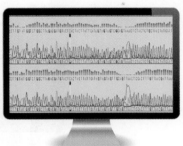
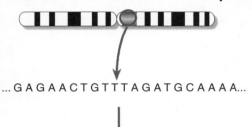

...GAGAACTGTTTAGATGCAAAA...

▸ A gene of interest is selected for analysis.

▸ High throughput 'Next-Gen' sequencing technologies allow the DNA sequence of the gene to be quickly determined.

Organism 1 ...G A G|A A|C T G T T T|A G|A T G C A A|A A...

Organism 2 ...G A G|A T|C T G T G T|A G|A T G C A|G A|A...

Organism 3 ...G A G|T T|C T G T G T|C G|A T G C A|G A|A...

Organism 4 ...G A G|T T|C T G T T T|C G|A T G C A|G A|G...

▸ Powerful computer software can quickly compare the DNA sequences of many organisms. Commonalities and differences in the DNA sequence can help to determine the evolutionary relationships of organisms. The blue boxes indicate differences in the DNA sequences.

▸ Once sequence comparisons have been made, the evolutionary relationships can be displayed as a **phylogenetic tree**. The example (right) shows the evolutionary relationships of the whales to some other land mammals.

▸ Bioinformatics has played an important role in determining the origin of whales and their transition from a terrestrial (land) form to a fully aquatic form.

▸ This phylogenetic tree was determined by comparing repetitive DNA fragments that are inserted into chromosomes after they have been reverse transcribed from an mRNA molecule. The locations of these repetitive fragments are predictable and stable, so they make reliable markers for determining species relationships. If two species have the same repeats in the same location, they are very likely to share a **common ancestor**.

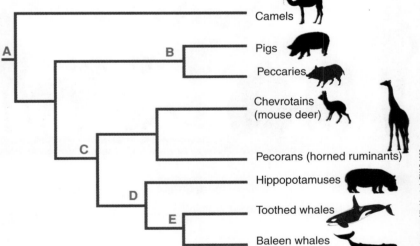

Camels
Pigs
Peccaries
Chevrotains (mouse deer)
Pecorans (horned ruminants)
Hippopotamuses
Toothed whales
Baleen whales

Data source: Nikaido et.al, PNAS 1996

1. The diagram above shows the relatedness of several mammals as determined by DNA sequencing of 10 genes:

   (a) Which land mammal are whales most closely related to? _____

   (b) Which letter shows where whales and the organism in (a) last shared a common ancestor?_____

   (c) Pigs were once considered to be the most closely related land ancestor to the whales. Use the phylogenetic tree above to describe the currently accepted relationship:

   _____

   _____

P LS4.A

229 Protein Evidence for Evolution

Key Question: How can protein homology be used to determine evolutionary patterns?

Protein homology

The amino acid sequence of proteins can be used to establish homologies (similarities) between organisms. Any change in the amino acid sequence reflects changes in the DNA sequence. As genetic relatedness increases, the number of amino acid differences due to mutation decreases.

Some proteins are common to many different species. These proteins are often highly conserved, meaning they mutate (change) very little over time. This is because they have critical roles, e.g. in cellular respiration, and mutations are likely to be detrimental to their function.

Evidence indicates that these highly conserved proteins are homologous and have been derived from a **common ancestor**. Because they are highly conserved, changes in the amino acid sequence are likely to represent major divergences between groups during the course of **evolution**.

The Pax-6 protein provides evidence for evolution

▸ The Pax-6 protein regulates eye formation during embryonic development.

▸ The Pax-6 gene is so highly conserved that the gene from one species can be inserted into another species, and still produce a normally functioning eye.

▸ This suggests the Pax-6 proteins are homologous, and the gene has been inherited from a common ancestor.

An experiment inserted mouse Pax6 gene into fly DNA and turned it on in a fly's legs. The fly developed fly eyes on its legs!

Hemoglobin homology

Hemoglobin is the oxygen-transporting blood protein found in most vertebrates. The beta chain hemoglobin sequences from different organisms can be compared to determine evolutionary relationships.

As genetic relatedness decreases, the number of amino acid differences between the hemoglobin chains of different vertebrates increases (below). For example, there are no amino acid differences between humans and chimpanzees, indicating they recently shared a common ancestor. Humans and frogs have 67 amino acid differences, indicating they had a common ancestor a very long time ago.

Human – chimpanzee 0
Gorilla 1
Gibbon 2
Rhesus monkey 8
Dog 15
Horse 25
Mouse 27
Kangaroo 38
Chicken 45
Frog 67

Increasing difference in amino acid sequence

Primates | Placental mammals | Marsupial | Non-mammalian vertebrates

1. (a) What is a highly conserved protein? _____

   _____

   (b) Why are highly conserved proteins good for constructing phylogenies? _____

   _____

   _____

2. Compare the differences in the hemoglobin sequence of humans, rhesus monkeys, and horses. What do these tell you about the relative relatedness of these organisms?

   _____

   _____

   _____

  LS4.A P

©2022 **BIOZONE** International
ISBN: 978-1-98-856692-4
Photocopying Prohibited

230 Developmental Evidence for Evolution

Key Question: How do similarities in the development of embryos, including the genetic control of development, provide strong evidence for evolution?

Developmental biology

Developmental biology studies the process by which organisms grow and develop. In the past, it was restricted to the morphology (physical appearance) of a growing fetus. Today, developmental biology focuses on the genetic control of development and its role in producing the large differences we see in the adult appearance of different species.

During development, vertebrate embryos pass through the same stages, in the same sequence, regardless of the total time period of development. This similarity is strong evidence of their shared ancestry. The stage of embryonic development is identified using a standardized system based on the development of structures, not by size or the number of days of development. The Carnegie stages (right) cover the first 60 days of development.

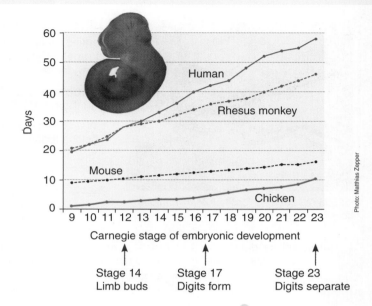

Carnegie stage of embryonic development

Stage 14
Limb buds

Stage 17
Digits form

Stage 23
Digits separate

Limb homology and the control of development

As we have seen, **homology**, e.g. in limb structure, is evidence of shared ancestry. How do these **homologous structures** become so different in appearance? The answer lies in the way the same genes are regulated during development.

All vertebrate limbs form as buds at the same stage of development. At first, the limbs resemble paddles, but apoptosis (programmed cell death) of the tissue between the developing bones separates the digits to form fingers and toes.

Like humans, mice have digits that become fully separated by apoptosis of the tissue between the bones during development. In bat forelimbs, this controlled destruction of the tissue between the forelimb digits is inhibited. The developmental program is the result of different patterns of expression of the same genes in the two types of embryos.

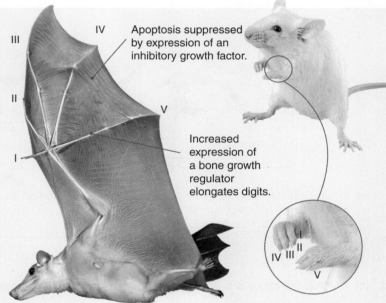

Apoptosis suppressed by expression of an inhibitory growth factor.

Increased expression of a bone growth regulator elongates digits.

Bat wings are highly specialized structures with unique features, such as elongated wrist and fingers (I-V) and membranous wing surfaces. The forelimb structures of bats and mice are homologous, but how the limb looks and works is quite different.

1. Describe a feature of vertebrate embryonic development that supports evolution from a common ancestor:

_____

_____

_____

2. Explain how different specialized limb structures can arise from a basic pentadactyl structure: _____

_____

_____

_____

231 Review Your Understanding

Key Question: How does scientific evidence allow us to continually build on our ideas of what dinosaurs looked like?

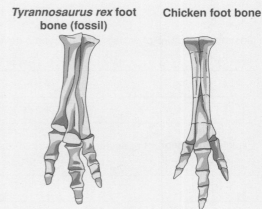

125 MYA feathered fossil of *Zhenyuanlong* fossil is a close relative of the velociraptor.

Junchang Lü and Stephen L. Brusatte Wikimedia (CC BY 4.0)

A dinosaur and a bird?

Scientific theory is developed through observation that is rigorously checked and repeated by the scientific community to confirm its validity. However, if new evidence arises, then the theory must be changed to accommodate it.

Birds are now classified by scientists as belonging to a group of dinosaurs called theropods. Birds have been on Earth for at least 150 million years.

1. What types of evidence could scientists have used to determine that birds belong to the (avian) dinosaur group?

_____

_____

_____

2. (a) Scientists identify the *Archaeopteryx* as a transitional species between birds and dinosaurs, but not likely a direct ancestor. Refer to the phylogenetic tree below and explain why this might be:

_____

_____

Birds

Velociraptor   Archaeopteryx

Crocodiles

Sauropods

Pterosauria   C   Dinosauria   D

A   B

E

F

Plotnick, Theodor & Holtz Jr., CC BY 4.0

(b) At what point in the phylogenetic tree would we expect to find the last common ancestor between birds and velociraptors? Explain your reasoning:

_____

_____

(c) Which of the above points are fossils likely to share with some features of birds? Explain your reasoning:

_____

_____

(d) Why is the ability to fly not an appropriate indicator of bird and dinosaur classification?

_____

_____

_____

Tyrannosaurus rex **foot bone (fossil)**   **Chicken foot bone**

3. The foot bones of *Tyrannosaurus rex* and a chicken look similar (right). What is a probable explanation for this?

_____

_____

_____

©2022 **BIOZONE** International
ISBN: 978-1-98-856692-4
Photocopying Prohibited

232 Summing Up

In 2004, a **fossil** of an unknown vertebrate was discovered in northern Canada and subsequently called *Tiktaalik roseae*. The *Tiktaalik* fossil was quite well preserved and many interesting features could be identified. These are shown in the photograph of the fossil below.

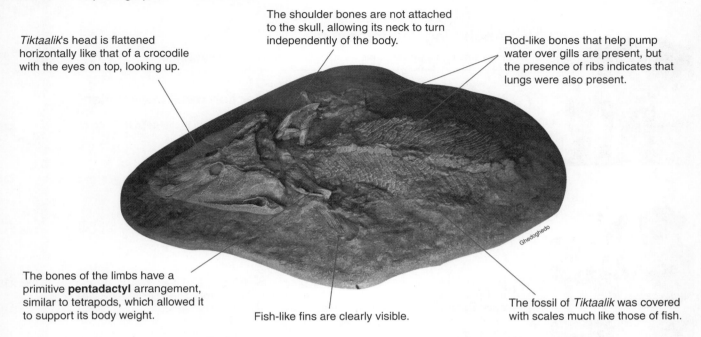

The shoulder bones are not attached to the skull, allowing its neck to turn independently of the body.

Tiktaalik's head is flattened horizontally like that of a crocodile with the eyes on top, looking up.

Rod-like bones that help pump water over gills are present, but the presence of ribs indicates that lungs were also present.

The bones of the limbs have a primitive **pentadactyl** arrangement, similar to tetrapods, which allowed it to support its body weight.

Fish-like fins are clearly visible.

The fossil of *Tiktaalik* was covered with scales much like those of fish.

1. Use the information above to place *Tiktaalik* on the time line of vertebrate evolution. Discuss the evidence for your decision.

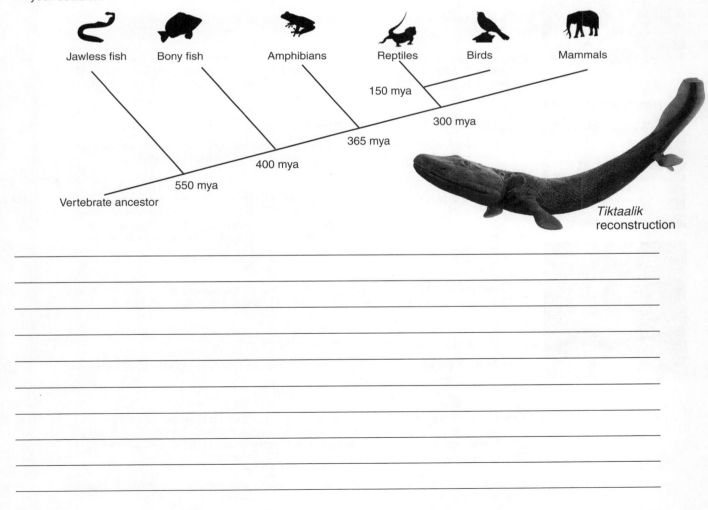

Jawless fish   Bony fish   Amphibians   Reptiles   Birds   Mammals

150 mya

300 mya

365 mya

400 mya

550 mya

Vertebrate ancestor

Tiktaalik reconstruction

_____
_____
_____
_____
_____
_____
_____
_____
_____
_____
_____

©2022 **BIOZONE** International
ISBN: 978-1-98-856692-4
Photocopying Prohibited

P   LS4.A

Natural Selection and Adaptation

Anchoring Phenomenon

How does an elephant lose its tusks? What is the link to poaching?

233 249

How does the process of natural selection result in evolution?

☐ 1 Use evidence from models to construct an explanation for how the genetic makeup of a population can change over time.

234

☐ 2 Simulate natural selection with a model to investigate changes occurring due to selection pressure in a population.

235

☐ 3 Interpret information to explain how adaptations can increase fitness in a species. Discuss the link between similar adaptations evolving to perform similar tasks.

236 237

☐ 4 Understand how adaptive radiation can provide evidence for evolution. Investigate and process beak depth data in finches, and coat color data in rock pocket mice. Develop evidence to support an argument that natural selection has occurred.

238 239

☐ 5 Use a computer simulation to model natural selection in deer mice.

240

☐ 6 Model insecticide resistance in insects, as an example of natural selection.

241

☐ 7 Investigate gene pool changes in rock pocket mice using a spreadsheet model to calculate the change due to natural selection. Interpret the data from the simulation to construct an explanation for the observations seen.

242

How do new species arise?

☐ 8 Define the term species, using dogs as a context. Use different models to develop an explanation for why dogs are all considered to belong to one single species.

243

☐ 9 Use models to understand how geographic isolation can create changes in the gene pool of two separate groups, leading to speciation. Identify features that act as geographic barriers. Identify and distinguish between divergent evolution, adaptive radiation, sequential evolution, phyletic gradualism, and punctuated equilibrium as patterns of evolution.

244 245

☐ 10 Investigate adaptive radiation in mammals. Use a case study of rodent biodiversity as an adaptive radiation example to link speciation to a range of different adaptations.

246

Why do species become extinct?

☐ 11 Analyze and interpret models and data of previous extinctions to explain why extinction is considered a natural phenomenon.

247

☐ 12 Use evidence from models and data to classify and provide reasons for extinction caused by human activity. Research several species that have become extinct due to human activity, examining evidence to discuss the precise cause.

248

☐ 13 Create a new species and discuss aspects about your species adaptations and survival ability in potential habitats.

250

233 How does an Elephant Lose its Tusks?

Key Question: How is poaching causing African elephants to be born without tusks?

Tusks and African elephants

▸ African elephants can be found across 23 countries in Africa and have adapted to a wide range of habitats.

▸ Both male and female African elephants typically have tusks, which are adapted from extended teeth. The tusks can vary in length.

▸ The tusks are made of ivory, a material much sought after by humans. Poaching (illegal hunting) of elephants has greatly reduced the total population, with around 14,000 individuals still being killed every year, mainly for their tusks.

▸ Tusklessness is a **phenotype** that is controlled by genes. Typically, around 2-4% of an African elephant population are born without tusks. Almost all are female.

▸ In some heavily poached areas, scientists have observed up to 60% of the elephants having the tuskless phenotype.

Carved ivory tusk

1. (a) What do you think might be the evolutionary advantage of having large tusks? Discuss in small groups, and write your ideas below:

_____

_____

_____

_____

(b) Why might that advantage now be a disadvantage to African elephants?_____

_____

_____

_____

_____

2. How do you think the term "Survival of the fittest" relates to the phenomenon of increasing numbers of African elephants being born without tusks?

_____

_____

_____

_____

_____

_____

234 How Evolution Occurs

Evolution

Evolution is the change in inherited characteristics in a population over generations. Evolution is the consequence of interaction between four factors: (1) The potential for populations to increase in numbers, (2) Genetic **variation** as a result of mutation and sexual reproduction, (3) competition for resources, and (4) proliferation of individuals with better survival and reproduction.

When better adapted organisms survive to produce a greater number of viable offspring, we call it **natural selection**. This has the effect of increasing their proportion in the population so that they become more common. This is the basis of Darwin's theory of evolution by natural selection.

We can demonstrate the basic principles of evolution using the analogy of a "population" of M&M's® candy.

In a bag of M&M's, there are many colors. This represents the variation in a population. As you and a friend eat through the bag of candy, you both leave the blue ones, which you both dislike, and return them to the bag.

The blue candy becomes more common...

Eventually, you are left with a bag of blue M&M's. Your selective preference for the other colors changed the make-up of the M&M's population. This is the basic principle of selection that drives evolution in natural populations.

Darwin's theory of evolution by natural selection

Darwin's theory of evolution by natural selection is outlined below. It is widely accepted by the scientific community today and is one of the founding principles of modern science.

Overproduction
Populations produce too many young: many must die.

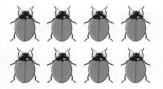

Populations generally produce more offspring than are needed to replace the parents. Natural populations normally maintain constant numbers. A certain number will die without reproducing.

Variation
Individuals show variation: some variations are more favorable than others.

Individuals in a population have different **phenotypes** and therefore, genotypes. Some **traits** are better suited to the environment, and individuals with these have better survival and reproductive success.

Natural selection
Natural selection favors the individuals best suited to the environment at the time.

Individuals in the population compete for limited resources. Those with favorable variations will be more likely to survive. Relatively more of those without favorable variations will die.

Inherited
Variations are inherited: the best suited variants leave more offspring.

©2022 **BIOZONE** International
ISBN: 978-1-98-856692-4

   LS4.B LS4.C CE

Variation, selection, and population change

Natural populations, like the ladybug population above, show genetic **variation**. This is a result of mutation, which creates new alleles, and sexual reproduction, which produces new combinations of alleles. Some variants are more suited to the environment of the time than others. These variants will leave more offspring, as described for the hypothetical population (right).

1. Variation through mutation and sexual reproduction:
In a population of brown beetles, mutations independently produce red coloration and 2 spot marking on the wings. The individuals in the population compete for limited resources.

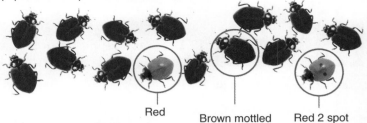

Red   Brown mottled   Red 2 spot

2. Selective predation:
Brown mottled beetles are eaten by birds but red ones are avoided.

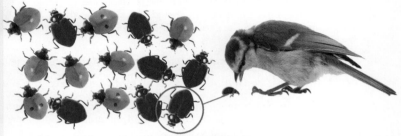

3. Change in the genetics of the population:
Red beetles have better survival and fitness and become more numerous with each generation. Brown beetles have poor fitness and become rare.

1. What produces the genetic variation in populations? _____

2. (a) Define evolution: _____
   _____

   (b) Identify the four factors that interact to bring about evolution in populations: _____
   _____
   _____
   _____
   _____

3. Using your answer to 2(b) as a basis, explain how the genetic make-up of a population can change over time: _____
   _____
   _____
   _____
   _____
   _____
   _____
   _____
   _____

©2022 **BIOZONE** International
ISBN: 978-1-98-856692-4
Photocopying Prohibited

235 Modeling Natural Selection

Key Question: How do adaptations provide an advantage when selection pressure is applied to a population?

Investigation 13.1 Investigating natural selection

See appendix for equipment list.

Selection in M&M's®

You can easily demonstrate natural selection using M&M's. M&M's have several colors, but they all taste the same. Therefore the only basis for selection is color. We call this selection pressure. We can see the effect of this when people act as M&M's "predators" and preferentially select their favorite color candy to eat.

1. Work in pairs. Open a bag of M&M's and pour them onto a clean surface.

2. Count the total "population" number of M&M's and record it here: _____

3. First, calculate the proportion of each color in the original population using:
Proportion = number of each color ÷ total "population" number. Enter these values below (row START).

4. Each person chooses 10 M&M's, <u>avoiding the blue ones</u>. Pick your favorite color first. If your favorite color is not available, e.g. blue, choose your second favorite. Eat (or discard) them.

5. Calculate the number of M&M's <u>remaining</u> and the proportion of each color in this population.
Proportion = number of each color ÷ total number remaining. Record these proportions below (Round 1).

6. Return the remaining M&M's to the bag (including the blue ones). From a second bag of M&M's restore the original "population" number using the proportions you calculated above. This represents the surviving M&M's reproducing. Calculate the number of each color of M&M's needed using the equation:
Proportion x number of M&M's you both ate (20). Round up or down to get whole numbers.

7. Repeat steps 3-5 three more times. Each round represents a generation.
Record the proportions left after each round (take photos after each round if you want).

	Blue	Green	Yellow	Orange	Red	Brown
START						
Round 1						
Round 2						
Round 3						
Round 4						

1. What happens to the blue M&M's over time? Explain: _____
_____
_____

2. How is this model similar to the process of **natural selection**? _____
_____
_____

3. How is this model different from the process of natural selection?_____
_____
_____
_____

      LS4.B LS4.C CE

©2022 **BIOZONE** International
ISBN: 978-1-98-856692-4
Photocopying Prohibited

236 Adaptation

Key Question: What are adaptations, and how are they classified?

Adaptation and fitness

▸ An **adaptation** is any heritable trait that equips an organism for its niche, enhancing its exploitation of the environment and contributing to its survival and successful reproduction.

▸ Adaptations are a product of **natural selection** and can be morphological (structural), physiological, or behavioral traits. Traits that are not helpful to survival and reproduction are not favored and will be lost.

▸ Adaptation is important in an evolutionary sense because adaptive features promote **fitness**. Fitness is a measure of an organism's reproductive success, i.e. its genetic contribution to the next generation.

Adaptive features of the North American beaver

North American beavers (*Castor canadensis*) are semi-aquatic and are able to remain submerged for up to 15 minutes. Their adaptations enable them to exploit both aquatic and terrestrial environments.

Beavers are strict herbivores and eat leaves, bark, twigs, roots, and aquatic plants. They do not hibernate. They live in domelike homes called lodges, which they build from mud and branches. Lodges are usually built in the middle of a pond or lake, with an underwater entrance, making it difficult for predators to attack.

Ears and nostrils

Valves in the ears and nose close when underwater. These keep water out.

Eyes

A clear eyelid protects the eye and allows the beaver to see while swimming.

Lips

Lips can close behind their front teeth.
This lets them carry objects and gnaw underwater, but keeps water out and stops them drowning.

Oxygen conservation

During dives, beavers slow their heartbeat and reduce blood flow to their extremities to conserve oxygen and energy.
This enables them to stay submerged for 15 minutes, even though they are not particularly good at storing oxygen in the tissues.

Front feet

Front paws are good at manipulating objects.
The paws are used in dam and lodge construction to pack mud and manipulate branches.

Thick insulating fat

A thick fat layer under the skin insulates the beaver from the cold water and helps keep it warm.

Teeth

Large, strong chisel-shaped incisors (front teeth) grow constantly.
These let beavers fell trees and branches for food and lodges.

Waterproof coat

A double-coat of fur (coarse outer hairs and short, fine inner hairs). An oil is secreted from glands and spread through the fur.
The underfur traps air against the skin for insulation and the oil acts as a waterproofing agent and keeps the skin dry in the water.

Large, webbed, hind feet

The webbing between the toes acts like a diver's swimming fins, and helps to propel the beaver through the water.

Large, flat paddle-like tail

The tail assists swimming and acts like a rudder. It is also used to slap the water in communication with other beavers, to store fat for the winter, and as a means of temperature regulation in hot weather because heat can be lost over the large, unfurred surface area.

 CE LS4.C LS4.B

Adaptations for diving in air-breathing animals

Air breathing animals that dive must cope with a lack of oxygen, which limits the length of the dive, and pressure, which limits the depth of the dive. Many different air-breathing vertebrates have diving representatives which have evolved from terrestrial ancestors and become adapted for an aquatic life. Diving air-breathers must maintain a supply of oxygen to the tissues during dives and can only stay underwater for as long as their oxygen supplies last. Their adaptations enable them to conserve oxygen and prolong their dive time.

Species for which there is a comprehensive fossil record, e.g. whales (right), show that adaptations for a diving lifestyle accumulated slowly during the course of the group's evolution.

Humpback whale

Penguin

Green turtle

Diving mammals

Dolphins, whales, and seals are among the most well adapted divers. They exhale before diving, so that there is no air in the lungs and nitrogen does not enter the blood. This prevents them getting the bends when they surface (a condition in which dissolved gases come out of solution at reduced pressures and form bubbles in the tissues).

During dives, oxygen is conserved by reducing heart rate dramatically, and redistributing blood to supply only critical organs. Diving mammals have high levels of muscle myoglobin, which stores oxygen, but their muscles also function efficiently using anaerobic metabolism.

Diving birds

Penguins show many of the adaptations typical of diving birds. During dives, a bird's heart rate slows, and blood is diverted to the head, heart, and eyes.

Diving reptiles

Sea turtles have low metabolic rates and their tissues are tolerant of low oxygen. These adaptations allow them to remain submerged for long periods and they surface only occasionally.

1. (a) What is an adaptation? _____
_____
_____

(b) How can an adaptation increase an organism's fitness? _____
_____
_____

2. The following list identifies some adaptations in a beaver which allow it to survive in its environment. Identify each adaptation as structural, physiological, or behavioral and describe its survival advantage:

(a) Large front teeth: _____
_____

(b) Lodge built in middle of pond: _____
_____

(c) Oil secreting glands in skin: _____
_____

3. (a) What restricts the amount of time diving animals can spend underwater? _____
_____

(b) How does reducing heart rate during a dive enable animals to stay underwater for longer? _____
_____
_____
_____

©2022 **BIOZONE** International
ISBN: 978-1-98-856692-4
Photocopying Prohibited

237 Similar Environments, Similar Adaptations

Key Question: Why do unrelated species often evolve similar adaptations to overcome the same environmental challenges?

Sometimes, genetically unrelated organisms evolve similar **adaptations** in response to the particular environmental challenges they face. The adaptations may result in different species having very similar appearances. Although the organisms are not closely related, **evolution** has produced similar solutions in order to solve similar ecological problems. This phenomenon is called convergent evolution.

Similar adaptations in unrelated plants

Cactus, N America)

Euphorbia

▸ The North American cactus and African *Euphorbia* species shown above are both xerophytes. They have evolved similar structural adaptations to conserve water and survive in a hot, dry, desert environment. Although they have a similar appearance, they are not related. They provide an excellent illustration of how unrelated organisms living in the same environment have independently evolved the same adaptations to survive.

▸ Their appearance is so similar at first glance that the *Euphorbia* is often mistaken for a cactus. Both have thick stems to store water and both have lost the presence of obvious leaves. Instead, they have spines or thorns to conserve water (a leafy plant would quickly exhaust its water reserves because of losses via transpiration). In cacti, spines are highly modified leaves. In *Euphorbia*, the thorns are modified stalks. It is not until the two flower that their differences are obvious.

Similar adaptations in unrelated animals

Lip Kee Yap cc2.0

Colugo (related to lemurs)

Flying squirrel (rodent)

The ability for mammals to glide between trees has evolved independently in unrelated animals. The characteristics listed below for the sugar glider are typically found in the gliding mammals shown above.

Sugar glider (marsupial)

Tail acts as a stabilizer and an air brake during flight.

The animals are nocturnal and have large eyes to help them see at night.

Skin is stretched between the front and back legs to form a wing-like flap. This allows them to glide up to 50 m between trees.

By moving its limbs, the animal has some control about the direction it flies in.

1. Explain why the North American cactus and African *Euphorbia* species have evolved such similar adaptations:

_____

_____

2. Suggest why gliding between trees (rather than walking) is an advantage to the gliding mammals described above:

_____

_____

3. Tenrecs, echidnas, and hedgehogs are examples of unrelated organisms that have all evolved spines. Explain the advantage of this adaptation to these animals and why it might have evolved:

_____

_____

CE P LS4.C LS4.B

238 Natural Selection in Finches

Key Question: How did studying Galápagos finch beaks provide evidence for evolution by natural selection?

- ▶ **Natural selection** acts on the **phenotypes** of a population. Individuals with phenotypes that increase their **fitness** produce more offspring, increasing the proportion of the genes corresponding to that phenotype in the next generation.

- ▶ Numerous population studies have shown that natural selection can cause phenotypic changes in a population relatively quickly. The effect of natural selection on a population can be verified by making quantitative measurements of phenotypic **traits**.

G.fortis

J Podos

Beak adaptations and feeding in Galápagos finches

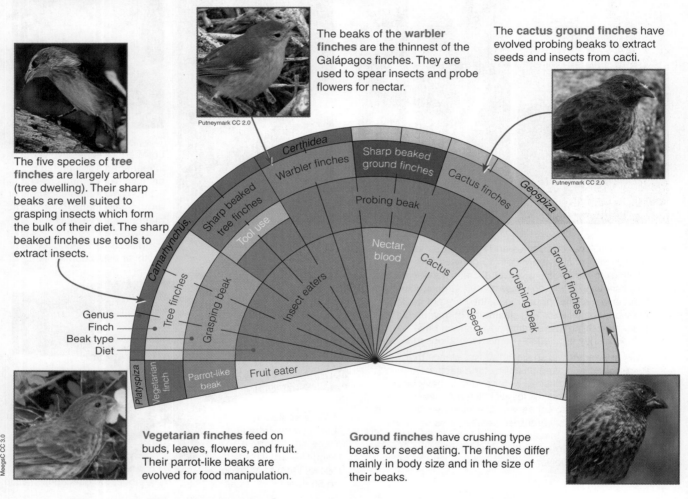

The beaks of the **warbler finches** are the thinnest of the Galápagos finches. They are used to spear insects and probe flowers for nectar.

Putneymark CC 2.0

The **cactus ground finches** have evolved probing beaks to extract seeds and insects from cacti.

Putneymark CC 2.0

The five species of **tree finches** are largely arboreal (tree dwelling). Their sharp beaks are well suited to grasping insects which form the bulk of their diet. The sharp beaked finches use tools to extract insects.

MeegsC CC 3.0

Vegetarian finches feed on buds, leaves, flowers, and fruit. Their parrot-like beaks are evolved for food manipulation.

Ground finches have crushing type beaks for seed eating. The finches differ mainly in body size and in the size of their beaks.

Diagram labels: Certhidea, Warbler finches, Sharp beaked ground finches, Cactus finches, Geospiza, Camarhynchus, Sharp beaked tree finches, Tool use, Probing beak, Nectar, blood, Cactus, Ground finches, Crushing beak, Seeds, Tree finches, Grasping beak, Insect eaters, Genus, Finch, Beak type, Diet, Platyspiza, Vegetarian finch, Parrot-like beak, Fruit eater

1. What are the main factors that contributed to the adaptive radiation of the Galápagos finches?

_____

_____

2. Given that the ancestral finch species radiated into so many different species, what might be said about the feeding and behaviour of the ancestral species?

_____

_____

3. Why is the adaptive radiation of the Galápagos finches important in providing evidence for evolution by natural selection?

_____

_____

      LS4.B LS4.C P CE

©2022 **BIOZONE** International
ISBN: 978-1-98-856692-4
Photocopying Prohibited

Adaptations are heritable

▶ Of all the Galápagos Island finches, the medium ground finch (*Geospiza fortis*) is particularly well studied. A population on the island Daphne Major had their beak depth measured shortly before the island experienced a severe drought. Researchers were interested in how the drought had affected the birds, and measured the beak depth of the survivors and their offspring.

▶ The finches on the Galápagos island (Darwin's finches) are famous as they are commonly used as examples of how evolution produces new species.

Medium ground finch

▶ In this activity, you will analyze data from the measurement of beak depths of the medium ground finch (*Geospiza fortis*) on the island of Daphne Major near the center of the Galápagos Islands. The measurements were taken in 1976 before a major drought hit the island, and in 1978 after the drought (survivors and survivors' offspring).

▶ Note: In table (right) No. = number of.

Beak depth (mm)	No. 1976 birds	No. 1978 survivors	Beak depth of offspring (mm)	Number of birds
7.30-7.79	1	0	7.30-7.79	2
7.80-8.29	12	1	7.80-8.29	2
8.30-8.79	30	3	8.30-8.79	5
8.80-9.29	47	3	8.80-9.29	21
9.30-9.79	45	6	9.30-9.79	34
9.80-10.29	40	9	9.80-10.29	37
10.30-10.79	25	10	10.30-10.79	19
10.80-11.29	3	1	10-80-11.29	15
11.30+	0	0	11.30+	2

4. Use the data above to draw two separate sets of histograms:

(a) On the left hand grid, draw side-by-side histograms for the number of 1976 birds per beak depth and the number of 1978 survivors per beak depth.

(b) On the right hand grid, draw a histogram of the beak depths of the offspring of the 1978 survivors.

NEED HELP?
See Activity
21

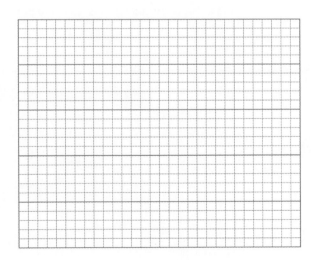

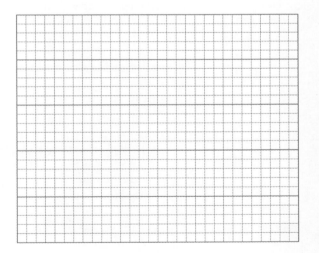

5. (a) On both of your graphs, mark the approximate mean beak depth.

(b) How much has the average moved from 1976 to 1978? _____

(c) Is beak depth heritable? What does this mean for the process of natural selection in the finches? _____

_____

_____

_____

239 Natural Selection in Rock Pocket Mice

Key Question: How does natural selection act upon the coat color of rock pocket mice?

Rock pocket mice are found in the deserts of southwestern United States and northern Mexico. They are nocturnal, foraging at night for seeds, while avoiding owls (their main predator). During the day, they shelter from the desert heat in their burrows. The coat color of the mice varies from light brown to very dark brown. Throughout the desert environment in which the mice live, there are outcrops of dark volcanic rock. The presence of these outcrops and the mice that live on them provide an excellent study in **natural selection**. The need to blend into their surroundings to avoid predation is an important selection pressure acting on the coat color of rock pocket mice.

▶ The coat color of the Arizona rock pocket mice is controlled by the Mc1r gene (a gene that in mammals is commonly associated with the production of the pigment melanin). Homozygous dominant (AA) and heterozygous mice (Aa) have dark coats, while homozygous recessive mice (aa) have light coats.
Coat color of mice in New Mexico is not related to the Mc1r gene.

▶ 107 rock pocket mice from 14 sites were collected and their coat color and the rock color they were found on were recorded by measuring the percentage of light reflected from their coat (low percentage reflectance equals a dark coat). The data are presented on the right:

Site	Rock type (V volcanic)	Percent reflectance (%) Mice coat	Percent reflectance (%) Rock
KNZ	V	4	10.5
ARM	V	4	9
CAR	V	4	10
MEX	V	5	10.5
TUM	V	5	27
PIN	V	5.5	11
AFT		6	30
AVR		6.5	26
WHT		8	42
BLK	V	8.5	15
FRA		9	39
TIN		9	39
TUL		9.5	25
POR		12	34.5

1. (a) What is the genotype(s) of the dark colored mice? _____

   (b) What is the genotype of the light colored mice? _____

2. Using the data in the table above and the grids below and on the facing page, draw column graphs of the percent reflectance of the mice coats and the rocks at each of the 14 collection sites.

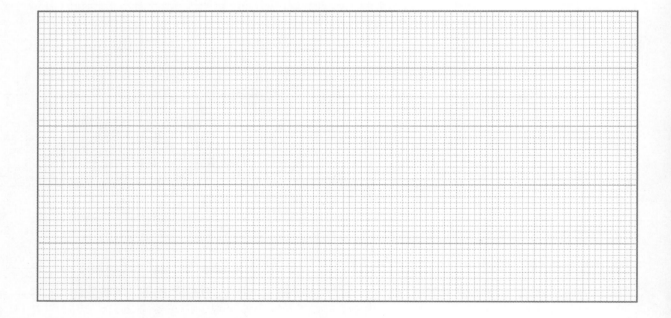

    LS4.B LS4.C P CE

©2022 **BIOZONE** International
ISBN: 978-1-98-856692-4
Photocopying Prohibited

3. (a) What do you notice about the reflectance of the rock pocket mice coat color and the reflectance of the rocks they were found on?

_____

_____

(b) Suggest a cause for the pattern in 3(a). How do the phenotypes of the mice affect where the mice live?

_____

_____

_____

_____

(c) What are two exceptions to the pattern you have noticed in 3(a)? _____

_____

_____

_____

_____

(d) How might these exceptions have occurred? _____

_____

_____

_____

4. The rock pocket mice populations in Arizona use a different genetic mechanism to control coat color than the New Mexico populations. What does this tell you about the evolution of the genetic mechanism for coat color?

_____

_____

_____

240 Natural Selection in Deer Mice

Key Question: How does the change in environmental selection pressure affect the coat color in a deer mouse population?

Computer simulations can be used to model natural selection

▸ **Natural selection** can be simulated using spreadsheets or computer programs. In this investigation, you will use Connectedbio Multi-Level Simulation (MLS), which will run on Mac or Windows platforms. It models change in phenotype (fur color) over time in a population.

▸ In this activity, you will model natural selection by adding a predator into a deer mouse population.

Access the MLS programme here:
https://short.concord.org/lm3

(you can also access the link via the **Resource Hub**)

▸ **Graph** displays data from investigation.
▸ **Predator** lets you add predators to the system.
▸ **Run** starts the investigation.

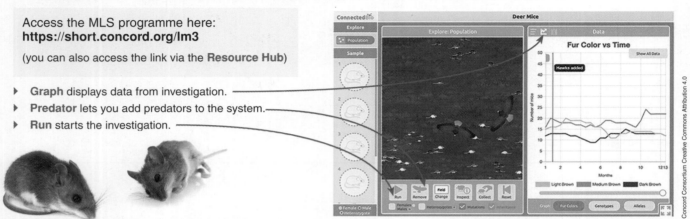

Investigation 13.2 Investigating natural selection in deer mice

This computer simulation can be used to model changes in gene pools of the genotypes in a population. Selection pressures can be easily manipulated and effects on the deer mice population can be analyzed from data displayed on graphs.

Access the simulation from the BIOZONE Resource Hub.

Setting up and running the simulation

1. Select beach habitat. Add predator.

2. Press run. Run simulation for 30 months.

3. Open data in right hand screen. Select "show all data". Observe data for fur colors, genotypes, and alleles.

1. (a) How did the proportion of deer mice with dark colored fur in the population change over the 30 months, compared to the light fur deer mice?

_____

_____

(b) Was this trend seen in fur color, genotype, and alleles of the population? If so, why might that be?

_____

_____

(c) What might happen to the dark brown fur allele if the simulation was allowed to run for a much longer time?

_____

_____

2. Reset and select field habitat. Repeat simulation. What differences did you observe in this second simulation?

_____

_____

_____

   LS4.B LS4.C P CE

©2022 **BIOZONE** International
ISBN: 978-1-98-856692-4
Photocopying Prohibited

241 Insecticide Resistance

Key Question: How does the application of insecticide act as a strong selection pressure on insects?

Selection for insecticide resistance

Insecticides are pesticides used to control insects considered harmful to humans, their livelihood, or environment. Insecticide use has increased since the use of man-made insecticides began in the 1940s.

▶ The widespread but often ineffective use of insecticides can lead to chemical resistance in insects. This means the same level of an insecticide no longer controls the pest. Effective control then requires a higher dosage rate or a different chemical.

▶ Random mutations (DNA changes) may also produce **traits** in individuals that are favored in the environment in which insecticides are being used.

▶ Ineffective application may include applying chemicals at the wrong dosage or at the wrong time, e.g. before rain, and applying sprays that insects may avoid by hiding under leaves.

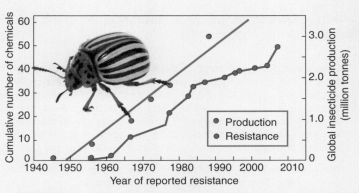

The Colorado potato beetle is a major potato pest that was originally found living on buffalo-bur in the Rocky mountains. Since man-made insecticides began to be produced, it has become resistant to more than 50 different types.

How does resistance become more common?

Insecticide use can act as a selective agent for chemical resistance in pest insects. Insects with a low natural resistance die from an insecticide application, but a few (those with a naturally higher resistance) will survive, particularly if the insecticide is not applied properly. These individuals will reproduce, giving rise to a new generation that will, on average, have a higher resistance to the insecticide.

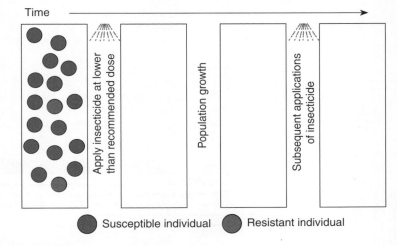

Mechanisms of resistance in insect pests

Insecticide resistance in insects can arise through a combination of mechanisms:

▶ Increased sensitivity to an insecticide will cause the pest to avoid a treated area.

▶ Certain genes confer stronger physical barriers, decreasing the rate at which the chemical penetrates the insect's cuticle.

▶ Detoxification by enzymes within the insect's body can render the insecticide harmless.

▶ Structural changes to the target enzymes make the insecticide ineffective.

1. Using the blue and red circles for susceptible and resistant individuals, draw a sequence in the boxes to the right to show what you think will happen to the proportion of susceptible and resistant individuals over time.

2. Why must farmers be sure that insecticides are applied correctly? _____

_____

_____

3. Describe two mechanisms that increase insecticide resistance in insects: _____

_____

_____

_____

©2022 **BIOZONE** International
ISBN: 978-1-98-856692-4
Photocopying Prohibited

CE | P | LS4.C | LS4.B

242 Gene Pool Simulation

Key Question: How can we use a computer model to simulate changes in the gene pool due to natural selection?

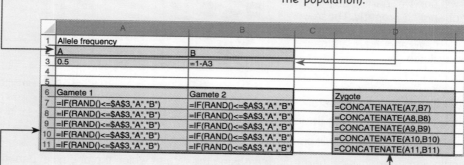

Light coat (aa)

Dark coat (AA) or (Aa)

Investigation 13.3 Investigating gene pool changes

Changes in gene pools are often modeled using physical representations of the genotypes in a population. This type of modeling is tedious and subject to human error. Modeling genotypic changes using a spreadsheet is quicker and allows the model to be changed to simulate different scenarios occurring in the gene pool. Some natural selection labs work by manually placing tokens representing alleles into a bag and withdrawing them randomly to make genotypes. A certain genotype is then selected against by not returning it to the bag and the next generation is drawn from the remaining alleles. This exercise is long and difficult to manipulate, and it reduces the population numbers over time so that an accurate simulation is not entirely possible.

These problems can be solved by using a spreadsheet to compute allele changes over time. Once the formulae are in place, the spreadsheet can be manipulated in different ways to produce a more accurate, yet still simple, simulation. Download the spreadsheet from the BIOZONE Resource Hub or use the notes and screenshots below to recreate the spreadsheet yourself.

PART 1: Setting Up the spreadsheet to investigate natural selection in rock pocket mouse coat colors

1. Open a new spreadsheet. First, switch off automatic calculation. This makes calculation of future allele frequencies simpler and under manual control, so that you can calculate them when you're ready. Each spreadsheet program will have slightly different ways of doing this. For Microsoft Excel, click on the **Formulas** tab then on the **Calculation Options** menu and click **Manual**. Calculations can then be made using the **Calculate Now** button beside the Calculation Options menu or by using the **F9** button.

2. The headings A and B represent the alleles A and a. This is necessary because the COUNTIF formula used later in the spreadsheet is not case sensitive (it does not recognise the difference between A and a).

3. 0.5 is the frequency of the A allele in the initial population (generation 0). The frequency of the B allele in the population is equal to 1–A. In our initial population, 50% of the alleles will be A and 50% will be B (A and a in the population).

	A	B	C	D
1	Allele frequency			
2	A	B		
3	0.5	=1-A3		
4				
5				
6	Gamete 1	Gamete 2		Zygote
7	=IF(RAND()<=$A$3,"A","B")	=IF(RAND()<=$A$3,"A","B")		=CONCATENATE(A7,B7)
8	=IF(RAND()<=$A$3,"A","B")	=IF(RAND()<=$A$3,"A","B")		=CONCATENATE(A8,B8)
9	=IF(RAND()<=$A$3,"A","B")	=IF(RAND()<=$A$3,"A","B")		=CONCATENATE(A9,B9)
10	=IF(RAND()<=$A$3,"A","B")	=IF(RAND()<=$A$3,"A","B")		=CONCATENATE(A10,B10)
11	=IF(RAND()<=$A$3,"A","B")	=IF(RAND()<=$A$3,"A","B")		=CONCATENATE(A11,B11)

4. The RAND formula produces a random number between 0 and 1 and compares it to the number in cell A3. If the random number is less than or equal to the number in cell A3, then an A is displayed in the cell. If the random number is greater, a B is displayed. The $ symbol tells the spreadsheet that cell A3 is a reference cell and must not change.

5. The CONCATENATE formula takes gametes A and B and puts them together to make the zygote.

6. Highlight cells and copy down all formulas to row 56 to produce 100 random gametes containing alleles A or B and 100 zygotes.

7. You now need to count up the number of AA, AB, and BB genotypes

	F	G	H	I	J
1	AA	AB	BB		Total Genotypes
2	=COUNTIF($D$7:$D$56,"AA")	=COUNTIF($D$7:$D$56,"AB")+COUNTIF($D$7:$D$56,"BA")	=COUNTIF($D$7:$D$56,"BB")		=SUM(F2:H2)
3					
4					

8. The COUNTIF formula counts up the number of AA, AB, and BB genotypes.

9. The SUM formula adds up the number of genotypes. It should add up to 50.

LS4.B   LS4.C   P   CE

©2022 **BIOZONE** International
ISBN: 978-1-98-856692-4
Photocopying Prohibited

10. Now you must calculate the number of A and B alleles present in this generation (Generation 1).In cell **F4,** type the heading **A** and in cell **G4,** type the heading **B**. In cell **I4,** type the heading **Total Alleles**.

11. Cell F5 adds up all the alleles from the AA genotype and the A alleles from the AB genotype. Cell **G5** adds up all the B alleles. Cell I5 adds up all the alleles. Click **Calculate Now** and you should see the number 100 appear.

12. Cells F9 and G9 calculate the frequency of As and Bs in Generation 1. Cell I9 adds up cell F9 and G9. This should add to 1.

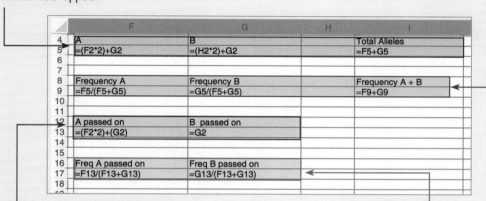

	F	G	H	I
4	A	B		Total Alleles
5	=(F2*2)+G2	=(H2*2)+G2		=F5+G5
6				
7				
8	Frequency A	Frequency B		Frequency A + B
9	=F5/(F5+G5)	=G5/(F5+G5)		=F9+G9
10				
11				
12	A passed on	B passed on		
13	=(F2*2)+(G2)	=G2		
14				
15				
16	Freq A passed on	Freq B passed on		
17	=F13/(F13+G13)	=G13/(F13+G13)		
18				

13. The **selection pressure** is against the recessive phenotype. The alleles in the recessive genotype (BB) will not be passed on, so the frequency of alleles in the population is different from the frequency of alleles that will be passed on. These cells calculate the **number** of alleles that will be passed on (excluding those in genotype BB).

14. These cells calculate the frequency of alleles that will be available to the next generation.

15. Finally, you must keep a record of each generation's allele frequencies before mating, i.e. before BB is excluded.

	F	G	H
20		A	a
21	Gen 0	0.5	=1-G21
22	Gen 1		=1-G22

16. Gen 0 was your starting population. Copy down to cell **F31** to get ten generations.

17. The frequency of A in Gen 0 was 0.5

18. The frequency of a is simply 1-A. Copy down to **H31**.

19. Click **Calculate Now**. Note the numbers that appear in cells **F9** and **G9**. Type the number in **F9** into **G22**. This is the frequency of the A allele in the first generation.

20. Now type the number in cell **F17** into cell **A3** and click **Calculate Now** to produce the second generation of alleles in cells F9 and G9. Again enter the number in **F9** into **G23** and the number in **F17** into **A3** before clicking **Calculate Now.**

21. Each time you do this, the spreadsheet calculates a new generation of genotypes and their alleles based in the number you enter into A3.

22. Save your spreadsheet.

PART 2: Natural Selection lab

Now that you have built the spreadsheet and are familiar with it, you can begin the natural selection lab.

1. To do this, you will select against the recessive light color coat phenotype (and hence the aa genotype, represented as BB in the spreadsheet). In this scenario, any BB individuals never get to breed (it is irrelevant what the phenotype is, simply that no BB individuals will enter their alleles into the next generation).

2. To start the lab, make sure **0.5** is entered into cell **A3**. Enter **0.5** into cell **G21** and make sure the cells below them are clear. Highlight cells **F17** and **G17** and under the **Format** menu click **Cells**, then click the **Number** category and set it to **2** decimal places. Click **OK**.

3. It is also worth tracking the numbers of AA, Aa and aa individuals before breeding. You can do this by simply recording the numbers on a new part of the spreadsheet, the same way as recording the A and a allele frequencies. Theoretically, Generation 0 will start as 12.5 AA, 25 Aa, and 12.5 aa, but because only whole numbers of individuals are allowed these will need to be rounded to the nearest whole number that still

produces a total of 50 (12, 26, 12). In cell **F35**, type the heading **Gen0**. **Highlight** the cell and copy it down to cell **F45**. In cell **G34**, type the heading **AA**, Cell **H34** type Aa and in **I34** type **aa**. Into cell **G35** type **12**, in **H35** type **26**, and in **I35** type **12**.

	F	G	H	I	
35	Gen 0	12	26	12	
36	Gen 1				
37	Gen 2				

4. Click **Calculate Now**.

5. Enter the results in cell **F9** into Generation 1 A (cell **G22**). Enter the numbers in **F2, G3**, and **H2** into **G36, H36**, and **I36**. Enter the number from **F17** into **A3** and click **Calculate Now** again.

6. Repeat this until you have ten generations of alleles.

PART 3: Graphing the data

1. You can now produce a graph of the results. Highlight the cells **F20** to **H31** and click **Insert** then click on a **line graph** with markers.

2. The graph should automatically produce two lines for A and a. Give the graph appropriate titles and axes labels by clicking **Add Chart Element** (depending on your spreadsheet program) and selecting **title** and **axes** labels.

3. Repeat this for the AA, Aa, and aa individuals.

4. Print the graphs and staple them to this page.

1. (a) What happens to the frequency of the a alleles in rock pocket mice over ten generations when the aa genotype is totally excluded from passing its alleles to the next generation?

_____

_____

_____

(b) What happens to the frequency of the A alleles in rock pocket mice over ten generations when the aa genotype is totally excluded from passing its alleles to the next generation?

_____

_____

_____

(c) Why do your observations from (a) and (b) happen? _____

_____

_____

(d) What is the effect on the rock pocket mouse coat color phenotypes over time? (Assume AA and Aa produce the same dominant dark coat phenotype and aa is the recessive light coat phenotype).

_____

_____

_____

(e) Predict what might happen if some of the aa genotype were able to breed: _____

_____

_____

_____

_____

©2022 **BIOZONE** International
ISBN: 978-1-98-856692-4
Photocopying Prohibited

243 What is a Species?

Key Question: How can we define a species?

The species is the basic unit of taxonomy. A biological species is defined as a group of organisms capable of interbreeding to produce fertile offspring.

▸ There can be difficulties in applying the biological species concept (BSC) in practice, as some closely related species are able to interbreed to produce fertile hybrids, e.g. species of *Canis*, which includes wolves, coyotes, domestic dogs, and dingoes.

▸ The BSC is more successfully applied to animals than to plants and organisms that reproduce asexually. Plants hybridize easily and can reproduce vegetatively. For some, e.g. cotton and rice, first generation hybrids are fertile but second generation hybrids are not.

▸ In addition, the BSC cannot be applied to extinct organisms. Some organisms, such as bacteria can also transfer genetic material to unrelated species. Increasingly, biologists are using DNA analysis to clarify relationships between closely related populations.

These five breeds of domestic dog may look and act differently, but they can all breed with each other and with the wolf (bottom right), a different species, to produce viable offspring.

Another way of defining a species is by using the phylogenetic species concept (PSC). Phylogenetic species are defined on the basis of their evolutionary history. This is determined on the basis of shared, derived characteristics. These are characteristics that evolved in an ancestor and are present in all its descendants.

▸ The PSC defines a species as the smallest group that all share a derived character state. It is useful in paleontology because biologists can compare both living and extinct organisms.

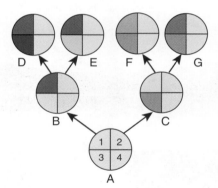

Species B and C are related to species A as they share three of four characteristics with it. However, they only share two characteristics with each other.

D and E share characteristics with B, while F and G share characteristics with C.

1. Define a biological species: _____

_____

_____

2. Why is it difficult to fully define a species in practice? _____

_____

_____

_____

3. How is the biological species different to the phylogenetic species? _____

_____

_____

4. There often appear to be greater differences between different breeds of dog than there are between different species of the *Canis* genus. Why are dogs all considered one species?

_____

_____

_____

P   LS4.C   LS4.B

244 How Species Form

Key Question: How do isolating mechanisms lead to the formation of new species?

Species formation

▸ Species evolve in response to selection pressures from the environment. These may be naturally occurring or caused by humans. The diagram below represents a possible sequence for the **evolution** of two hypothetical species of butterfly from an ancestral population. As time progresses (from top to bottom of the diagram below), the amount of genetic difference between the populations increases, with each group becoming increasingly isolated from the other. Gene flow is reduced when populations are separated. Continual reduction in gene flow by isolating mechanisms may eventually lead to the formation of new species.

▸ The isolation of two gene pools from one another may begin with geographical barriers. This may be followed by isolating mechanisms that occur before the production of a zygote, e.g. behavioral changes, and isolating mechanisms that occur after a zygote is formed, e.g. hybrid sterility. As the two **gene pools** become increasingly isolated and different from each other, they are progressively labelled: population, race, and subspecies. Finally, they attain the status of separate species.

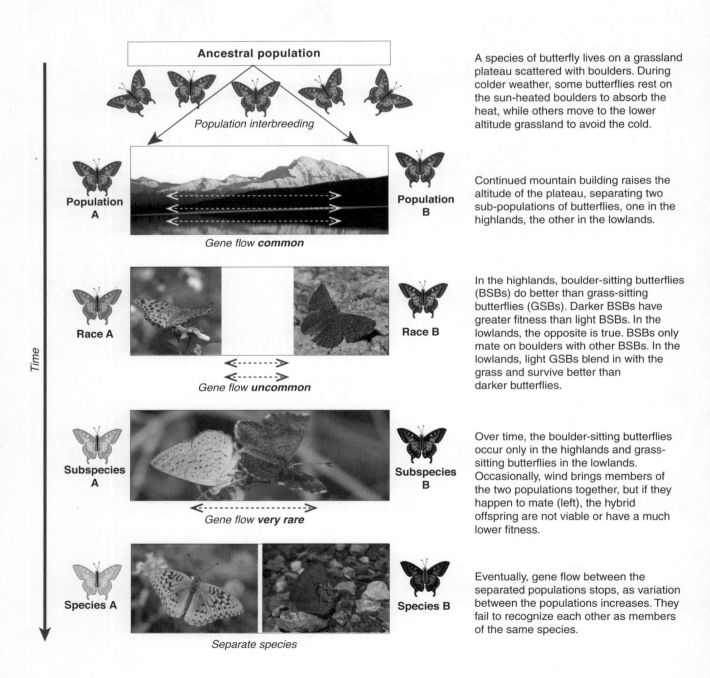

Ancestral population

Population interbreeding

Population A — **Population B**
Gene flow **common**

Race A — **Race B**
Gene flow **uncommon**

Subspecies A — **Subspecies B**
Gene flow **very rare**

Species A — **Species B**
Separate species

Time

A species of butterfly lives on a grassland plateau scattered with boulders. During colder weather, some butterflies rest on the sun-heated boulders to absorb the heat, while others move to the lower altitude grassland to avoid the cold.

Continued mountain building raises the altitude of the plateau, separating two sub-populations of butterflies, one in the highlands, the other in the lowlands.

In the highlands, boulder-sitting butterflies (BSBs) do better than grass-sitting butterflies (GSBs). Darker BSBs have greater fitness than light BSBs. In the lowlands, the opposite is true. BSBs only mate on boulders with other BSBs. In the lowlands, light GSBs blend in with the grass and survive better than darker butterflies.

Over time, the boulder-sitting butterflies occur only in the highlands and grass-sitting butterflies in the lowlands. Occasionally, wind brings members of the two populations together, but if they happen to mate (left), the hybrid offspring are not viable or have a much lower fitness.

Eventually, gene flow between the separated populations stops, as variation between the populations increases. They fail to recognize each other as members of the same species.

LS4.B LS4.C   P

©2022 **BIOZONE** International
ISBN: 978-1-98-856692-4
Photocopying Prohibited

Geographic isolation

▸ Geographic isolation describes the isolation of a species population (gene pool) by some kind of physical barrier, e.g. mountain range, water body, desert, or ice sheet. Geographic isolation is a frequent first step in the subsequent reproductive isolation of a species.

▸ An example of geographic isolation leading to speciation is the large variety of cichlid fish in the rift lakes of East Africa (right). Geologic changes to the lake basins have been important in the increase of cichlid fish species.

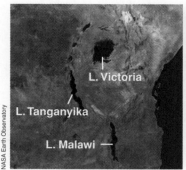

Reproductive isolating mechanisms

▸ Reproductive isolating mechanisms (RIMs) are reproductive barriers that are part of a species' biology and therefore do not include geographic isolation. They prevent interbreeding and therefore gene flow between species. Single barriers may not completely stop gene flow, so most species commonly have more than one type of barrier. Some RIMs (prezygotic - before fertilization) include differences in breeding season, differences in mating behaviors, and differences in copulatory structures. Other RIMS (postzygotic - after fertiization) involve a mismatch of chromosomes in the zygote, and include hybrid sterility, e.g. mule (right), hybrid inviability, and hybrid breakdown.

The white-tailed antelope squirrel (left) and the Harris's antelope squirrel (right) in the southwestern United States and northern Mexico, are separated by the Grand Canyon (center) and have evolved to occupy different habitats.

Mules are a cross between a male donkey and a female horse. The donkey contributes 31 chromosomes while the horse contributes 32, making 63 chromosomes in the mule. This produces sterility in the mule as meiosis cannot produce gametes with an even number of chromosomes.

1. Identify some geographical barriers that could separate populations: _____

_____

_____

2. Why is a geographical barrier not considered a reproductive isolating mechanism? _____

_____

_____

3. Identify the two types of reproductive isolating mechanisms, and explain the difference between them:

_____

_____

_____

_____

_____

4. Why is more than one reproductive isolation barrier needed to completely isolate a species? _____

_____

_____

_____

245 Patterns of Evolution

Key Question: How do populations diverge from their common ancestor and form new species?

▸ The diversification of an ancestral group into two or more species in different habitats is called divergent evolution. This is shown on the right, where two species diverge from a common ancestor. Note that another species (W) arose but became extinct.

▸ When divergent evolution involves the formation of a large number of species to occupy different niches, it is called adaptive radiation.

▸ The evolution of species may not necessarily involve branching. A species may accumulate genetic changes that, over time, result in a new species. This is known as sequential evolution.

▸ **Evolution** of species does not always happen at the same pace. Two models describe the pace of evolution.

Phyletic gradualism proposes that populations diverge slowly by accumulating adaptive features in response to different selective pressures.

Punctuated equilibrium proposes that most of a species' existence is spent in stasis and evolutionary change is rapid. The stimulus for evolution is a change in some important aspect of the environment, e.g. mean temperature.

It is likely that both mechanisms operate at different times for different taxonomic groups.

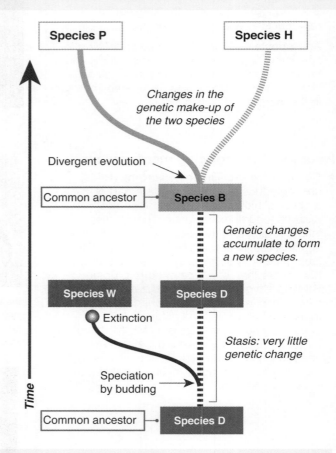

Species P   Species H

Changes in the genetic make-up of the two species

Divergent evolution

Common ancestor → Species B

Genetic changes accumulate to form a new species.

Species W   Species D

Extinction

Stasis: very little genetic change

Speciation by budding

Common ancestor → Species D

Time

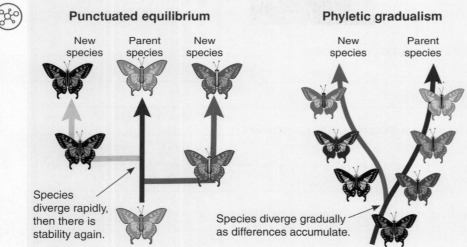

Punctuated equilibrium

New species   Parent species   New species

Species diverge rapidly, then there is stability again.

Phyletic gradualism

New species   Parent species

Species diverge gradually as differences accumulate.

The ginkgo tree (*Gingko biloba*) is an example of a living fossil. It is almost identical to a fossil species living 270 million years ago. The fossil record suggests it evolved by very slow phyletic gradualism.

Joe Schneid, Louisville, Kentucky

1. In the hypothetical example of divergent evolution illustrated at the top of the page:

   (a) Identify the type of evolution that produced species B from species D: _____

   (b) Identify the type of evolution that produced species P and H from species B: _____

   (c) Name all species that evolved from: Common ancestor D: _____   Common ancestor B: _____

2. When do you think punctuated evolution is most likely to occur, and why? _____

   _____

   _____

©2022 **BIOZONE** International
ISBN: 978-1-98-856692-4
Photocopying Prohibited

246 Evolution and Biodiversity

Key Question: How did adaptive radiation of mammals increase the biodiversity of the group?

▸ The mammals underwent a spectacular adaptive radiation following the sudden extinction of the non-avian dinosaurs 66 million years ago, which left many niches vacant. Most modern groups of mammals appeared very quickly. Placental mammals have become dominant over other groups of mammals throughout the world, except in Australia where marsupials remain dominant. In Australia, marsupials became isolated after the breakup of the supercontinent, Gondwana. Placental mammals (bats) reached Australia 15 million years ago, followed by rodents, 10 million years ago.

▸ The diagram below shows a simple evolutionary tree for the mammals. The width of the bars indicates the number of species in each of the three mammalian groups.

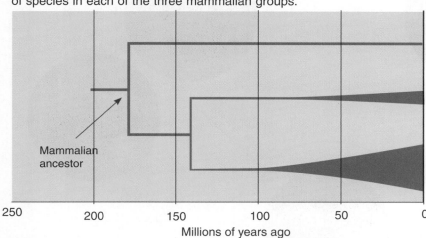

Monotremes
(platypus, echidna)
5 species

Marsupials
(e.g. kangaroo)
324 species

Eutherians
(Placental mammals)
5010 species

Mammalian ancestor

250   200   150   100   50   0
Millions of years ago

Rodents

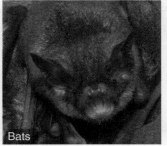
Bats

Primates

Marsupials

The placental mammals make up the largest group of mammals. Embryos are carried inside the mother's body and obtain nutrients via the placenta. The young are born live and well developed. The group consists of 18 orders, although some orders contain many more species than others. Rodents are the largest order, making up nearly 40% of mammalian species, followed by bats. Primates also make up a significant proportion of the mammals.

The five species of monotremes lay eggs, whereas marsupials give birth to very undeveloped young that develop in a pouch.

1. (a) When did the mammals first evolve? _____

    (b) When did placental mammals split from the marsupials? _____

    (c) Suggest why the mammals underwent such a major adaptive radiation after the dinosaurs died out: _____

    _____

    _____

    (d) Suggest why marsupials remained the dominant mammalian group in Australia after the split between the marsupial and placental mammals:

    _____

    _____

    _____

    _____

©2022 **BIOZONE** International
ISBN: 978-1-98-856692-4
Photocopying Prohibited

P   LS4.C   LS4.B

Rodent biodiversity

Rodents make up 40% of mammalian species, making them easily the most successful of the mammalian groups. They have adapted to a huge number of habitats, from deserts to forest. All rodents have upper and lower incisor teeth that grow continuously. Fossils with distinctive rodent features first appeared about 66 million years ago.

Squirrel-like rodents

Squirrels are found on many continents. Their lifestyles include tree dwelling, ground dwelling, and gliding forms. Like most rodents, they are social, with prairie dogs forming large communities called towns.

Porcupine-like rodents

South American capybaras are the largest of all rodents. They occupy habitats from forests to savannahs. Porcupines are found throughout the Old and New Worlds. Their spines make an almost impenetrable defense against predators. The group also includes guinea pigs, which are popular as pets.

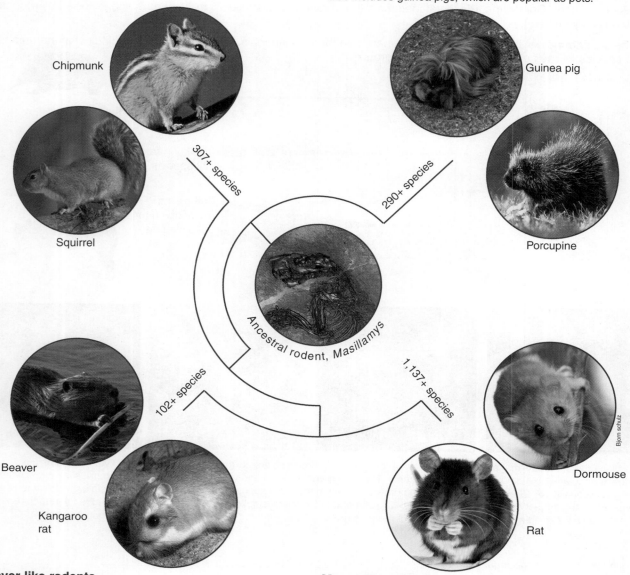

Chipmunk

Guinea pig

307+ species

290+ species

Squirrel

Porcupine

Ancestral rodent, *Masillamys*

102+ species

1,137+ species

Beaver

Dormouse

Kangaroo rat

Rat

Beaver-like rodents

Beavers are one of the larger types of rodents. They live near rivers, streams and lakes, chewing through small trees to build dams across streams and to make lodges to live in. Gophers live in burrows, while kangaroo rats are so well adapted to the desert, they virtually never need to drink.

Mouse-like rodents

Rats and mice are found in virtually every part of the world, thanks to their generalist adaptations and human assisted travel. There are at least 100 species of rats and mice alone. The group also includes voles, lemmings, jerboas, and dormice.

2. What anatomical feature do all rodents have?_____

3. (a) Describe some of the habitats rodents have occupied: _____

_____

(b) Describe some adaptations of rodents (you may use extra paper and attach it to this page): _____

_____

_____

©2022 **BIOZONE** International
ISBN: 978-1-98-856692-4
Photocopying Prohibited

247 Extinction is a Natural Process

Key Question: How does the natural process of extinction affect biodiversity?

Extinction is the death of an entire species; no individuals are left alive. Extinction is an important and natural process in **evolution** and describes the loss of a species forever. Extinction provides opportunities, in the form of vacant niches, for the evolution of new species. More than 98% of species that have ever lived are now extinct, most of these before humans were present.

▸ Extinction is the result of a species being unable to adapt to an environmental change. Either it evolves into a new species, and the ancestral species becomes effectively extinct, or it and its lineage becomes extinct.

▸ A mass extinction describes the widespread and rapid (in geologic terms) decrease in life on Earth and involves not only the loss of species, but the loss of entire families, which are made up of many genera and species. Such events are linked to major climate shifts or catastrophic events. There have been five previous mass extinctions.

▸ The diagram below shows how the diversity of life has varied over the history of life on Earth and aligns this with major geologic and climatic events.

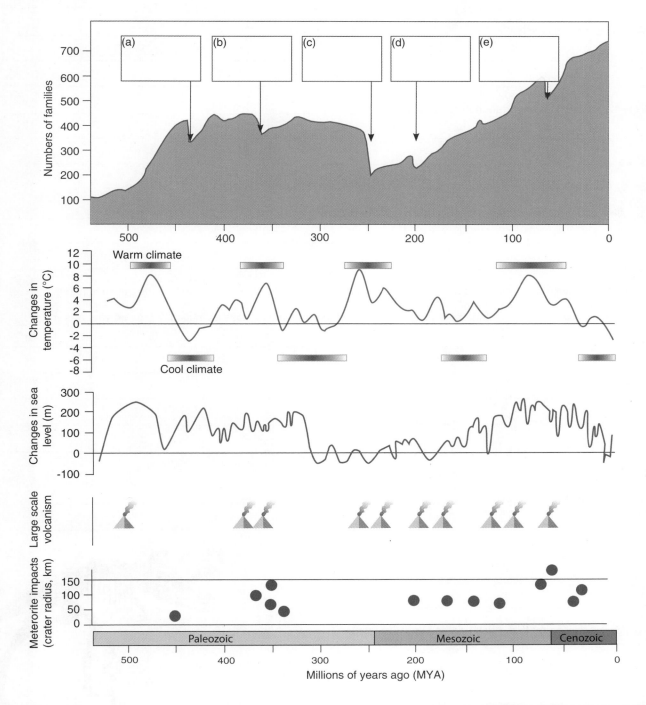

CE   LS4.C

Graptolite 0.5 cm

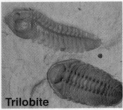
Trilobite

Coral

Conodont (reconstruction)

Dinosaur

Ordovician extinction (458-440 MYA). Second largest extinction of marine life: >60% of marine invertebrates died. One of the coldest periods in Earth's history.

Devonian extinction (375-360 MYA). Marine life affected, especially brachiopods, trilobites, and reef building organisms.

Permian extinction (252 MYA). Nearly all life on Earth perished. 57% of families and 83% of genera were wiped out. 96% of marine species became extinct.

Triassic extinction (201.3 MYA). At least half of the species present became extinct, vacating niches and ushering in the age of the dinosaurs.

Cretaceous extinction (66 MYA). Marked by the extinction of nearly all dinosaur species (their descendants, the birds, survive).

1. What is the general cause of extinction? _____
   _____
   _____

2. What is a mass extinction? _____
   _____
   _____

3. Why is extinction an important natural process? _____
   _____
   _____

4. Study the data on the opposite page carefully and answer the following questions:

   (a) What is the evidence that extinction is a natural process? _____
   _____
   _____

   (b) Label the five mass extinctions and their approximate starting dates on the top graph on the previous page:

   (c) Is there reason to believe that there is any one cause for any of these mass extinctions? Explain your answer:
   _____
   _____
   _____
   _____
   _____

   (d) What happened to the diversity of life soon after each mass extinction?
   _____
   _____
   _____

5. Which of the five extinctions appears to be the most severe?
   _____

6. Which extinctions mark the rise and end of the dinosaurs?
   _____
   _____

248 Humans and Extinction

Key Question: How has human activity been directly responsible for a number of extinctions?

How have humans affected extinction rates?

▶ Human activity is the cause of many recent **extinctions**. One very famous extinction was the extermination of the dodo, within 70 years of its first sighting on the island of Mauritius, in 1598. Another, was the extinction of the passenger pigeon, which went from an estimated 3 billion individuals before the arrival of Europeans in North America, to extinction by 1914. More recent extinctions include the Yangtze River dolphin in 2006, and the Pinta Island tortoise, a giant tortoise, in 2012 (with the death of the last individual, named Lonesome George).

▶ Human activities, such as hunting and destruction of habitat by pollution, and land clearance for agriculture and urbanization, have caused many recent extinctions. The effects of climate change on habitats and life histories, e.g. breeding times, may drive many other vulnerable species to extinction.

The golden toad (left) from Costa Rica was officially declared extinct in 1989.

▶ Species become extinct naturally at an estimated rate of about one species per one million species per year. This is called the background extinction rate. Since the 1500s, at least 412 vertebrate or plant species have become extinct (below). Proportionally, birds have been affected more than any other vertebrate group, although a large number of frog species are in rapid decline.

Organism	Total number of species (approx)*	Known extinctions (since ~1500 AD)*
Mammals	5487	87
Birds	9975	150
Reptiles	10,000	22
Amphibians	6700	39
Plants	300,000	114

* These numbers vastly underestimate the true numbers because so many species are undescribed.

Natural History Museum CC 2.5

The dodo (skeleton left) has become synonymous with human-caused extinction ("dead as a dodo"). The dodo was endemic to the island of Mauritius in the Indian Ocean and became extinct by 1662, just 64 years after its discovery. Its extermination was so rapid that we are not even completely sure what it looked like and know very little about it.

1. (a) What is the total number of known extinctions over the last 500 years?_____

   (b) Why is this number probably an underestimate? _____

   _____

2. (a) There are approximately 10,000 living or recently extinct types of bird. Assuming a background extinction rate, how many bird species should be becoming extinct per year? Show your working:

   _____

   _____

   (b) Since 1500 AD, 150 birds species are known to have become extinct. How many times greater is this rate of extinction than the background rate for birds (assume 1500 AD to 2015)?

   _____

   _____

3. Identify three reasons for an increase in extinct rates over the last 515 years: _____

   _____

   _____

   _____

CE   LS4.C

Threatened and endangered

The International Union for Conservation of Nature (IUCN) began its Red List in 1964. It is the most comprehensive inventory of the conservation status of the Earth's numerous biological species. By continually updating the list, it is possible to track the changes in species' populations and therefore their risk of extinction.

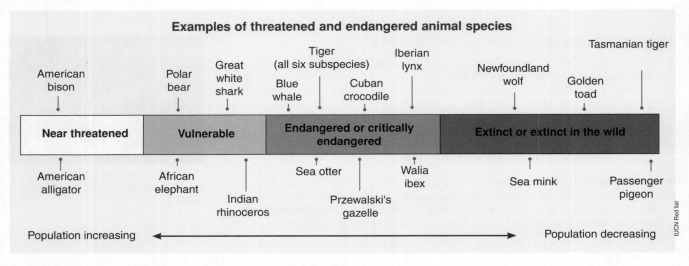

Examples of threatened and endangered animal species

4. The IUCN has established a Red List Index (RLI) for four taxonomic groups: reef forming corals, amphibians, birds, and mammals. This index focuses on the genuine status of changes. An RLI of 1.0 equates to all species qualifying as Least Concern (unlikely to become extinct in the near future). An RLI of 0 means that all species have become extinct. The figure (right) shows the trends in risk for the four taxonomic groups currently completed.

(a) Which group is moving most rapidly towards extinction risk?

_____

(b) Which group is, on average, the most threatened?

_____

(c) Why would an index like this be useful and how could it help to highlight environmental issues of concern?

_____

_____

_____

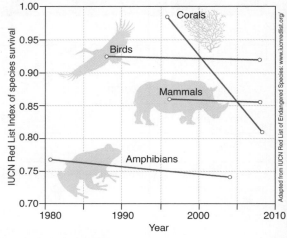

5. Humans are not always directly responsible for the extinction of a species, but many extinctions can be indirectly attributed to humans. Discuss the reasons for this:

_____

_____

_____

_____

_____

_____

6. Carry out some research to identify two species, other than those mentioned earlier, that have recently become extinct. Identify the date of their extinction, and possible reasons for their extinction:

_____

_____

_____

_____

_____

249 Review Your Understanding

Key Question: How is poaching causing the African elephants to be born without tusks?

1. What do you think is wrong with this statement, "An elephant is more likely to lose its tusks if poaching is occurring"?

_____

_____

2. Tuskless females will pass the tuskless genes to 50% of their daughters. The tuskless gene is carried on the x (female) chromosome. Why is the increase in tuskless African elephants over time an example of natural selection?

_____

_____

_____

_____

3. (a) A long term study of a Mozambique population of African elephants showed changing proportions of 2,1, and no tusk elephants. What most likely caused an increase in tuskless elephants from 1970-2000?

_____

_____

_____

_____

_____

_____

Population size vs presence of tusks in elephants

Number of tusks: 2 1 0

(*n* = 54) 18.5% ... no data available ... (*n* = 108) 50.9% ... (*n* = 91) 33%

9.3% 68.5% ... 8.3% 40.7% ... 7.7% 58.4%

Population size axis: 0, 500, 1000, 1500, 2000, 2500

Year axis: 1970, 1980, 1990, 2000, 2010

Data from Campbell-Station et al. (2021)

_____

_____

(b) Three factors are relevant to observations seen in the study of the Mozambique African elephant population:
   i. The presence of the tuskless gene is fatal to males, hence the lack of tuskless male elephants in the population.
   ii. Tusks are used by elephants to perform many tasks and are an important evolutionary adaptation.
   iii. Conservation efforts were made after 2000 that reduced poaching of the elephants.
   Relate the factors above to the changes in the proportion of tuskless elephants in the Mozambique population since 2000, and link them to the process of evolution by natural selection:

_____

_____

_____

_____

_____

_____

_____

_____

_____

©2022 **BIOZONE** International
ISBN: 978-1-98-856692-4
Photocopying Prohibited

250 Summing Up

In this assessment task, you will design a living organism and explain how its adaptations help it survive:

You are to think of a new kind of living organism, it can be anything, but not something already living. Your new organism could be a combination of other living things, e.g. a cog, which has some features of a cat and some of a dog. You also need to think of the environment that the organism usually lives in, e.g tropical rainforest, high humidity, etc.

1. Describe your new organism, including one specific adaptation that helps the organism survive (e.g. thick fur coat helps the organism to …). You will also need to describe the environment that your organism lives in:

Hypothetical "Cog"

_____

_____

_____

_____

_____

_____

_____

_____

_____

_____

_____

_____

_____

_____

_____

_____

2. The adaptations of living organisms vary, usually in a normal distribution, e.g. claws may vary in length through the population. Explain how the specific adaptation for your organism varies through the population, from the most extreme forms to the median form:

_____

_____

_____

_____

_____

_____

3. Describe how the variation in the adaptation affects the ability of the organism to survive in its normal environment, assuming there is no selective pressure for change:

_____

_____

_____

_____

LS4.B LS4.C CE

©2022 **BIOZONE** International
ISBN: 978-1-98-856692-4
Photocopying Prohibited

4. Imagine now, that a change in the environment has introduced a slight directional selection pressure on the adaptation of your organism.

(a) Which extreme of your organism's adaptation is negatively affected? _____

_____

(b) Explain why: _____

_____

_____

_____

(c) If the selection pressure remains for many generations, describe how your organism will change over time:

_____

_____

_____

5. Imagine now, that the selection pressure changes to act upon the median form of your organism's adaptation.

(a) Describe how the population of the organism will be affected over many generations:

_____

_____

_____

(b) Will this affect the ability of the organisms at the extreme ends of the range to breed with one another?

_____

(c) If so, why, and how will this affect the species? _____

_____

_____

_____

_____

6. Now, imagine that a mutation in your organism affects one extreme of the phenotypic range so that the organism's fitness increases, i.e. it increases the organism's chance of surviving and reproducing.

(a) What is the effect of the mutation on the adaptation, i.e. what change is there and how does this affect the organism for the better?

_____

_____

_____

(b) Imagine that the mutation only increases chances of survival under certain circumstances, e.g. a mutation from brown to white fur enhances survival in the snow.

i) What is the circumstance? _____

_____

_____

ii) How will this limited enhancement of survival affect the evolution of the species over many generations?

_____

_____

Biodiversity

Activity number

Anchoring Phenomenon

Can't see the wood for the trees: How can we retain ecological biodiversity?   251 260

How can we measure biodiversity?

☐ 1 Recall the definition of biodiversity. Mathematically compare the biodiversity of two modelled ecosystems. Calculate the biodiversity of different ecosystems using Simpson's Index of Diversity. Contrast between the terms species richness and species diversity.   252

☐ 2 Investigate the impact of human activities on biodiversity. Use a computer tool to calculate Simspon's Index of Diversity of a simulated ecosystem. Use a simulation model to test different possible solutions to sustaining biodiversity.   252

☐ 3 Understand the importance and range of ecosystem services available to humans. Link the level of biodiversity and availability of ecosystem services.   253

☐ 4 Identify the main biodiversity hotspots indicated on a map. Research and summarize the main characteristics of one hotspot in greater detail.   254

How does human activity impact biodiversity?

☐ 5 Understand the link between climate change and human activity. Make use of data to discuss how human-induced climate change and agricultural land use have impacted biodiversity.   255

What are some solutions can we use to conserve biodiversity?

☐ 6 Define conservation, specifically ex-situ conservation, in the context of biodiversity solutions. Discuss the advantages and disadvantages of ex-situ conservation. Evaluate the roles of zoos, aquaria, and seed banks as methods of increasing biodiversity.   256

☐ 7 Define in-situ conservation methods in the context of American bison diversity. Analyze American bison populations to support an argument on bison recovery successes and risks. Research American bison to discuss recovery of species and genetic diversity.   257

☐ 8 Define conservation genetics and discuss its relevance in the context of the Florida panther recovery project. Research the Florida panther to discuss recovery of species and genetic diversity.   258

☐ 9 Use information and data from the impacts of farming and conservation efforts in the Maasai Mara National Reserve to discuss how conservancies helped to reduce conflict between humans and wildlife.   259

☐ 10 Design, and present, a solution to halt endangered Indiana bat population decline and ultimately restore their numbers.   261

251 Can't see the Wood for the Trees

Key Question: How has human activity affected a biodiversity hotspot, and what are some possible solutions for restoring it, or preventing more loss?

▶ The Madrean Pine-Oak Woodlands is an area of around 500,000 km² situated in Mexico, with patches extending into Southern United States. The ecosystem has a very high biological diversity, but is threatened by human activity.

The Madrean Pine-Oak Woodlands contains over 5000 species of flowering plants, many endemic (found nowhere else) to the area, as well as 500 bird, 380 reptile, 200 amphibian, and 330 mammal species.

Biodiversity loss in the Madrean Pine-Oak Woodlands is occurring due to commercial logging, deforestation (clearing forests) to provide agricultural land, climate change, and poor forest management leading to fires.

1. Why do you think having many endemic species makes it extra important to conserve the biodiversity in this area?

_____

_____

_____

2. Each of the factors listed below affects the Madrean Pine-Oak Woodlands ecosystem. Work in small groups to discuss each of these and then write what you think some of those effects might be:

(a) Commercial logging: _____

_____

_____

_____

(b) Deforestation for agriculture: _____

_____

_____

_____

_____

(c) Climate change: _____

_____

_____

_____

_____

(d) Uncontrolled forest fires: _____

_____

_____

_____

_____

252 Biodiversity

Key Question: How is biodiversity measured in an ecosystem?

What is biodiversity?

Biodiversity is the amount of biotic variation within a given group. It could be the number of species in a particular area or the amount of genetic diversity in a species. Ecosystem biodiversity refers to the number of ecosystems in a given region and is usually correlated with species diversity. To conserve biodiversity, conservation of habitats is important.

Biodiversity tends to be clustered in certain parts of the world, called **hotspots**, where species diversity is high. Tropical forests and coral reefs (above) are some of the most diverse ecosystems on Earth.

Biodiversity of Earth	
Type of organism	**Estimated number of species**
Protozoa	36,400
Brown algae, diatoms	27,500
Invertebrates	7.6 million
Plants	298,000
Fungi	611,000
Vertebrates	60,000

The latest estimate of the number of eukaryotic species on Earth is 8.7 million, of which only 1.2 million have been formally described. Prokaryotic species are so variable and prolific, the number of species is virtually inestimable and could be hundreds of millions.

Measuring biodiversity

Biodiversity is measured for a variety of reasons, e.g. to assess the success of conservation work or to measure the impact of human activity. One measure of biodiversity is to simply count all the species present (the species richness) although this may give an imprecise impression of the ecosystem's biodiversity. Species evenness gives a measure of relative abundance of species, i.e. how close the numbers of each species in an environment are.

Two ecosystems with quite different biodiversity.

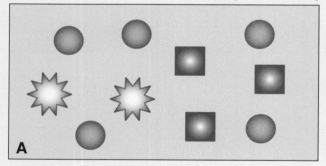

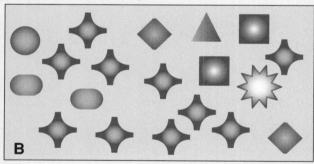

Species richness is a simple method of estimating biodiversity in which the number of species is counted. This does not show if one species is more abundant than others. Species evenness is a method in which the proportions of different species in an ecosystem are estimated.

Diversity Indices use mathematical formulae based on the species abundance and the number of each species to describe the biodiversity of an ecosystem. Many indices produce a number between 0 and 1 to describe biodiversity, 1 being high diversity and 0 being low diversity.

1. Calculate the number and percentage of eukaryotic species that are still to be formally described: _____

   _____

2. Describe in words the species richness and species evenness of ecosystem A and B above:

   A _____

   B _____

       LS4.C LS4.D CE

Comparing the biodiversity of different ecosystems

▶ The Simpson's Index of Diversity: $D = 1 - (\sum(n/N)^2)$ is a method of quantitatively measuring diversity, where n = number of individuals of each species in the sample and N = total number of individuals (of all species) in the sample. It generates a value between 0 and 1. The closer the value is to 1, the higher the biodiversity.

▶ At student observed that, in a conifer plantation, there seemed to be only a few different invertebrate species and wondered if more would be found in a nearby oak woodland. They carried out an investigation and the results are tabled below. The invertebrates are not drawn to scale.

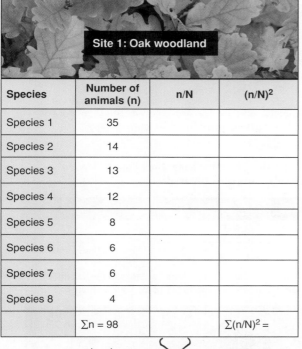

Site 1: Oak woodland

Species	Number of animals (n)	n/N	$(n/N)^2$
Species 1	35		
Species 2	14		
Species 3	13		
Species 4	12		
Species 5	8		
Species 6	6		
Species 7	6		
Species 8	4		
	$\sum n = 98$		$\sum(n/N)^2 =$

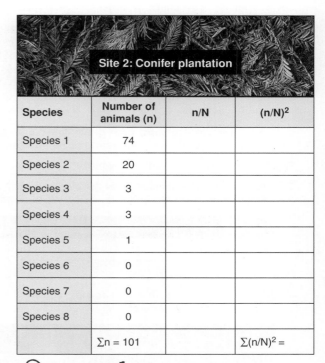

Site 2: Conifer plantation

Species	Number of animals (n)	n/N	$(n/N)^2$
Species 1	74		
Species 2	20		
Species 3	3		
Species 4	3		
Species 5	1		
Species 6	0		
Species 7	0		
Species 8	0		
	$\sum n = 101$		$\sum(n/N)^2 =$

| Species 1 Mite | Species 2 Ant | Species 3 Earwig | Species 4 Woodlice | Species 5 Centipede | Species 6 Longhorn beetle | Species 7 Small beetle | Species 8 Pseudoscorpion |

3. (a) Complete the two tables above by calculating the values for n/N and $(n/N)^2$ for the student's two sampling sites:

(b) Calculate the Simpson's Index of Diversity for site 1: _____

(c) Calculate the Simpson's Index of Diversity for site 2: _____

(d) Compare the diversity of the two sites and suggest any reasons for it: _____

_____

_____

_____

_____

4. (a) Species richness is a measure of the number of different species in an area. Which of the two areas sample above has the greatest species richness?

_____

(b) Why would measuring species richness not be as informative as measuring species diversity?_____

_____

_____

_____

Investigation 14.1 Investigating biodiversity and human impacts

Biodiversity changes in an ecosystem can be modelled using a simulation. Different species in an ecosystem can be represented by different colored tokens or beads. Individuals in an ecosystem can be removed or added to represent different effects, including effects of human activity. These can be overpopulation, overexploitation, adverse habitat alterations, pollution, invasive species, and changes in climate.

There are many ways to quantify biodiversity. For the purpose of this activity, you will use **Simpson's Index of Diversity**, once more. A simple online calculator can be found at https://www.easycalculation.com/statistics/simpson-diversity-index.php and also accessed through BIOZONE's Resource Hub.

PART 1: Setting Up the simulation

1. Teachers mix up enough colored tokens into a large bag or bowl so that every group has enough for 2 or 3 handfuls. Provide a ratio of Red:4 Blue:3 Green:2 Yellow:1 for every 10 beads or tokens.

2. Working in small groups, each group takes a handful of randomly mixed different colored beads or tokens from a large container that your teacher has already prepared. Spread your handful onto a tray.

PART 2: Measuring initial biodiversity

3. Add up the total number of each color and record in the table. Use the **Simpson's Index of Diversity** calculator to calculate initial diversity. Write down the **Simpson's Index of Diversity**

Species	Red	Blue	Green	Yellow	Simpson's Index of Diversity
Total number					

PART 3: Human impacts

4. After each of these following rounds, add number of each color/species and calculate **Simpson's Index of Diversity**. Develop and perform at least 2 more scenarios and calculate **Simpson's Index of Diversity**

Scenario	Red	Blue	Green	Yellow	Simpson's Index of Diversity
A. Invasive species: Species Red has risen to plague proportions. Add 2 additional red tokens for every red you have in your tray.					
B. Climate change: Species yellow has been impacted significantly and cannot adapt to temperature increase. Remove all yellow tokens.					
C.					
D.					

Biodiversity in an ecosystem can be improved using a wide range of solutions, such as preventing the cutting of trees, restoring the habitats and replanting, banning animal hunting or capture, making areas protected, re-introducing species.

5. (a) How does the simulation represent changes in biodiversity? _____

_____

_____

    (b) What trend are you seeing in biodiversity due to human impacts, giving an example from above? _____

_____

_____

_____

©2022 **BIOZONE** International
ISBN: 978-1-98-856692-4
Photocopying Prohibited

PART 4: Biodiversity solutions

5. In your groups, develop at least four different solutions and the action that the solution will create for the different colored tokens, i.e. remove, take away, add a new color. Start with a new handful of tokens.

Species	Red	Blue	Green	Yellow	Simpson's Index of Diversity
Total number					

6. Simulate each solution with your tokens and calculate the **Simpson's Index of Diversity** after each.

Scenario	Red	Blue	Green	Yellow	Simpson's Index of Diversity
A.					
B.					
C.					
D.					

6. (a) How would we know if a solution was effective at increasing biodiversity? _____
_____
_____

(b) Which of your solutions was the most effect at increasing biodiversity and why do you think this might be? _____
_____
_____

7. Selecting one of your solutions from above, what might be any possible constraints of cost, safety, and reliability as well as cultural, and environmental impacts if it was actioned as a real-world solution?
_____
_____
_____
_____
_____

8. How well do you consider that your biodiversity solution is represented by the simulation? and what are the limitations of the simulation?
_____
_____
_____

253 Humans Depend on Biodiversity

Key Question: What are the ecosystem services that humans depend upon?

Ecosystems provide services

▸ Humans depend on Earth's ecosystems for the services they provide. These **ecosystem services** include resources such as food and fuel, as well as processes such as purification of the air and water. These directly affect human health.

▸ The **biodiversity** of an ecosystem affects its ability to provide these services.

▸ Biologically diverse and resilient ecosystems that are managed in a sustainable way are better able to provide the ecosystem services on which we depend.

▸ The UN has identified four categories of ecosystem services: supporting, provisioning, regulating, and cultural.

▸ Regulating and provisioning services are important for human health and security (security of resources and security against natural disasters).

▸ Cultural services are particularly important to the social fabric of human societies and contribute to well being. These are often things we cannot value in monetary terms.

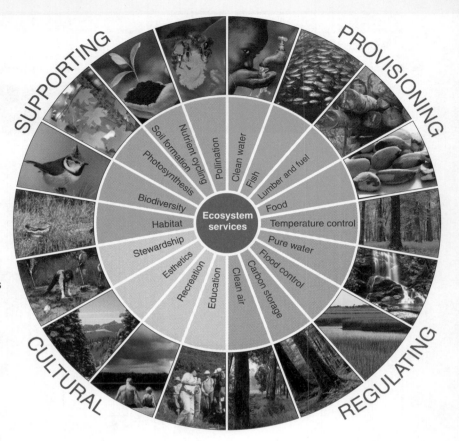

Biodiversity is important in crop development, e.g. promoting disease resistance. Many medical breakthroughs have come from understanding the biology of wild plants and animals.

High biodiversity creates buffers between humans and infectious diseases, e.g. Lyme disease, and increases the efficiency of processes such as water purification.

Biodiversity and ecosystem health are essential for reducing the effects of human activities (e.g. pollution) and the effects of environmental disasters (e.g. eruptions and landslides).

1. What are ecosystem services, and why are they important to humans? _____

_____

_____

_____

2. What is the relationship between biodiversity and the ability of an ecosystem to provide essential ecosystem services?

_____

_____

_____

  LS4.D CE

©2022 **BIOZONE** International
ISBN: 978-1-98-856692-4
Photocopying Prohibited

254 Biodiversity Hotspots

Key Question: What and where are Earth's biodiversity hotspots?

▸ Biodiversity is not distributed evenly on Earth. It tends to be clustered in certain parts of the world, called **biodiversity hotspots**. These regions are biologically diverse and ecologically distinct regions under the greatest threat of destruction from human activity. They are identified on the basis of the number of species present, the amount of endemism (species unique to a specific geographic location), and the extent to which the species are threatened.

▸ Biodiversity hotspots make up less than 2% of Earth's land surface but support nearly 60% of the world's plant and vertebrate species. Their conservation is considered central to securing global biodiversity.

▸ Habitat destruction and human-induced climate change are major threats to biodiversity hotspots. The introduction of invasive or predatory species can also place the biodiversity of these regions in danger.

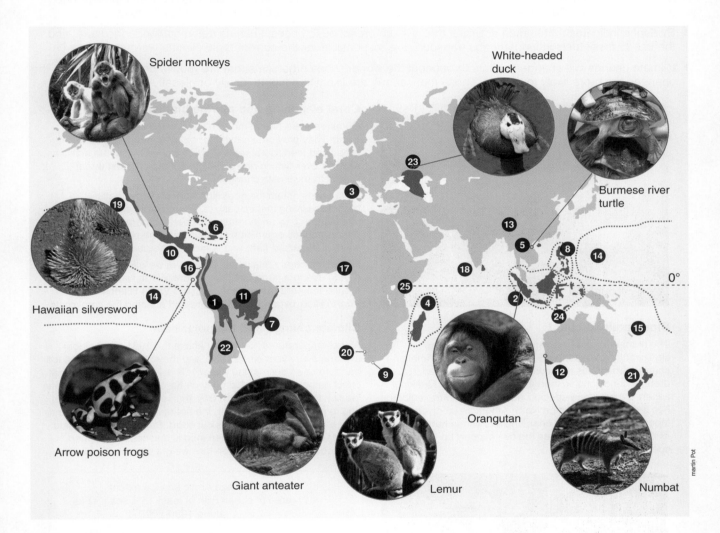

1. Looking at the map, where are most of the hotspots concentrated?_____

_____

2. Many of the biodiversity hotspots coincide with regions of very high human population density. How does high population density create greater risk of biodiversity loss in these regions?

_____

_____

3. Use your research tools (including the BIOZONE **Resource Hub**) to identify each of the 25 biodiversity hotspots illustrated in the diagram above. For one region that interests you, summarize the characteristics that have resulted in it being identified as a biodiversity hotspot. Attach your summary to this page.

©2022 **BIOZONE** International
ISBN: 978-1-98-856692-4
Photocopying Prohibited

255 How Humans Affect Biodiversity

Key Question: How is human activity affecting biodiversity on Earth?

▸ The activities of an expanding human population are contributing to an increase in **extinction** rates above the natural level, and local and global reductions in **biodiversity**.

▸ As human demand for resources increases, increasing pressure is placed on habitats and their natural populations. The effects of these pressures are often detrimental to ecosystem health and biodiversity.

▸ A decline in biodiversity reduces the ability of ecosystems to resist change and to recover from disturbance. Humans depend both directly and indirectly on healthy ecosystems, so a loss of biodiversity affects us too.

How does climate change affect biodiversity?

▸ **Climate change** is the result of increasing amounts of greenhouse gases, such as carbon dioxide, being added into the atmosphere, which traps more heat from the Sun and leads to a continuous rise in the average temperature of the Earth's surface. Climate change is a significant contributor to loss of biodiversity.

▸ Evidence indicates that human activities that release greenhouse gases, such as deforestation, agriculture, and the use of fossil fuels in industry and transport, are responsible for the current rapid climate warming.

▸ Climate change will change habitats throughout the world. Those organisms that are mobile or able to adapt to the changes are more likely to survive. Those that cannot, are likely to become locally or globally extinct.

Coral bleaching and marine biodiversity

Coral reefs are one of the most biodiverse ecosystems on Earth. They provide a habitat for many marine species to breed and feed. Coral reefs are threatened by human activities such as over-fishing and pollution, but the greatest threat comes from climate change.

Most corals obtain their nutrition from photosynthetic organisms living in their tissues. When coral becomes stressed, e.g. by an increase in ocean temperature, the photosynthetic organisms are expelled. The coral becomes white or bleached (left) and is more vulnerable to disease.

Half of the coral reefs in the Caribbean were lost in one year (2005) due to a large bleaching event. Warm waters centered around the northern Antilles, near the Virgin Islands and Puerto Rico, expanded southward and affected the coral reefs.

Biodiversity on land

The number of quaking aspen trees has declined significantly across the US in the last decade. Climate change, which has reduced rainfall and increased in drought conditions, is thought to be the cause. Quaking aspen is a keystone species, so its loss in some regions has a significant effect on North American biodiversity. Moose, elk, deer, black bear, and snowshoe hare browse its bark, and aspen groves (below) support up to 34 species of birds.

Climate change and the polar bear

Arctic temperatures have been above the 1981-2022 global average every year since 1988, and the extent of Arctic sea ice has also been decreasing (below). This is having a detrimental effect on polar bears, which rely on the sea ice to hunt their prey (seals). In Canada's Hudson Bay, the sea ice is melting earlier and forming later. As a result, the bears must swim further to hunt and their hunting time is cut short. Survival and breeding success during summer, when food is inaccessible, is then reduced because they put on less weight through the winter.

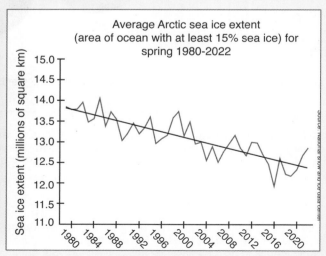

Average Arctic sea ice extent (area of ocean with at least 15% sea ice) for spring 1980-2022

©2022 **BIOZONE** International
ISBN: 978-1-98-856692-4
Photocopying Prohibited

How land use affects biodiversity

Natural grasslands (top, right) are diverse and productive ecosystems. Ancient grasslands may have contained 80-100 plant species, in contrast to currently cultivated grasslands, which may contain as few as three species. Unfortunately, many of the management practices that promote grassland species diversity conflict with modern farming methods. Appropriate management can help to conserve grassland ecosystems while still maintaining the land's viability for agriculture and food production.

Demand for food increases as the population grows. Modern farming techniques favor monocultures (bottom, left) to maximize yield and profit. However, monocultures, in which a single crop type is grown year after year, are low diversity systems and food supplies are vulnerable if the crop fails. Rice, maize, and wheat alone make up two thirds of human food consumption and are staples for more than 4 billion people. This creates issues of food security for the human population, currently, and in the future.

▸ Human activities have had major effects on the biodiversity of Earth.

▸ Nearly 40% of the Earth's land surface is devoted to agricultural use. In these areas, the original biodiversity, a polyculture of plants and animals, has been severely reduced. Many of these areas are effectively monocultures, where just one type of plant is grown.

▸ The graph (right) shows that, as the land is more intensively used, the populations and variety of plants and animals fall.

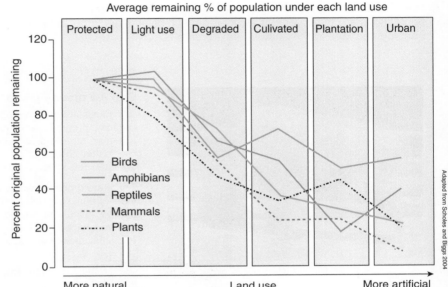

Average remaining % of population under each land use

Adapted from Scholes and Biggs 2004

1. How does coral bleaching affect marine biodiversity? _____
   _____
   _____
   _____

2. (a) Describe the trend in Arctic sea ice extent since the 1980s: _____
   _____
   _____

   (b) Reduced sea ice means polar bears need to swim longer distances to hunt seals. What effect is this likely to have on the polar bear population numbers?
   _____
   _____
   _____

3. Study the land use graph above and describe the effect of land development on biodiversity: _____
   _____
   _____
   _____
   _____
   _____

256 *Ex-Situ* Conservation

Key Question: How do *ex-situ* conservation methods assist critically endangered species?

▸ **Conservation** aims to maintain the biodiversity of a particular ecosystem. This is achieved by managing species in their habitats (*in-situ*) or by employing conservation programs away from the natural environment (*ex-situ*).

▸ *Ex-situ* means off site conservation. Individuals of an endangered species are removed from their natural habitat and placed in a new, secure location where their needs for survival are met. Zoos, aquaria, and botanical gardens are common sites for *ex-situ* conservation programs. They house and protect specimens for breeding and, where possible, they reintroduce captive-bred organisms into the wild to restore natural populations. Breeding is carefully recorded and managed to maintain **genetic diversity**, i.e. the diversity of genes and alleles within a species.

▸ *Ex-situ* conservation methods are particularly useful when species are critically endangered, i.e. have a very high risk of extinction in the wild. However, *ex-situ* conservation is expensive and labor intensive. Also, because the populations are often very small, their genetic diversity is very limited and the species may be unable to adapt to ongoing changes in their natural environments.

Captive breeding and relocation

Individuals are captured and bred under protected conditions. If breeding programs are successful and there is suitable habitat available, captive-bred individuals may be relocated to the wild where they can establish natural populations. Many zoos now have an active role in captive breeding.

There are problems with captive breeding. Individuals are inadvertently selected for fitness in a captive environment and their survival in the wild may be compromised. This is especially so for marine species. However, for some taxa such as reptiles, birds, and small mammals, captive rearing is very successful.

The important role of zoos and aquaria

As well as their role in captive breeding programs and as custodians of rare species, zoos have a major role in public education. They raise awareness of the threats facing species in their natural environments and gain public support for conservation work. Modern zoos tend to concentrate on particular species and are part of global programs that work together to help retain genetic diversity in captive bred animals.

Above: Captive breeding at the Bronx Zoo in the early 1900s was important for the recovery of American bison in the wild.

Right: A puppet "mother" shelters a takahe chick. Takahe, a rare species native to New Zealand, were brought back from the brink of extinction through a successful captive breeding program.

Right: The okapi is an endangered species of rare forest antelope related to giraffes. Okapi are only found naturally in the Ituri Forest, in the northeastern rainforests of the Democratic Republic of Congo, Africa, an area at the front line of an ongoing civil war. Successful breeding programs have been developed in a number of zoos in the US, including San Diego zoo.

1. What is the purpose of conservation? _____

_____

_____

2. What is *ex-situ* conservation? _____

_____

©2022 **BIOZONE** International
ISBN: 978-1-98-856692-4
Photocopying Prohibited

The role of botanic gardens

Botanic gardens use their expertise and resources to play a critical role in plant conservation. They maintain seed banks, nurture rare species, maintain a living collection of plants, and help to conserve indigenous plant knowledge. They also have an important role in plant research and public education.

In the US, 30% of plant species are threatened. The US Botanic Garden (Washington, DC) houses more than 500 rare plant species. They are involved in the conservation of endangered species by maintaining live specimen collections, studying wild plants at risk, banking seeds of rare plants, and introducing rare plants to the commercial growers.

Seed banks and gene banks

Seed banks and gene banks help preserve the genetic diversity of species. A seed bank (above) stores seeds as a source for future planting in case seed reserves elsewhere are lost. The seeds may be from rare species whose genetic diversity is at risk, or be the seeds of crop plants. In some cases, the seeds are of ancient varieties no longer in commercial production.

3. Describe the key features of *ex-situ* conservation methods: _____

_____

_____

_____

4. Describe some challenges or disadvantages associated with *ex-situ* conservation: _____

_____

_____

_____

_____

5. Describe three key roles of zoos and aquaria and explain the importance of each:

(a) _____

_____

(b) _____

_____

(c) _____

_____

6. Explain the role of gene and seed banks in the conservation of endangered species: _____

_____

_____

_____

_____

_____

257 *In-Situ* Conservation

Key Question: How can *in-situ* (on site) conservation methods manage ecosystems?

What is *in-situ* conservation?

▶ ***In-situ* conservation** means the conservation of a species in its natural environment.

▶ *In-situ* methods focus on ecological preservation or restoration (cleaning up the ecosystem) and often involves removing predators or invasive species.

▶ *In-situ* conservation protects more species at once, including unknown species.

▶ *In-situ* methods have several disadvantages:

- Ecological restoration is a long term process. It involves collaboration between local communities and institutions with scientific expertise.

- Populations may continue to decline during restoration, either because of lag in response or because the population has become critically low.

- Illegal poaching can be difficult to control.

In the US, the Endangered Species Act (ESA) protects species and the ecosystems on which they depend. The act is administered by the US Fish and Wildlife Service and the National Oceanic and Atmospheric Administration. As a result of this and other efforts, species such as the snowy egret (above right), white tailed deer, and wild turkey (above left), which were all critically endangered, are common again.

Returning bison to the wild

The American bison, also called buffalo, was hunted nearly to **extinction** in the 1800s. Since the 1900s, both *ex-situ* and *in-situ* conservation methods have been used successfully to boost bison numbers, although their recovery still presents several challenges. Reduced **genetic diversity**, low numbers, and interbreeding with domestic cattle mean that, even today, the numbers of "wild bison" are still quite low. American bison consists of two subspecies, the plains bison and the wood bison. In this activity, bison refers to both subspecies.

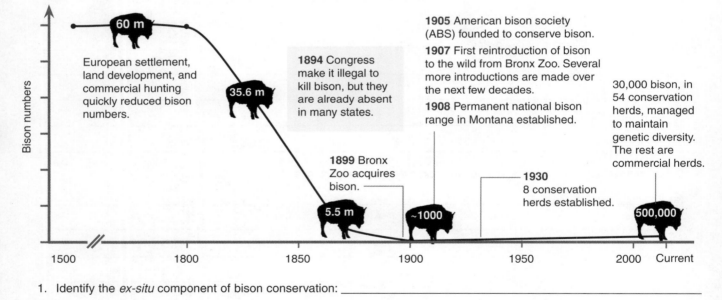

Bison numbers

60 m

European settlement, land development, and commercial hunting quickly reduced bison numbers.

35.6 m

1894 Congress make it illegal to kill bison, but they are already absent in many states.

1899 Bronx Zoo acquires bison.

5.5 m

1905 American bison society (ABS) founded to conserve bison.

1907 First reintroduction of bison to the wild from Bronx Zoo. Several more introductions are made over the next few decades.

1908 Permanent national bison range in Montana established.

~1000

1930 8 conservation herds established.

30,000 bison, in 54 conservation herds, managed to maintain genetic diversity. The rest are commercial herds.

500,000

1500 1800 1850 1900 1950 2000 Current

1. Identify the *ex-situ* component of bison conservation: _____

_____

2. How have *in-situ* conservation methods and legislation contributed to the success of bison recovery in North America?

_____

_____

3. Identify a possible risk for bison conservation today: _____

_____

4. Use the information on this page and the reports located on the **Resource Hub** page to create or revise a projection for recovery of species and genetic diversity in American bison. Write a brief report and attach it to this page.

      LS4.C LS4.D CE

©2022 **BIOZONE** International
ISBN: 978-1-98-856692-4
Photocopying Prohibited

258 Conservation and Genetic Diversity

Key Question: How can conservation methods help to maintain genetic diversity?

Conservation genetics involves many branches of science and uses genetic methods to restore **genetic diversity** in a declining species.

One of the biggest problems occurring when a species' population declines is loss of genetic diversity (the loss of gene diversity within a species). This increases the relatedness between individuals because there are fewer individuals to breed with. Decreased genetic diversity and increased relatedness can result in inbreeding depression (the reduced fitness of individuals as a result of inbreeding), and can dramatically reduce population viability.

The Florida panther (right) is an example of how conservation genetics has been used to restore genetic diversity in an endangered population.

▸ In the late 1970s, the Florida panther population had become critically low and occupied just 5% of its historical range. Population models showed it would be extinct within a few decades. Individuals often had several abnormalities including kinked tails, heart defects, and sperm defects. It was determined that these were due to inbreeding depression.

▸ In 1995, eight female panthers were translocated from Texas to increase genetic diversity. Since then, there has been an increase in the population growth rate and an improvement in the survival and health of individuals (graph, right).

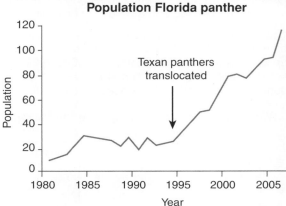

Population Florida panther

Texan panthers translocated

Conservation genetics is an important concept for conservation and breeding programs. By keeping detailed breeding records of which individuals are bred together and monitoring the genetic relatedness of populations, conservation scientists can make sure endangered species avoid inbreeding depression and maintain their genetic diversity.

Conservation genetics were used to rebuild the Illinois prairie chicken population.

Critically endangered birds, such as the New Zealand kakapo, have lost much of their genetic diversity.

1. What is conservation genetics? _____

_____

2. (a) Why were panthers from Texas translocated to Florida? _____

_____

_____

(b) Why is it important to maintain genetic diversity in populations? _____

_____

_____

3. Use the reports located on the **Resource Hub** page to create or revise a projection for recovery of the Florida panther. One report provides in-depth information for gifted and talented students. Write a brief report and attach it to this page.

259 Maasai Mara Case Study

Key Question: How can the Maasai Mara case study be used to show the need to balance human and environmental needs in conservation?

How farming affected the Maasai Mara region

The Maasai Mara National Reserve is a region in south western Kenya covering 1500 km². It is part of the much larger Mara-Serengeti ecosystem, which covers around 25,000 km². Considerable change has occurred in the Maasai Mara region since the start of last century. Early in the 20th century, the region was much less populated and the land was used mainly for nomadic agriculture (raising cattle). European settlers forced many of the Maasai off their traditional lands. In 1945, more land was turned into reserves and the Maasai land placed in Trust. As a result of changes in governments and ideals, Trust land was redesignated as group ranches. This encouraged the Maasai to subdivide the land to acquire individual titles in order to secure legal rights to lands, rather than risk losing them outright. Privatization led to an increase in mechanized farming and a reduction in wildlife.

The changes to the land use has had a drastic affect on wildlife (below). Wildebeest numbers dropped from approximately 150,000 in 1977 to 40,000 in 2010. Water buffalo numbers dropped from 40,000 to 5000. Livestock numbers increased (cattle, sheep, and goats).

Wildlife and livestock changes in the Maasai Mara

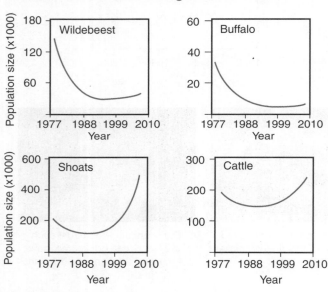

The positive effect of conservancies

Over time, it was realized that modern farming methods limited the range of wildlife and also grazing options for livestock. In 2005, many landowners in the northern Maasai Mara began consolidating their land into conservancies (land set aside for conservation), aiming to generate income through tourism. This included establishing partnerships with tourism operators. The success of this approach has seen a rapid expansion of conservancies.

Although the formation of conservancies has benefited the wildlife and many people, those who do not own any land have been no better off. In addition, livestock are only allowed into the conservancies during certain times, e.g. drought, which has led to higher stocking rates outside the park and conservancies.

The development of conservancies has had benefits for the community. Payments by tour operators (above) for use of the land are made directly to the land owners, reducing loss of income through bureaucratic handling and corruption. Today, there are 8 conservancies representing 92,000 ha. Around $3.6 million is paid to the conservancies each year.

1. (a) How has land use in Maasai Mara region changed since the early 20th century? _____

_____

   (b) What effect did farming have on wildebeest and buffalo in the Maasai Mara region? _____

_____

_____

2. In what way did conservancies help reduce conflict between humans and wildlife? _____

_____

_____

_____

  LS4.D CE

©2022 **BIOZONE** International
ISBN: 978-1-98-856692-4
Photocopying Prohibited

260 Review Your Understanding

Key Question: How has human activity affected a biodiversity hotspot, and what are some possible solutions for restoring it, or preventing more loss?

1. The Madrean Pine-Oak Woodlands area is considered a hotspot. What does this term mean? _____

_____

_____

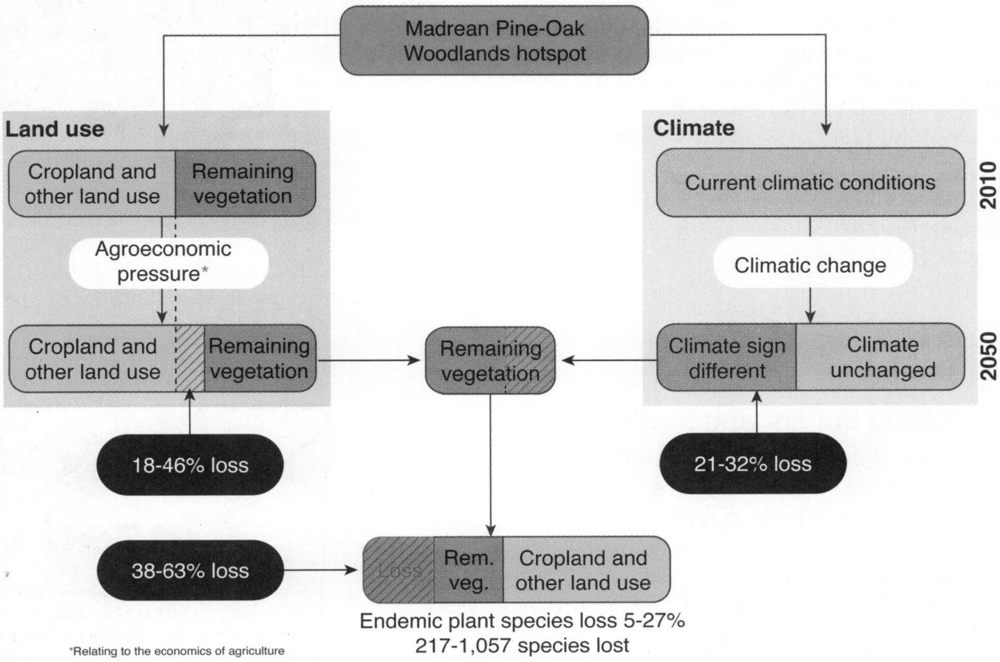

*Relating to the economics of agriculture

Endemic plant species loss 5-27%
217-1,057 species lost

Adapted from Habel, JC, Rasche, L, Schneider, UA, et al. Final countdown for biodiversity hotspots. 2019; 12:e12668. https://doi.org/10.1111/conl.12668

2. Using information from the diagram above, what are the key factors that will affect plant biodiversity loss in the Madrean Pine-Oak Woodlands hotspot?

_____

_____

_____

3. Research an endangered species found in your local area and present a one page report of your findings, including features about the species, its habitat and niche, the population size, and how human activity has affected it. What potential conservation action(s) might be effective in reducing the loss of your selected species? Attach your report to this page.

_____

_____

_____

_____

_____

_____

_____

_____

261 Summing Up

About the Indiana bat

The Indiana bat (*Myotis sodalis*) is an insectivorous (insect eating) bat native to North America. It is found in the eastern US, from New Hampshire to northern Florida, and west to Iowa, Missouri, and Oklahoma. In winter, Indiana bats hibernate in caves. In spring, they migrate north and make their home in tree cavities or under loose tree bark.

Winter hibernation allows the bats to conserve energy when there are too few insects available as food. The bats do not eat during hibernation, but live off stored fat laid down before winter when there is ample insect food available. However, each time a bat is woken during hibernation, energy stores are used up more rapidly. If they are disturbed too often, they may not have enough fat reserves to survive the winter.

The Indiana bat has an important ecological role. It is a major predator of night-flying insects and contributes energy to the cave food web through guano (excrement) and decomposition of its body when it dies.

The Indiana bat population is estimated to be 244,000. Very few caves provide suitable hibernation conditions, so the majority of the population hibernates in only a few caves. About 23% hibernate in caves in Indiana. The bats hibernate in tight clusters (above).

Why is the Indiana bat endangered?

Population numbers of the Indiana bat have declined by more than 50% in the last 10 years and they are now listed as endangered. Several factors have contributed to this.

▸ Although they have a large range, they have very few suitable winter hibernation sites. They must roost in cool, humid caves with temperatures above freezing but below 10°C.

▸ Changes in cave structure, e.g. blocking an entrance, can change the temperature in a cave, making it unsuitable for hibernating bats.

▸ Human disturbance can rouse the bats from hibernation. Some caves are fitted with gates to stop people getting in, but a poor gate design also stops the bats entering the cave.

▸ White nose syndrome (WNS) is a fungal disease and grows around the mouth and wings of hibernating bats. The disease causes abnormal behaviors in the bats, such as daytime flights during winter. Nearly 6 million bats (of different species) have been killed by WNS since 2007.

▸ WNS is spread by direct contact with other infected bats or with infected material in the cave, e.g. guano. The fungus can be infectious for a very long time in the cave and can be spread to new caves by people and bats.

▸ Increased pesticide use kills insects that the bats eat. Indiana bats only eat insects and have no other food sources.

1. A number of factors have contributed to the severe reduction in the Indiana bat population in the last decade. Using the information above, and in the **Resource Hub**, your task is to design a solution to halt their population decline and ultimately restore their numbers so that they are no longer listed as endangered. You should consider the effect of biological factors and human activities on bat conservation and use both scientific and engineering solutions in your conservation plan.

   Once you have established your plan, you should present the information as an educational resource, e.g. a pamphlet, poster, or slide presentation, designed to teach the general public about the ecological importance of the Indiana bat and how they can help protect it.

©2022 **BIOZONE** International
ISBN: 978-1-98-856692-4

Appendix: English/Spanish Glossary

Each glossary term is first provided in English (black text) with the Spanish translation directly beneath (blue text).

abiotic factor: Non-living, physical features in an ecosystem, including temperature, humidity, and rainfall.

factor abiótico: Características físicas no vivas en un ecosistema, incluida la temperatura, la humedad y la lluvia.

---

accuracy: The correctness of a measurement; how close a measured value is to the true value.

exactitud: La exactitud de una medición; qué tan cerca está un valor medido del valor verdadero.

---

adaptation: The process by which populations become more suited to their environments.

adaptación: El proceso por el cual las poblaciones se adaptan más a sus entornos.

---

aerobic: A biological process that requires oxygen.

aerobio: Un proceso biológico que requiere oxígeno.

---

allele: Any of the alternative versions of a gene that may produce distinguishable phenotypes.

alelo: Cualquiera de las versiones alternativas de un gen que puede producir fenotipos distinguibles.

---

altruism: The activity of an organism whose behavior benefits other organisms, at a personal cost to itself.

altruismo: La actividad de un organismo cuyo comportamiento beneficia a otros organismos, a un costo personal para sí mismo.

---

anaerobic: A biological process that does not require the presence of oxygen.

anaeróbico: Un proceso biológico que no requiere la presencia de oxígeno.

---

anthropogenic change: A change that results from the influence of human beings on the natural world.

cambio antropogénico: Un cambio que resulta de la influencia de los seres humanos en el mundo natural.

---

assumption: A statement that is assumed to be true but is not (or cannot be) tested.

presunción: Una afirmación que se supone que es verdadera pero que no se prueba (o no se puede probar).

---

base-pairing rule: The rule that states that nucleotides are specific in their pairing activity, i.e. A always pairs with T, and G always pairs with C.

regla de emparejamiento de bases: La regla que establece que los nucleótidos son específicos en su actividad de emparejamiento, es decir, A siempre se empareja con T, y G siempre se empareja con C.

---

biodiversity: The amount of biological variation present in a region (includes genetic, species, and habitat diversity).

biodiversidad: La cantidad de variación biológica presente en una región (incluye genética, especies y diversidad de hábitat).

---

biodiversity hotspot: A region that contains significant numbers of species that are found nowhere else in the world.

hotspot de biodiversidad: Una región que contiene un número significativo de especies que no se encuentran en ningún otro lugar del mundo.

---

bioinformatics: The use of computer science, mathematics, and information theory to organize and analyze complex biological data.

Bioinformática: El uso de las ciencias computacionales, las matemáticas y la teoría de la información para organizar y analizar datos biológicos complejos.

---

biological drawing: An illustration that visually communicates the structure of a subject being studied, showing specific details.

dibujo biológico: Una ilustración que comunica visualmente la estructura de un tema que se está estudiando, mostrando detalles específicos.

---

biotic factor: Relating to the living factors in an ecosystem, including distribution and abundance.

factor biótico: Relacionado con los factores vivos en un ecosistema, incluida la distribución y la abundancia.

---

carbon cycle: The process by which carbon is exchanged between living organisms, the earth and its atmosphere.

ciclo del carbono: El proceso por el cual el carbono se intercambia entre los organismos vivos, la tierra y su atmósfera.

---

carrying capacity: The maximum number of organisms that can be sustained by a specific environment.

capacidad de carga: El número máximo de organismos que pueden ser sostenidos por un entorno específico.

---

catalyst: A substance that modifies and increases the rate of a chemical reaction without being consumed in the process.

catalizador: Sustancia que modifica y aumenta la velocidad de una reacción química sin ser consumida en el proceso.

---

cell: The smallest biological unit that can survive on its own. It is the base unit of all living organisms.

celda: La unidad biológica más pequeña que puede sobrevivir por sí sola. Es la unidad base de todos los organismos vivos.

---

cell cycle: The cycle of stages that occur in a cell as it grows and divides to produce new daughter cells.

célula: La unidad biológica más pequeña que puede sobrevivir por sí sola. Es la unidad base de todos los organismos vivos.

---

cell differentiation: The process by which a cell changes from one cell type to another, usually to a more specialized type of cell.

diferenciación celular: El proceso por el cual una célula cambia de un tipo de célula a otro, generalmente a un tipo de célula más especializada.

---

cell division: The cycle of stages that occur in a cell as it grows and divides to produce new daughter cells.

división celular: El ciclo de etapas que ocurren en una célula a medida que crece y se divide para producir nuevas células hijas.

---

cellular respiration: The series of metabolic reactions that oxidize organic molecules to produce ATP.

respiración celular: La serie de reacciones metabólicas que oxidan las moléculas orgánicas para producir ATP.

English/Spanish Glossary

chloroplast: An organelle within the cells of plants and green algae that contains chlorophyll and is the site of photosynthesis.

cloroplasto: Un orgánulo dentro de las células de las plantas y algas verdes que contiene clorofila y es el sitio de la fotosíntesis.

chromatin: A complex of DNA and proteins, making up the chromosomes.

cromatina: Un complejo de ADN y proteínas, que componen los cromosomas.

chromosome: A cellular structure consisting of one DNA molecule and associated protein molecules.

cromosoma: Estructura celular que consiste en una molécula de ADN y moléculas de proteínas asociadas.

climate change: A change in the patterns of climate such as temperature and rainfall that is attributed to human activity, including the use of fossil fuels.

cambio climático: Un cambio en los patrones del clima, como la temperatura y las precipitaciones, que se atribuye a la actividad humana, incluido el uso de combustibles fósiles.

common ancestor: An ancestor that two or more descendant species have in common.

ancestro común: Un ancestro que dos o más especies descendientes tienen en común.

competition: Interaction within or between species in which individuals attempt to access the same limited resource.

competición: Interacción dentro o entre especies en la que los individuos intentan acceder al mismo recurso limitado.

conservation: Activities designed to protect biological diversity in a specific area.

conservación: Actividades destinadas a proteger la diversidad biológica en una zona específica.

consumer: An organism that feeds on producers, other consumers, or non-living organic material.

consumidor: Un organismo que se alimenta de productores, otros consumidores o material orgánico no vivo.

control: A component of an experiment that is isolated from the effects of the independent variable. It demonstrates that any change to the dependent variable must come from the independent variable.

control: Componente de un experimento aislado de los efectos de la variable independiente. Demuestra que cualquier cambio en la variable dependiente debe provenir de la variable independiente.

controlled variable: Another word for independent variable. The variable being controlled or changed by the experimenter.

variable controlada: Otra palabra para variable independiente. El variable es controlado o cambiado por el experimentador.

cooperative behavior: An evolutionary response to reduce the competition between individuals of the same species.

comportamiento cooperativo: Una respuesta evolutiva para reducir la competencia entre individuos de la misma especie.

crossing over: The reciprocal exchange of genetic material between non-sister chromatids during prophase 1 of meiosis.

cruzando: El intercambio recíproco de material genético entre cromátidas no hermanas durante la profase 1 de la meiosis.

data: A set of values of qualitative or quantitative variables, collected through observation.

datos: Conjunto de valores de variables cualitativas o cuantitativas, recogidos a través de la observación.

deforestation: The removal of forests by cutting, burning or other large scale activity by humans, usually to make way for crops or monoculture plantations.

deforestación: La eliminación de bosques mediante la tala, quema u otra actividad a gran escala por parte de los seres humanos, generalmente para dar paso a cultivos o plantaciones de monocultivos.

density: The number of individuals per unit area or volume.

densidad: El número de individuos por unidad de área o volumen.

dependent variable: The variable being tested and measured in an experiment, whose value depends on that of the independent variable.

variable dependiente: La variable que se está probando y medido en un experimento, cuyo valor depende del de la variable independiente.

descriptive statistics: Also called summary statistics, these are brief descriptors that help summarise features of data.

estadística descriptiva: También llamadas estadísticas resumidas, estas son descriptores breves que ayudan a resumir las características de los datos.

diabetes mellitus: A disease in which the body is unable to produce or respond to the hormone, insulin, to maintain optimum levels of glucose in the blood.

diabetes mellitus: Una enfermedad en la que el cuerpo es incapaz de producir o responder a la hormona, la insulina, para mantener niveles óptimos de glucosa en la sangre.

dihybrid: An organism that is heterozygous, with two different alleles at a genetic location.

dihíbrido: Un organismo que es heterogótico, con dos alelos diferentes en una ubicación genética.

distribution: The spatial arrangement of organisms.

distribución: La disposición espacial de los organismos.

DNA: A large molecule composed of two polynucleotide chains that carries the genetic code and enables cells to function.

ADN: La duplicación de una molécula de ADN, produciendo dos copias idénticas de una molécula de ADN original.

DNA replication: The duplication of a DNA molecule, producing two identical copies from one original DNA molecule.

Replicación del ADN: La duplicación de una molécula de ADN, produciendo dos copias idénticas de una molécula de ADN original.

DNA sequencing: A technique used to determine the base sequence (As, Ts, Cs, and Gs) in a DNA molecule.

Secuenciación del ADN: Técnica utilizada para determinar la secuencia de bases (As, Ts, Cs y Gs) en una molécula de ADN.

---

ecological pyramid: A graphical representation of the relationship between different organisms in an ecosystem, and their trophic levels.

pirámide ecológica: Representación gráfica de la relación entre los diferentes organismos de un ecosistema y sus niveles tróficos.

---

ecosystem: All the organisms in a given area as well as the abiotic factors with which they interact.

ecosistema: Todos los organismos en un área determinada, así como los factores abióticos con los que interactúan.

---

ecosystem services: The services that ecosystems provide to humans in various forms including food, fuel, air and water purification.

servicios ecosistémicos: Los servicios que los ecosistemas proporcionan a los seres humanos en diversas formas, incluyendo alimentos, combustible, aire y purificación de agua.

---

enzyme: Globular proteins that act as biological catalysts for specific reactions.

enzima: Proteínas globulares que actúan como catalizadores biológicos para reacciones específicas.

---

eukaryotic cell: An animal cell in which the genetic material is contained within a distinct, membrane-bound nucleus.

célula eucariota: Célula animal en la que el material genético está contenido dentro de un núcleo distinto unido a la membrana.

---

evo-devo: Evolutionary developmental biology. A scientific area of research that compares the developmental processes of different organisms.

evo-devo: Biología evolutiva del desarrollo. Un área científica de investigación que compara los procesos de desarrollo de diferentes organismos.

---

evolution: The change in the heritable characteristics of populations over successive generations.

evolución: El cambio en las características hereditarias de las poblaciones a lo largo de generaciones sucesivas.

---

ex-situ **conservation**: Activities that take place to protect and conserve a species outside of its natural environment, e.g. zoos, aquariums and botanic gardens.

conservación *ex situ*: Actividades que se llevan a cabo para proteger y conservar una especie fuera de su entorno natural, por ejemplo, zoológicos, acuarios y jardines botánicos.

---

extinction: No living members of a species are left in the wild or captivity.

extinción: Ningún miembro vivo de una especie queda en la naturaleza o en cautividad..

---

feedback mechanisms: A process by which the body detects changes to physiological states and either brings the body back to the normal state (negative) or away from the normal state (positive).

mecanismos de retroalimentación: Un proceso por el cual el cuerpo detecta cambios en los estados fisiológicos y devuelve al cuerpo al estado normal (negativo) o se aleja del estado normal (positivo).

---

fermentation: An anaerobic metabolic process by which a carbohydrate, such as starch or a sugar, is converted into an alcohol or an acid.

fermentación: Un proceso metabólico anaeróbico por el cual un carbohidrato, como el almidón o un azúcar, se convierte en un alcohol o un ácido.

---

fitness: An organism's ability to survive to reproductive age and produce offspring. A mathematical measure of an organism's genetic contribution to the next generation.

aptitud: La capacidad de un organismo para sobrevivir hasta la edad reproductiva y producir descendencia. Una medida matemática de la contribución genética de un organismo a la próxima generación.

---

flocking: Activity undertaken by birds in which members group together and each individual aligns its movements to the other individuals around it.

bandada: Actividad subestimada por las aves en la que los miembros se agrupan y cada individuo alinea sus movimientos con los otros individuos a su alrededor.

---

food chain: A model that is used to demonstrate the feeding relationships between organisms in an ecosystem.

cadena alimentaria: Un modelo que se utiliza para demostrar las relaciones de alimentación entre los organismos en un ecosistema.

---

food web: The combination of all the food chains in a particular ecosystem.

red trófica: La combinación de todas las cadenas alimentarias en un ecosistema en particular.

---

fossil: The preserved remains, found in the earth's crust, of animal or plants that lived a long time ago.

fósil: Los restos conservados, encontrados en la corteza de la tierra, de animales o plantas que vivieron hace mucho tiempo.

---

fossil record: The history of life as documented by fossils.

registro fósil: La historia de la vida documentada por los fósiles.

---

gene: A unit of hereditary information consisting of a specific nucleotide sequence in DNA.

genotipo: Unidad de información hereditaria que consiste en una secuencia específica de nucleótidos en el ADN.

---

gene expression: The transcription and translation of a gene.

expresión génica: La transcripción y traducción de un gen.

---

gene pool: The collective genetic information within a population of interbreeding organisms.

reserva genética:La información genética colectiva dentro de una población de organismos que se cruzan.

---

genetic diversity: The differences in the genetic makeup of individuals in a population.

diversidad genética: Las diferencias en la composición genética de los individuos en una población.

English/Spanish Glossary

genotype: The genetic makeup of an organism.

genotipo: La composición genética de un organismo.

---

global warming: The gradual heating of the earth, attributed to human activities such as the burning of fossil fuels.

Calentamiento global: El calentamiento gradual de la tierra, atribuido a actividades humanas como la quema de combustibles fósiles.

---

glucose: A simple sugar that functions as the main source of metabolic energy in living things.

glucosa: Un azúcar simple que funciona como la principal fuente de energía metabólica en los seres vivos.

---

graph: A diagram that is used to present scientific data, usually with two variables on an x and y axis. It allows relationships between variables to be clearly visualized.

gráfico: Un diagrama que se utiliza para presentar datos científicos, generalmente con dos variables en un eje x e y. Permite visualizar claramente las relaciones entre variables.

---

habitat: The natural environment in which an organism lives, including all of the biotic and abiotic factors.

hábitat: El medioambiente natural en el que vive un organismo, incluidos todos los factores bióticos y abióticos.

---

heat energy: The energy that results from the movement of particles such as individual atoms moving against each other.

energía térmica: La energía que resulta del movimiento de partículas como átomos individuales que se mueven uno contra el otro.

---

herding: Activity undertaken by grazing mammals in which members group together and each individual aligns its movements to the other individuals around it. Usually, one or more members keeps guard while the others graze.

pastoreo: Actividad de mamíferos de pastoreo en la que los miembros se agrupan y cada individuo alinea sus movimientos con los otros individuos a su alrededor. Por lo general, uno o más miembros mantienen la guardia mientras los demás pastan.

---

heterozygous: Having two different alleles for any hereditary characteristic.

heterocigótico: Tener dos alelos diferentes para cualquier característica hereditaria.

---

histone: A small basic protein found in the nucleus of eukaryotic cells that organizes DNA strands to form chromatin.

histona: Una pequeña proteína básica que se encuentra en el núcleo de las células eucariotas que organiza las hebras de ADN para formar cromatina.

---

histone modification: Changes to histones, most commonly by methylation or acetylation.

modificación de histonas: Cambios en las histonas, más comúnmente por metilación o acetilación.

---

home range: The area normally inhabited by a specific organism. The size of the home range is dependent on the resources available to the organism within that area.

gama de inicio: El área normalmente habitada por un organismo específico. El tamaño del rango de hogar depende de los recursos disponibles para el organismo dentro de esa área.

---

homeostasis: The steady-state physiological condition of the body.

homeostasis: La condición fisiológica de estado estacionario del cuerpo.

---

homologous structure: Structures found in different organisms that result from a common ancestor, e.g. forelimb of a seal and wing of a bird.

estructura homóloga: Estructuras que se encuentran en diferentes organismos que resultan de un ancestro común, por ejemplo, la extremidad anterior de una foca y el ala de un ave.

---

homology: Similarity between two different species of organisms due to shared ancestry.

homología: Similitud entre dos especies diferentes de organismos debido a la ascendencia compartida.

---

homozygous: Where both chromosomes possess identical alleles for a gene, at a specific locus.

homocigótico: Donde ambos cromosomas poseen alelos idénticos para un gen, en un locus específico.

---

hydrologic cycle: Also known as the water cycle. Describes that way in which water circulates between the earth and its atmosphere.

ciclo hidrológico: También conocido como el ciclo del agua. Describe esa forma en que el agua circula entre la tierra y su atmósfera.

---

hypothesis: A tentative explanation, proposition, or set of propositions capable of being tested by scientific experimentation.

hipótesis: Una explicación tentativa, proposición o conjunto de proposiciones capaces de ser probadas por experimentación científica.

---

in-situ conservation: A method of conserving a species within its own environment.

conservación in situ: Un método para conservar una especie dentro de su propio entorno.

---

independent assortment: With reference to inheritance, describing how alleles for separate traits are passed to the gametes independently of one another.

surtido independiente: Con referencia a la herencia, describiendo cómo los alelos para rasgos separados se pasan a los gametos independientemente unos de otros.

---

independent variable: The variable being set by the experimenter which is assumed to have a direct effect upon the dependent variable.

variable independiente: La variable establecida por el experimentador que se supone que tiene un efecto directo sobre la variable dependiente.

---

interphase: The period in the cell cycle when the cell is not dividing, which accounts for about 90% of the cell cycle. During interphase, cellular metabolic activity is high and cell size may increase.

interfase: El período en el ciclo celular cuando la célula no se está dividiendo, que representa aproximadamente el 90% del ciclo celular. Durante la interfase, la actividad metabólica celular es alta y el tamaño celular puede aumentar.

---

interspecific competition: Competition for resources between different species.

competencia interespecífica: Competencia por los recursos entre diferentes especies.

intraspecific competition: Competition for resources between member of the same species.

competencia intraespecífica: Competencia por los recursos entre miembros de la misma especie.

invasive species: A species that is not native to that ecosystem.

especies invasoras: Una especie que no es nativa de ese eccosistema.

keystone species: A species that occupies an essential role in an ecosystem and on which most or all of the other species in an ecosystem depend, directly or indirectly.

especies clave: Una especie que ocupa un papel esencial en un ecosistema y de la que dependen, directa o indirectamente, la mayoría o la totalidad de las demás especies de un ecosistema.

kin selection: Altruistic behavior that involves close relatives (see altruism).

selección de parentesco: Comportamiento altruista que involucra a parientes cercanos (ver altruismo).

mean: The sum of the data divided by the number of data entries; a measure of central tendency in a normal distribution.

promedio: La suma de los datos dividida por el número de entradas de datos; una medida de tendencia central en una distribución normal.

median: The middle number in an ordered sequence of numbers. For an odd number of values, it is the average of the two middle numbers.

mediana: El número medio en una secuencia ordenada de números. Para un número impar de valores, es el promedio de los dos números medios.

meiosis: The process of double nuclear division in sexually reproducing organisms which results in cells with half the original number of chromosomes (haploid).

meiosis: El proceso de doble división nuclear en organismos que se reproducen sexualmente que da como resultado células con el medio número eoriginal de cromosomas (haploide).

migration: Seasonal movement of animals from one region or habitat to another.

migración: Movimiento estacional de animales de una región o hábitat a otro.

mitochondria: organelles in eukaryotic cells that serves as the site of cellular respiration.

mitocondria: orgánulos en células eucariotas que sirven como sitio de respiración celular.

mitosis: The phase of the cell cycle resulting in nuclear division.

mitosis: La fase del ciclo celular que resulta en la división nuclear.

mode: The value that occurs most often in a data set.

modo: El numero que se produce con mayor frecuencia en un conjunto de datos.

model: A conceptual, mathematical or physical representation of a real-world phenomenon.

modelo: Una representación conceptual, matemática o física de un fenómeno del mundo real.

monohybrid: An organism that is homozygous, with two alleles that are the same, at a genetic location.

monohíbrido: Un organismo que es homocigoto, con dos alelos que son iguales, en una ubicación genética.

mutation: A change in the nucleotide sequence of an organism's DNA (or RNA).

mutación: Un cambio en la secuencia de nucleótidos del ADN (o ARN) de un organismo.

mutualism: Biological interaction between (usually two) species that benefits both parties.

mutualismo: Interacción biológica entre (generalmente dos) especies que beneficia a ambas partes.

natural selection: The differential survival and reproduction of favourable phenotypes.

selección natural: La supervivencia diferencial y la reproducción de fenotipos favorables.

negative feedback: The differential survival and reproduction of favourable phenotypes.

selección natural: La supervivencia diferencial y la reproducción de fenotipos favorables.

niche: The position occupied by an organism within its specific environmental conditions.

nicho: La posición que ocupa un organismo dentro de sus condiciones ambientales específicas.

nitrogen cycle: The processes by which nitrogen, in different forms, is cycled between living and non-living things in marine, terrestrial and atmospheric ecosystems.

ciclo del nitrógeno: Los procesos por los cuales el nitrógeno, en diferentes formas, se recicla entre seres vivos y no vivos en ecosistemas marinos, terrestres y atmosféricos.

nucleotide: An organic molecule that is the building block of DNA and RNA. Consist of a sugar molecule (ribose in RNA or deoxyribose in DNA) attached to a phosphate group and a nitrogen-containing base.

nucleótido: Una molécula orgánica que es el bloque de construcción del ADN y el ARN. Consiste en una molécula de azúcar (ribosa en ARN o desoxirribosa en ADN) unida a un grupo fosfato y una base que contiene nitrógeno.

observation: The activity of watching or recording what is happening in a given, often experimental, setting.

observación: La actividad de observar o registrar lo que está sucediendo en un entorno dado, a menudo experimental.

organ system: A group of organs in the body that work together to perform a particular function, e.g. the circulatory system involves the heart and the different blood vessels.

sistema de órganos: Un grupo de órganos en el cuerpo que trabajan juntos para realizar una función particular, por ejemplo, el sistema circulatorio involucra el corazón y los diferentes vasos sanguinios.

English/Spanish Glossary

organelle: A subcellular structure with one or more specific jobs to perform in the cell.

orgánulo: Estructura subcelular con uno o más trabajos específicos para realizar en la célula.

---

overfishing: The practice of removing fish from an area at a rate higher than that which allows the population can replace itself.

sobrepesca: La práctica de retirar peces de un área a un ritmo superior al que permite a la población puede reemplazarse a sí misma.

---

oxygen cycle: The movement of oxygen between biotic and abiotic components of ecosystems.

ciclo del oxígeno: El movimiento del oxígeno entre los componentes bióticos y abióticos de los ecosistemas.

---

parasitism: Biological interaction in which one organism, the parasite, benefits at the expense of the other, the host.

parasitismo: Interacción biológica en la que un organismo, el parásito, se beneficia a expensas del otro, el huésped.

---

pentadactyl limb: A limb with a specific arrangement of bones, containing five digits. It is used to provide evidence that all organisms possessing it derived from a common ancestor.

extremidad pentadactyl: Una extremidad con una disposición específica de huesos, que contiene cinco dígitos. Se utiliza para proporcionar evidencia de que todos los organismos que lo poseen derivan de un ancestro común.

---

phenotype: The observable physical and physiological traits of an organism, which are determined by its genetic makeup, environment and epigenetic factors.

fenotipo: Los rasgos físicos y fisiológicos observables de un organismo, que están determinados por su composición genética, entorno y factores epigenéticos.

---

photosynthesis: A process used by green plants, algae, and some bacteria to convert light energy into chemical energy (carbohydrate).

fotosíntesis: Un proceso utilizado por plantas verdes, algas y algunas bacterias para convertir la energía de la luz en energía química (carbohidratos).

---

phylogenetic tree: A branching diagram showing evolutionary relationships among organisms. Can be based on cladistic analysis, in which case it is called a cladogram.

árbol filogenético: Un diagrama de ramificación que muestra las relaciones evolutivas entre los organismos. Puede basarse en el análisis cladístico, en cuyo caso se denomina cladograma.

---

polynucleotide: A molecule consisting of nucleotides bound together by chemical bonds, e.g. DNA and RNA.

polinucleótido: Una molécula que consiste en nucleótidos unidos entre sí por enlaces químicos, por ejemplo, ADN y ARN.

---

population: A group of individuals of the same species living in a given area at a given time.

población: Un grupo de individuos de la misma especie que viven en un área determinada en un tiempo specifico.

---

population growth: Change in the size of a population over time.

crecimiento de la población: Cambio en el tamaño de una población a lo largo del tiempo.

---

positive feedback: In physiology, a control mechanism in which a change in a variable triggers a response that reinforces or amplifies the change.

retroalimentación positiva: En fisiología, un mecanismo de control en el que un cambio en una variable desencadena una respuesta que refuerza o amplifica el cambio.

---

precision: How close repeated measurements are to each other, i.e. repeatability.

precisión: Qué tan cerca están las mediciones repetidas entre sí, es decir, la repetibilidad.

---

predation: Biological interaction in which one organism, the predator, kills and eats another, its prey.

predación:

Interacción biológica en la que un organismo, el predador, mata y se come a otro, su presa.

---

prediction: What is expected to happen if the hypothesis of an experiment or scenario is true.

predicción: Lo que se espera que suceda si la hipótesis de un experimento o escenario es cierta.

---

producer: An organism that produces its own food using materials from inorganic sources (also known as an autotroph).

productor: Un organismo que produce su propio alimento utilizando materiales de fuentes inorgánicas (también conocido como autótrofo).

---

prokaryotic cell: Bacterial cells that lack any membrane-bound organelles or nucleus. And contain a simple chromosome of DNA.

célula procariota: Células bacterianas que cualquier orgánulo o núcleo unido a la membrana. Y contienen un simple cromosoma de ADN.

---

protein: A biologically functional molecule consisting of one or more polypeptides folded into a specific three-dimensional structure.

proteína: Una molécula biológicamente funcional que consiste en uno o más polipéptidos plegados en una estructura tridimensional específica.

---

qualitative data: Non-numerical data that describes qualities or characteristics.

datos cualitativos: Datos no numéricos que describen cualidades o características.

---

quantitative data: Numerical data expressing a certain quantity, amount, or range.

datos cuantitativos: Datos numéricos que expresan una determinada cantidad, monto o rango.

raw data: Experimental data that is collected in the field or laboratory and has not yet been processed.

datos brutos: datos experimentales que se recogen en el archivo o laboratorio y que aún no han sido procesados.

---

recombination: The process by which genes are exchanged between different chromosomes to produce new combinations of alleles.

recombinación: El proceso por el cual los genes se intercambian entre diferentes cromosomas para producir nuevas combinaciones de alelos.

---

resilience: The property of ecosystems or populations to recover from disturbances.

resiliencia: La propiedad de los ecosistemas o poblaciones para recuperarse de las perturbaciones.

---

schooling: Activity undertaken by fish in which members group together and each individual aligns its movements to the other individuals around it.

escolarización: Actividad precedida por los peces en la que los miembros se agrupan y cada individuo alinea sus movimientos con los otros individuos a su alrededor.

---

scientific method: The processes applied to the way in which scientists discover how the universe works. It involves a specific way of asking questions, observing, measuring, and interpreting data to formulate hypotheses and theories.

método científico: Los procesos aplicados a la forma en que los científicos descubren cómo funciona el universo. Implica una forma específica de hacer preguntas, observar, medir e interpretar datos para formular hipótesis y teorías.

---

semi-conservative replication: The normal mechanism of DNA replication, where each strand acts as a template for a new double helix.

replicación semiconservadora: El mecanismo normal de replicación del ADN, donde cada hebra actúa como una plantilla para una nueva doble hélice.

---

social group: A socially ordered group of organisms in which individuals have specific roles.

grupo social: Un grupo socialmente ordenado de organismos en el que los individuos tienen roles específicos.

---

specialized cell: A cell that has developed the characteristics needed to perform particular functions.

célula especializada: Una célula que ha desarrollado las características necesarias para realizar funciones particulares.

---

stem cell: A cell that has the ability to develop into many other different cell types.

célula madre: Una célula que tiene la capacidad de desarrollarse en muchos otros tipos de células diferentes.

---

sustainability: The ability of the earth to maintain itself without depletion of non-renewable resources.

sostenibilidad: La capacidad de la tierra para mantenerse sin agotamiento de recursos no renovables.

---

table: A way of presenting data in a structured format that allows relationships and trends to be easily recognized.

tabla: Una forma de presentar los datos en un formato estructurado que permite reconocer fácilmente las relaciones y tendencias.

---

thermoregulation: The maintenance of internal body temperature within a tolerable range.

termorregulación: El mantenimiento de la temperatura corporal interna dentro de un rango tolerable.

---

tissue: An integrated group of cells with a common structure, function, or both.

tejido: Un grupo integrado de células con una estructura, función o ambas comunes.

---

trait: A specific variant of a phenotype, controlled by one or more genes, e.g. flower colour in plants or tongue rolling ability in humans.

rasgo: Una variante específica de un fenotipo, controlada por uno o más genes, por ejemplo, el color de las flores en las plantas o la capacidad de rodar la lengua en los seres humanos.

---

transcription: The process of copying a segment of DNA into a strand of mRNA.

transcripción: El proceso de copiar un segmento de ADN en una hebra de ARNm.

---

transitional fossil: A fossil that exhibits traits common to both an ancestral group and its derived descendants.

fósil de transición: Un fósil que exhibe rasgos comunes tanto a un grupo ancestral como a sus descendientes derivados.

---

translation: The process of decoding a strand of mRNA to produce a sequence of amino acids.

traducción: El proceso de decodificación de una hebra de ARNm para producir una secuencia de aminoácidos.

---

transpiration: The evaporative loss of water from a plant.

transpiración: La pérdida evaporativa de agua de una planta.

---

trophic level: A level or position in a food chain, food web or ecological pyramid. An organism's trophic level is determined by its feeding behavior.

nivel trófico: Un nivel o posición en una cadena alimentaria, red alimentaria o pirámide ecológica. El nivel trófico de un organismo está determinado por su comportamiento de alimentación.

---

variable: A measurable property that changes over time or can take on different values.

variable: Una propiedad medible que cambia con el tiempo o puede tomar diferentes valores.

---

variation: The diversity of phenotypes and genotypes within a population or species.

variación: La diversidad de fenotipos y genotipos dentro de una población o especie.

---

zygote: A fertilised egg.

cigoto: Un óvulo fertilizado.

Appendix: Equipment list

2: Cell Structure

INVESTIGATION 2.1
Preparing an onion slide

Per student/pair
Light microscope
Onion/onion leaf
Glass microscope slides
Coverslips
Scalpel or razor
Iodine stain
Filter paper/tissue paper

INVESTIGATION 2.2
Simple diffusion across a membrane

Per student/pair
200 mL beaker
1 mL pipette
Glucose dipsticks
Lugol's indicator
4 x test tubes
Dialysis tubing
Thread or nylon line
Distilled water
1% starch solution
10% glucose solution
Timer or watch

INVESTIGATION 2.3
Estimating osmolarity

Per student/pair
6 x 500 mL beakers
Balance and equipment to weigh sugar
Table sugar or lab sucrose
Potato
Cork borer or scalpel
Paper towels
Marker pen

INVESTIGATION 2.4
How cell shapes affect diffusion

Per student/pair
Phenolphthalein infused agar
200 mL beaker
0.1 mol/L NaOH
Tongs
Paper towels
Razor/scalpel

INVESTIGATION 2.5
The effect of temperature on membrane permeability

Per student/pair
15 x 10 mL test tubes
200 mL beaker
Beetroot
Cork borer (internal diameter 4 mm)
Five water baths at 0° (ice), 20°, 40°, 60°, 90°C
Paper towels
Colorimeter set to 530 nm
Distilled water
Marker pen
Timer or watch

INVESTIGATION 2.6
Extracting DNA

Per pair
5 - 6 strawberries
1 large zip-lock bag
100 mL water
5 mL detergent
pinch of salt
1 x filter paper
1 x glass filter funnel
1 x 250 mL glass beaker
1 x glass rod
~100 mL ethanol (for rinsing)
2 x centrifuge tubes
Centrifuge

INVESTIGATION 2.7
Modeling protein structure

Per student/pair/group
Pipe cleaners (2 white, 2 pink, 2 purple, 4 blue)
Sticky tape
2 x binder clips or paper clips

INVESTIGATION 2.8
Effect of temperature on enzyme activity

Per group/temperature
1 x spotting plate/reaction plate
1 x test tube
1 x plastic pipette
Water bath
Timer
0.1 M iodine solution (I_2KI)
2 mL 1% amylase solution
1 mL buffer solution (pH 7.0)
1 mL 1% starch solution

INVESTIGATION 2.9
Forearm movements

No equipment needed

3: Feedback Mechanisms

INVESTIGATION 3.1
Modeling the effect of insulation

Per pair/group/test material
2 x 250 mL beakers
2 x 100 mL beakers
2 x thermometers
2 x larger containers (to fit beakers)
Insulating material (fat/lard, feathers, wool, cotton balls)
Weights or tape (optional)
Timer
2 x 100 mL warm tap water (~45°C)
Iced water

INVESTIGATION 3.2
Investing body shape and temperature regulation

Per group
Aluminum foil
Scissors
1 x thermometer (or datalogger)
1 x heat lamp

INVESTIGATION 3.3
Investigating effect of exercise on heart rate.

1 x stopwatch per group

INVESTIGATION 3.4
Investigating effect of exercise on breathing rate.

Equipment depends on group method

INVESTIGATION 3.5
Investigating plant transpiration

Per pair/group
250 mL conical flask with rubber bung
Petroleum jelly
1 cm^3 pipette
Clamp stand
Leafy plant shoot
Water
Cooking oil (for optional set up)
Timer or watch
Lamp, or plastic bag and water spray bottle, or fan
A4 or graph paper

4: Growth and development

INVESTIGATION 4.1
Modeling mitosis

Per student/pair
4 x pipe-cleaners (2 colors) cut in half
Yarn or string
A3 sheet of paper
Marker

5: Energy in Living Systems

INVESTIGATION 5.1
Measuring bubble production in Cabomba

Per pair/group
1.0 g *Cabomba aquatica*
Balance
Scissors
Water
1 x large beaker (large enough to hold the glass funnel)
1 x glass funnel
0.2 mol/L sodium hydrogen carbonate solution (enough to cover the plant)
1 x test tube
1 x lamp with a 60W bulb
Lux meter
Timer
1 x ruler or tape measure

INVESTIGATION 5.2
Measuring respiration in germinating seeds

Per group
3 x boiling tubes
Marker pen
6 x cotton balls
15% KOH solution
2 x eye dropper or plastic pipette
3 x gauze pieces
Germinated bean seeds (enough to fill one quarter of the boiling tube)
Ungerminated bean seeds (enough to fill one quarter of the boiling tube)
Glass beads (enough to fill one quarter of the boiling tube)
3 x 2-hole tube stoppers
3 x bent glass tubes or pipettes
3 x tubes (must be able to be clamped shut)
3 x screw clips
A few drops of colored liquid
3 x syringes (must fit tube with screw clamp attached)
3 x clamp stands or rack
Water bath (25°C)
Ruler
Timer

INVESTIGATION 5.3
Modeling photosynthesis and cellular respiration

Per Individual, pair, or group
Scissors

6: Interdependence in Ecosystem

INVESTIGATION 6.1
Investigating carrying capacity

Per student
Teaspoon
Cup
Per class
2 trays of shallow bowls
100 dried beans, peas, or beads etc.

INVESTIGATION 6.2
Creating a model of logistic growth

Per student/pair
Computer
Spreadsheet application e.g. Excel

INVESTIGATION 6.3
Density independent growth

Per student/pair
Computer and online access to
Populus software (free)
https://cbs.umn.edu/populus/overview

INVESTIGATION 6.4 and 6.5
Density dependent growth

Per student/pair
Computer and online access to
Populus software (free)
https://cbs.umn.edu/populus/overview

7: Energy Flow and Nutrient Cycles

INVESTIGATION 7.1
Exploring biomass pyramids

Per student/pair
Computer and online access to
HHMI interactive module
https://www.biointeractive.org/classroom-resources/exploring-biomass-pyramidsl

INVESTIGATION 7.2
A model of the carbon cycle

Per group
1 x 2 L clear soda bottle with lid
1 scoop aquarium gravel
Several dead leaves
1 x aquatic plant (e.g. Cabomba)
3-4 small pond snails
2 L filtered pond water

8: The Dynamic Ecosystem

INVESTIGATION 8.1
Pathways for toxins in food webs

Per group
Tape (for writing on)
Marker pen
Bag of small colored beads
1 x syringe or pipette capable of delivering 0.5 mL
10 x small test tubes
3 x medium test tubes
1 x large test tube
1 x 50 mL beaker
1 x 100 mL beaker
Test tube rack(s)

INVESTIGATION 8.2
A model of human impacts on fish stocks

Per student
Small cup
Teaspoon
Per group:
2 bowls
120 red beads, 80 yellow beads, 80 blue beads
Chopsticks
Spatula
Stopwatch

11: Variation of Traits

INVESTIGATION 11.1
Phenotypic variation in your class

No equipment required

INVESTIGATION 11.2
Modeling meiosis using iceblock sticks

Per pair
8 x Popsicle sticks
8 x sticky dots
Colored pencils or markers
Marker pen

INVESTIGATION 11.3
Modeling antibiotic resistance

Per student/pair
Computer
Spreadsheet application e.g. Excel

13: Natural selection, adaptation, and evolution

INVESTIGATION 13.1
Modeling selection with M&M's®

Per group
100 M&Ms®
1 x lidded container
1 x plate

INVESTIGATION 13.2
Phenotypic variation in your class

Per student/pair
Computer
Access https://short.concord.org/lm3 (free programme)

INVESTIGATION 13.3
Investigating gene pool changes

Per student/pair
Computer
Spreadsheet application e.g. Excel

14: Biodiversity

INVESTIGATION 14.1
Biodiversity and human impacts

Per student group
40 x red plastic tokens or brick
30 x blue plastic tokens or bricks
20 x plastic tokens or bricks
10 x plastic tokens or bricks
1 x plastic tub
1 x plastic tray

Credits

We acknowledge the generosity of those who have provided photographs for this edition: • Louisa Howard and Katherine Connolly, Dartmouth College Electronic Microscope Facility • PASCO for photographs of probeware • D. Dibenski for the photo of the flocking auklets • Stephen Moore for his photos of aquatic invertebrates.

We also acknowledge the photographers that have made their images available through Wikimedia Commons under Creative Commons Licences 2.0, 2.5. or 3.0: • James Hedberg • Matthias Zepper • Temsabulta • CDC: Dr Lucille K. Georg • Mnolf • Yaminchhipa10 • Jakob Suckale • SubtleGuest • Oxlamb • Jerald E. Dewey USDA • Mike Baird • Ansgar Walk • Bjørn Christian Tørrissen • Photaro • Paul Whippey • Wendy Kaveney • Gina Mikel • Alex Wild Public Domain • Jeffmock • Alastair Rae • Rocky Mountain Laboratories, NIAID, NIH • Xiangyux (PD) • Putney Mark • Allan and Elaine Wilson • UtahCamera • Dario • Jpbarrass • Bill Rhodes • Brocken Inaglory • dsworth Center-New York State Department of Health • NPS: Dan Richards• Dr Graham Beards • Kristian Peters • "Mike" Michael L. Baird • JesseW900 • Al Aumuller. NY World telegram and the Sun, Public Domain • Joe Schneid, Louisville, Kentucky • Pengo • Lusb • Suseno • Kaldari PD • dsworth Center: New York State Department of Health • Dr. Graham Beards • Cgoodwin • Steve Garvie • Luc Viatour www.Lucniz.be • NJR ZA • Biglaci • BirdPhotos.com • Cephas • 350z33 • Temsabuita • Haplochromis • Wallombi • CSIRO • Wipeter • U.S. Bureau of Reclamation • BS Thurner Hof • Matt Reinbold • Scott Edhardt, Public Domain • USDA/ Scott Bauer • Derek Quinn • Nicholls H • Bruno de Giusti • BS Thurner Hof • Scott Ehardt • ATamari • Takahashi • Adrian A. Smith • Yasunori Koide • Marc King • it:Utente:Cits • Ghedoghedo • NY State Dept of Health • J Podos • Lip Kee Yap • Putneymark• Mousa Direct Ltd • Fred Wierum • Bjorn schulz • Wilson44691 • Moussa Direct Ltd. • Heinz-Josef Lücking • Natural History Museum • Karl Magnacca • Famartin • Reverend Edward Brain, D.D. • D. Eason (DOC) • Raul654 • Ingfbruno • R.C Johnson • Dept of National Resources, Illinois • Junchang Lü and Stephen L. Brusatte • Plotnick, Theodor & Holtz Jr • Nobu • Tamura Geoff Gallice • Christian Ziegle • Richard Ling • Heikki Valv • Fir0002 • Martin Po • Famartin

Contributors identified by coded credits are: BF: Brian Finerran (University of Canterbury), BH: Brendan Hicks (Uni. of Waikato), CDC: Centers for Disease Control and Prevention, Atlanta, USA, CSIRO: The Commonwealth Scientific and Industrial Research Organisation, DoC: NZ Department of Conservation, EII: Education Interactive Imaging, FAOUN: Food and Agricultural Organization of the United Nations, KP: Kent Pryor, MPI: Max Planck Institute, NASA: National Aeronautics and Space Administration, NPS: National Park Services, NIH: National Institutes of Health, NOAA: National Oceanic and Atmospheric Administration www.photolib.noaa.gov, NYSDEC: New York State Dept of Environmental Conservation, RCN: Ralph Cocklin, RA: Richard Allan, USGS: United States Geological Survey, WBS: Warwick Silvester (Uni. of Waikato), WMU: Waikato Microscope Unit, USDA: United States Department of Agriculture, USFW: United States Fish and Wildlife Service, USGS: United States Geological Survey, NPS: National Park Service, DoC: New Zealand Department of Conservation.

Royalty free images, purchased by BIOZONE International Ltd, are used throughout this workbook and have been obtained from the following sources: Corel Corporation from their Professional Photos CD-ROM collection; IMSI (Intl Microcomputer Software Inc.) images from IMSI's MasterClips® and MasterPhotos™ Collection, 1895 Francisco Blvd. East, San Rafael, CA 94901-5506, USA; ©1996 Digital Stock, Medicine and Health Care collection; © 2005 JupiterImages Corporation www.clipart.com; ©Hemera Technologies Inc, 1997-2001; ©Click Art, ©T/Maker Company; ©1994., ©Digital Vision; Gazelle Technologies Inc.; PhotoDisc®, Inc. USA, www.photodisc.com. • TechPool Studios, for their clipart collection of human anatomy: Copyright ©1994, TechPool Studios Corp. USA (some of these images were modified by Biozone) • Totem Graphics, for their clipart collection • Corel Corporation, for use of their clipart from the Corel MEGAGALLERY collection • 3D images created using Bryce, Vue 6, Poser, and Pymol • iStock images • Art Today.

Index